Methods in Molecular Biology™

Series Editor
John M. Walker
School of Life Sciences
University of Hertfordshire
Hatfield, Hertfordshire, AL10 9AB, UK

For further volumes:
http://www.springer.com/series/7651

Mass Spectrometry of Glycoproteins

Methods and Protocols

Edited by

Jennifer J. Kohler

Department of Biochemistry
University of Texas Southwestern Medical Center
Dallas, TX, USA

Steven M. Patrie

Department of Pathology
University of Texas Southwestern Medical Center
Dallas, TX, USA

Humana Press

Editors
Jennifer J. Kohler
Department of Biochemistry
University of Texas Southwestern Medical Center
Dallas, TX, USA

Steven M. Patrie
Department of Pathology
University of Texas Southwestern Medical Center
Dallas, TX, USA

ISSN 1064-3745 ISSN 1940-6029 (electronic)
ISBN 978-1-4939-6203-7 ISBN 978-1-62703-146-2 (eBook)
DOI 10.1007/978-1-62703-146-2
Springer New York Heidelberg Dordrecht London

Softcover re-print of the Hardcover 1st edition 2013

Printed on acid-free paper

Humana Press is a brand of Springer
Springer is part of Springer Science+Business Media (www.springer.com)

Preface to Mass Spectrometry of Glycoproteins

Glycosylation is the most abundant post-translational modification of proteins. Estimates vary widely, but a common assessment is that upwards of 50% of eukaryotic proteins are modified by some type of glycan. Indeed, the difficulties associated with accurately assessing the glycosylation status of intra- and extracellular proteins are the primary motivations for this volume. Over the past 30 years, insight into the biological roles of glycan modifications has grown dramatically, yet this field has often struggled due to the inadequacies of accessible analytical methods. Fortunately, simultaneous to the recent expansion of knowledge in glycobiology, a similar transformation has occurred in the field of glycoproteomics. New enrichment techniques, novel ionization methods, mass spectrometry technologies, the expanding role of high-performance liquid chromatography, and improved informatic resources have transformed niche characterization of discrete glycoproteins into a powerful "omics" toolset that can simultaneously characterize diverse glycoproteins and the glycans they carry. Although proteomics approaches have been applied to broad classes of post-translational modifications, glycoproteins represent a particularly challenging case due to the heterogeneity of glycan structures, the lability of glycosidic bonds, the isobaric nature of many monosaccharides, and the difficulties associated with determining the unique structure of a branched molecule from compositional analysis. As presented in this volume, the latest glycoproteomics tools are meeting these challenges, providing unprecedented information about the structure and diversity of glycoproteins. It is an exciting time to be both a glycobiologist and mass spectrometrist.

Our contributing authors have highlighted the key aspects of most glycoproteomics workflows, including: the robust sample preparation techniques; the advanced chromatographic strategies for improving dynamic range; the advanced mass spectrometry instrumentation and associated ionization and fragmentation methods; and informatics tools used for identifying glycoproteins and characterizing the associated glycans. As in all experiments, sample preparation is paramount for successful glycoprotein characterization and thus the first seven chapters of this volume provide detailed descriptions of methods that reliably enrich glycosylated proteins, glycopeptides, and glycans from complex samples. Similarly, since mass spectrometry analysis is most often performed at the peptide level, Chaps. 8 and 9 are included to provide the reader with detailed protocols for best laboratory practices during the digestion of glycoproteins. Related to the emphasis on sample preparation, mass spectrometry workflows are facilitated by limiting the complexity of samples; therefore, state-of-the-art chromatographic separations tools that often interface directly to the mass spectrometer are detailed in Chaps. 10, 11, and 12. Chapters 13, 14, 15, and 16 build upon the techniques from the previous chapters while emphasizing emerging quantification strategies for both glycoproteins and associated glycans, the latter being essential since these modifications are typically substoichiometric. Since the improved duty cycle of modern technologies enables data collection at an unprecedented rate, Chaps. 17 and 18 provide details on the glyco-specific computational tools that are essential allies to all mass spectrometry workflows. Finally, we close the volume with four "case studies." These protocols detail the implementation of sample preparation, mass spectrometry, and data analysis in the study of real-world samples, including specimens from human diseased tissues and from the biologics industry.

We hope that *Mass Spectrometry of Glycoproteins* will serve as an essential resource for those who work at the interface of glycobiology and mass spectrometry. We envision that these protocols will serve as a critical foundation for collaborative efforts that rely on specialized knowledge in these two fields. On that note, we gratefully acknowledge the contributions of all of the authors who provided protocols for this volume. Their expertise in the application of mass spectrometry to glycobiology problems is unique and essential. We are also indebted to Dr. John Walker and the staff at Humana Press and Springer Science+Business Media for their support and encouragement in the preparation of this book.

Dallas, TX, USA ***Jennifer J. Kohler***
Dallas, TX, USA ***Steven M. Patrie***

Contents

Contributors

KIYOKO F. AOKI-KINOSHITA • *Department of Bioinformatics, Soka University, Aliso Viejo, CA, USA*
DAMARIS BAUSCH-FLUCK • *Institute of Molecular Systems Biology, Swiss Federal Institute of Technology (ETH) Zurich, Zurich, Switzerland; National Center of Competence in Research (NCCR) Neural Plasticity and Repair—Center for Proteomics, UZH/ETH Zurich, Zurich, Switzerland*
ROGER W. BEUERMAN • *Singapore Eye Research Institute, Singapore, Singapore*
ANDRÉ M. DEELDER • *Department of Parasitology, Leiden University Medical Center, Leiden, The Netherlands*
HEATHER DESAIRE • *Department of Chemistry, University of Kansas, Lawrence, KS, USA*
PENELOPE DRAKE • *Department of Obstetrics, Gynecology and Reproductive Sciences, University of California, San Francisco, CA, USA*
LARISSA S. FENN • *Department of Chemistry, Vanderbilt University, Nashville, TN, USA*
SUSAN FISHER • *Department of Obstetrics, Gynecology and Reproductive Sciences, University of California, San Francisco, CA, USA*
CORINA FLANGEA • *Department of Chemical and Biological Sciences, "Aurel Vlaicu" University of Arad, Timisoara, Romania*
BRAD GIBSON • *Department of Obstetrics, Gynecology and Reproductive Sciences, University of California, San Francisco, CA, USA*
EDEN P. GO • *Department of Chemistry, University of Kansas, Lawrence, KS, USA*
HOWARD C. HANG • *The Laboratory of Chemical Biology and Microbial Pathogenesis, The Rockefeller University, New York, NY, USA*
JINTANG HE • *Department of Surgery, University of Michigan Medical Center, Ann Arbor, MI, USA*
ANDREAS HOFMANN • *Institute of Molecular Systems Biology, Swiss Federal Institute of Technology (ETH) Zurich, Zurich, Switzerland; Program in Molecular Life Sciences, University of Zurich (UZH)/ETH Zurich, Zurich, Switzerland; National Center of Competence in Research (NCCR) Neural Plasticity and Repair—Center for Proteomics, UZH/ETH Zurich, Zurich, Switzerland*
PETER HØJRUP • *Department of Biochemistry and Molecular Biology, University of Southern Denmark, Odense, Denmark*
HYEJIN HWANG • *Department of Pathology, University of Washington School of Medicine, HMC, Seattle, WA, USA*
TOSHIAKI ISOBE • *Department of Chemistry, Tokyo Metropolitan University, and CREST, Japan Science and Technology Agency (JST), Tokyo, Japan*
OLE NØRREGAARD JENSEN • *Department of Biochemistry and Molecular Biology, University of Southern Denmark, Odense, Denmark*
PIA HØNNERUP JENSEN • *Institute of Molecular Medicine, University of Southern Denmark, Odense, Denmark*
HIROYUKI KAJI • *Department of Chemistry, Tokyo Metropolitan University, Tokyo, Japan*
DAVID KATES • *Analytical Development and Support, Global Biological Development, Bayer Healthcare LLC, Berkeley, CA, USA*

CAROLIEN A.M. KOELEMAN • *Department of Parasitology, Leiden University Medical Center, Leiden, The Netherlands*
JENNIFER J. KOHLER • *Department of Biochemistry, University of Texas Southwestern Medical Center, Dallas, TX, USA*
GÖRAN LARSON • *Department of Clinical Chemistry and Transfusion Medicine, Institute of Biomedicine, the Sahlgrenska Academy at the University of Gothenburg, Gothenburg, Sweden*
JUN LIU • *Analytical Development and Support, Global Biological Development, Bayer Healthcare LLC, Berkeley, CA, USA*
YASHU LIU • *Department of Surgery, University of Michigan Medical Center, Ann Arbor, MI, USA*
DAVID M. LUBMAN • *Department of Surgery, University of Michigan Medical Center, Ann Arbor, MI, USA*
HAOJIE LU • *Department of Chemistry and Institute of Biomedical Sciences, Fudan University, Shanghai, China*
ZHI YUAN MA • *Department of Biochemistry and Molecular Biology, Drexel University, Philadelphia, PA, USA*
ANOOP MAYAMPURATH • *School of Informatics and Computing, Indiana University, Bloomington, IN, USA*
JOHN A. MCLEAN • *Department of Chemistry, Vanderbilt University, Nashville, TN, USA*
SIMON MYSLING • *Department of Biochemistry and Molecular Biology, University of Southern Denmark, Odense, Denmark*
JONAS NILSSON • *Department of Clinical Chemistry and Transfusion Medicine, Institute of Biomedicine, the Sahlgrenska Academy at the University of Gothenburg, Gothenburg, Sweden*
RON ORLANDO • *Departments of Biochemistry and Molecular Biology, and Chemistry, University of Georgia, Athens, GA, USA*
STEVEN M. PATRIE • *Department of Pathology, University of Texas Southwestern Medical Center, Dallas, TX, USA*
MATTHEW R. PRATT • *Department of Chemistry, University of Southern California, Los Angeles, CA, USA; Department of Molecular and Computational Biology, University of Southern California, Los Angeles, CA, USA*
KATHRYN R. REBECCHI • *Department of Chemistry, University of Kansas, Lawrence, KS, USA*
ANDREAS M. RIZZI • *Institute of Analytical Chemistry and Food Chemistry, University of Vienna, Vienna, Austria*
MICHAEL J. ROTH • *Department of Pathology, University of Texas Southwestern Medical Center, Dallas, TX, USA*
NICOLE SAMUELS • *Analytical Development and Support, Global Biological Development, Bayer Healthcare LLC, Berkeley, CA, USA*
BIRGIT SCHILLING • *Buck Institute for Age Research, Novato, CA, USA*
ALINA SERB • *Department of Biochemistry, "Victor Babes" University of Medicine and Pharmacy, Timisoara, Romania*
JOANNE SEVERS • *Analytical Development and Support, Global Biological Development, Bayer Healthcare LLC, Berkeley, CA, USA*

MIN SHI • *Department of Pathology, University of Washington School of Medicine, HMC, Seattle, WA, USA*
EUGEN SISU • *Department of Biochemistry, "Victor Babes" University of Medicine and Pharmacy, Timisoara, Romania; Chemistry Institute of the Romanian Academy, Timisoara, Romania*
YULIYA SKOROBOGATKO • *Department of Biochemistry and Molecular Biology, Drexel University, Philadelphia, PA, USA*
STEPHANIE H. STALNAKER • *Complex Carbohydrate Research Center, University of Georgia, Athens, GA, USA*
HAIXU TANG • *School of Informatics and Computing, Indiana University, Bloomington, IN, USA*
KEITH VOSSELLER • *Department of Biochemistry and Molecular Biology, Drexel University, Philadelphia, PA, USA*
YOSHINAO WADA • *Osaka Medical Center and Research Institute for Maternal and Child Health, Osaka, Japan*
LANCE WELLS • *Complex Carbohydrate Research Center, University of Georgia, Athens, GA, USA; Department of Chemistry, University of Georgia, Athens, GA, USA; Department of Biochemistry and Molecular Biology, University of Georgia, Athens, GA, USA*
BERND WOLLSCHEID • *Institute of Molecular Systems Biology, Swiss Federal Institute of Technology (ETH) Zurich, Zurich, Switzerland; National Center of Competence in Research (NCCR) Neural Plasticity and Repair—Center for Proteomics, UZH/ETH Zurich, Zurich, Switzerland*
MANFRED WUHRER • *Department of Parasitology, Leiden University Medical Center, Leiden, The Netherlands*
YAWEI XU • *Department of Chemistry and Institute of Biomedical Sciences, Fudan University, Shanghai, China*
CHUAN-YIH YU • *School of Informatics and Computing, Indiana University, Bloomington, IN, USA*
ALINA D. ZAMFIR • *Department of Chemical and Biological Sciences, "Aurel Vlaicu" University of Arad, Timisoara, Romania*
BALYN W. ZARO • *Department of Chemistry, University of Southern California, Los Angeles, CA, USA*
GERHILD ZAUNER • *Department of Parasitology, Leiden University Medical Center, Leiden, The Netherlands*
ANA-MARIA ZAGREAN • *Neuroscience Laboratory, "Carol Davila" University of Medicine and Pharmacy, Bucharest, Romania*
JING ZHANG • *Department of Pathology, University of Washington School of Medicine, HMC, Seattle, WA, USA*
LIJUAN ZHANG • *Department of Chemistry and Institute of Biomedical Sciences, Fudan University, Shanghai, China*
PENG ZHAO • *Complex Carbohydrate Research Center, University of Georgia, Athens, GA, USA; Department of Chemistry, University of Georgia, Athens, GA, USA*
LEI ZHOU • *Singapore Eye Research Institute, Singapore, Singapore*

Chapter 1

Introduction to Glycosylation and Mass Spectrometry

Steven M. Patrie, Michael J. Roth, and Jennifer J. Kohler

Abstract

Glycosylation is increasingly recognized as a common and biologically significant post-translational modification of proteins. Modern mass spectrometry methods offer the best ways to characterize the glycosylation state of proteins. Both glycobiology and mass spectrometry rely on specialized nomenclature, techniques, and knowledge, which pose a barrier to entry by the nonspecialist. This introductory chapter provides an overview of the fundamentals of glycobiology, mass spectrometry methods, and the intersection of the two fields. Foundational material included in this chapter includes a description of the biological process of glycosylation, an overview of typical glycoproteomics workflows, a description of mass spectrometry ionization methods and instrumentation, and an introduction to bioinformatics resources. In addition to providing an orientation to the contents of the other chapters of this volume, this chapter cites other important works of potential interest to the practitioner. This overview, combined with the state-of-the-art protocols contained within this volume, provides a foundation for both glycobiologists and mass spectrometrists seeking to bridge the two fields.

Key words: Proteomics, Glycomics, Nomenclature, Mass spectrometry, Ionization methods, Glycosylation, Glycoproteins, Bioinformatics, Biomarker, Electrospray ionization, Matrix-assisted laser desorption/ionization, Liquid chromatography

The field of *proteomics* studies protein networks by rationalizing the composition and dynamics of translated gene products (1). A critical aspect of these investigations includes the characterization of protein post-translational modifications (PTMs) (e.g., glycosylation, phosphorylation, ubiquitination, acetylation, methylation, etc.), which are important stabilizers of protein–protein interactions and modulators of signal cascades. The glycosylation of proteins is the most common PTM and is an essential determinant of protein activity and function. It is estimated that >50% of mammalian proteins are glycosylated so it is not surprising to find that glycoproteins are involved in the coordination of most intra- and intercellular processes (e.g., immune function, cellular division/migration/adhesion, host–pathogen interactions, enzyme catalysis) (2, 3). Glycosylation is also recognized as an important element of disease pathophysiology (e.g., cancer, autoimmune, diabetes, Alzheimer's, hematologic

Jennifer J. Kohler and Steven M. Patrie (eds.), *Mass Spectrometry of Glycoproteins: Methods and Protocols*, Methods in Molecular Biology, vol. 951, DOI 10.1007/978-1-62703-146-2_1, © Springer Science+Business Media, LLC 2013

disorders, and allergies) and numerous investigators and organizations like the Human Disease Glycomics/Proteome Initiative (HGPI) (4) seek to translate global glycosylation profiles into diagnostic biomarkers. Mass spectrometry (MS), combined with sample preparation, chromatography, and informatics technologies, has emerged as an effective platform to determine the structure of glycans (e.g., sugar composition, antennary profiles), their sites of attachment to the polypeptide, and glycoform stoichiometry (5, 6). These tools also enable the billion-dollar biologics industry which seeks to meet the FDA's regulatory guidelines (e.g., ICH.Q6A or ICH.Q6P (7)) that specify physicochemical and biological activity criteria for "human-like" therapeutic proteins to ensure their pharmacokinetic, pharmacodynamic, and immunogenic properties (8).

This volume, written for biochemists, chemists, and glycobiologists, seeks to provide life science investigators with state-of-the-art protocols and benchmarks for glycoprotein analysis. We describe methods for enrichment, separation, and preparation of glycoprotein and glycan samples for MS analysis (Parts I–III), quantification of glycan and protein population changes by modern MS (Part IV), and interpretation of MS data with bioinformatics tools (Part V). To provide laboratory benchmarks for the reader, nearly all our chapters include representative examples of the practical uses of MS in characterizing glycoprotein samples. We also highlight several case studies that demonstrate the utility of glycosylation analysis in biomarker development and intact recombinant glycoprotein characterization (Part VI).

Throughout this book are references to glycobiology, equipment, and techniques that are commonly used in glycoproteomics. To aid the reader, here we briefly describe the basics of glycosylation and MS workflows.

Nomenclature: Glycan structures are annotated via convenient symbol and text nomenclature designed for glycan annotation in a mass spectrum (Fig. 1). To better understand this nomenclature we direct readers to the second edition of the textbook *Essentials of Glycobiology*, which prescribes rules to standardize the symbol nomenclature for figures (both color and black and white), as well as details on textual nomenclature for branched linkages written in either linear or two-dimensional formats (2, 9).

Nomenclature for tandem MS (MS/MS or MS^n) (*v.i.*) spectra of peptides or proteins describe N- and C-terminal fragment ions with designations of the polypeptide backbone bond cleavage site (Fig. 2a) (10). The most common cleavage occurs at either the amide bonds for ergodic dissociation methods, denoted as "b" and "y" ions, or the backbone N-Cα bonds for non-ergodic techniques, denoted "c" and "z" ions. Similarly, carbohydrate MS/MS nomenclature follows the conventions outlined by Costello and Domon (Fig. 2b) (11). Fragment ions that contain the unreduced

a

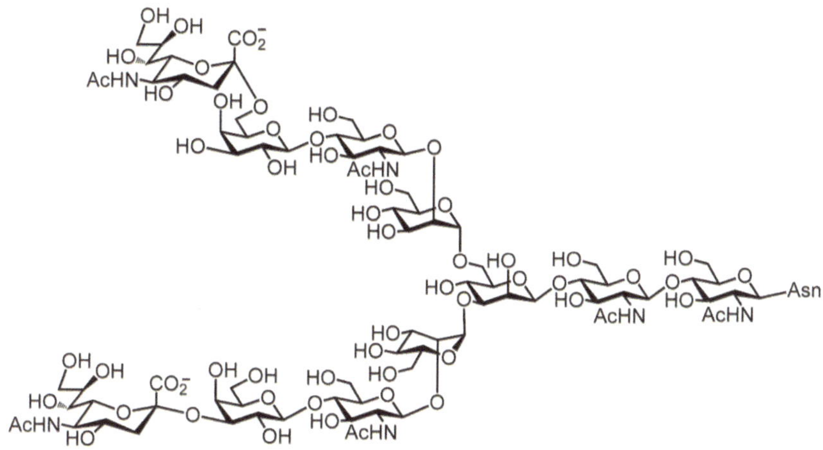

b

Siaα6Galβ4GlcNAcβ2Manα$_6$

Manβ4GlcNAcβ4GlcNAcβ-Asn

Siaα3Galβ4GlcNAcβ2Manα^3

c

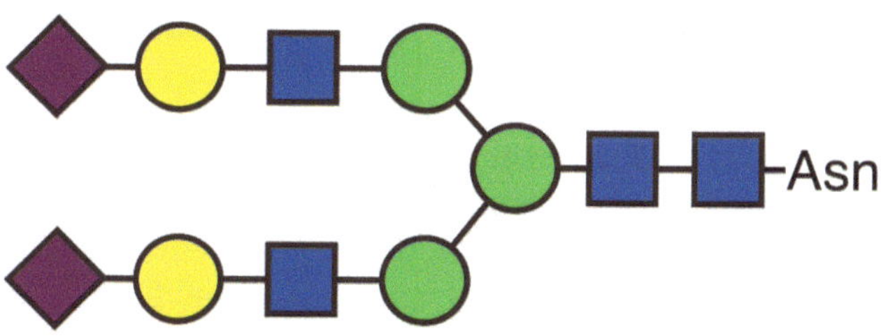

Fig. 1. Glycan representations. The same N-linked glycan is depicted in three ways: (**a**) chemical structure; (**b**) two-dimensional text format; (**c**) symbol nomenclature. Simplified nomenclatures are useful shorthand but may fail to specify important information. For example, the simple symbol nomenclature shown here does not indicate the regiochemistry or stereochemistry of the sugar linkages.

termini are designated by "A," "B," and "C" while the reduced termini are "X," "Y," and "Z." The B, C, Y, and Z fragment ions denote dissociation at glycosidic bonds while A and X ions represent interring dissociation sites with superscripts that designate the two ring bonds cleaved. The subscripts designate the position relative to the respective termini.

Glycosylation: Glycosylation is a common PTM: an estimated 50% of proteins are glycoproteins (3). post-translational glycosylation of proteins occurs in all three domains of life, although archaeal and bacterial glycosylation remain less well-characterized than eukaryotic glycosylation (12, 13). This book focuses on analysis of

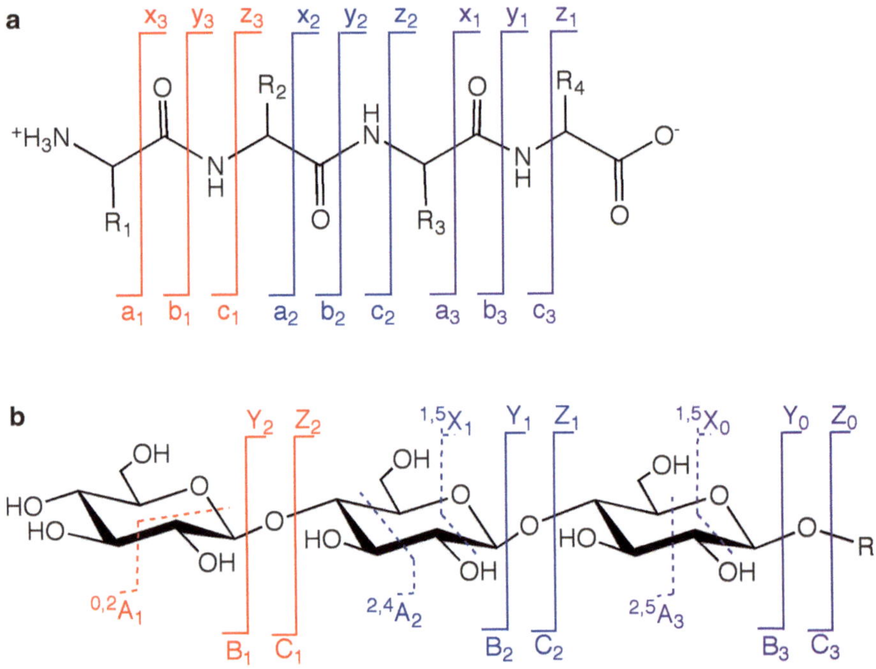

Fig. 2. Typical nomenclature used to describe MS/MS fragment ions for (**a**) polypeptides or (**b**) oligosaccharides denoting location of bond dissociation for various fragment ion types.

eukaryotic glycoproteins, which are produced in the secretory pathway, as well as in the cytoplasm and nucleus.

Enzymes residing in the secretory pathway—the endoplasmic reticulum (ER) and the Golgi—are responsible for the biosynthesis of multiple classes of glycoproteins including asparagine- or N-linked glycoproteins, mucin-type O-linked glycoproteins, proteoglycans, O-fucose glycoproteins, O-mannose glycoproteins, and collagen. Golgi- and ER-resident glycosyltransferases transfer sugars from nucleotide sugar donors to glycoprotein substrates as they traffic through the secretory pathway. Glycosyltransferases are membrane proteins that localize to specific subsites within the secretory pathway. The localization of these enzymes dictates the order in which glycosylation events occur. In this way, the secretory pathway serves as an assembly line for glycoprotein biosynthesis. Unlike the production of nucleic acid and peptide polymers, glycan biosynthesis is not template directed. The nontemplated nature of glycan biosynthesis results in a key characteristic of glycosylation: heterogeneity. Heterogeneity occurs both at the level of the occupancy of potential glycosylation sites as well as the diversity of structures present at each site. The term "glycoforms" refers to different isoforms of a protein that vary with respect to the number or structure of attached glycans. Since different glycoforms of a protein may differ dramatically in their physical and biological properties (14), mass spectrometric methods that provide quanti-

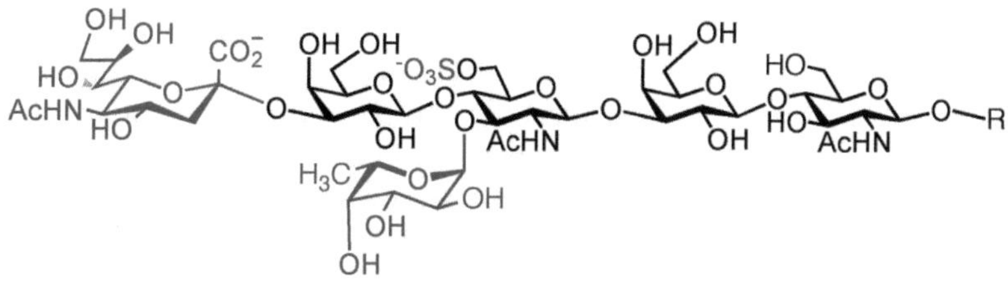

Fig. 3. Modified LacNAc repeats. The structure shown consists of two repeats of the LacNAc disaccharide (*black*). The leftmost LacNAc is modified with sialic acid, sulfate, and fucose (*contrasting color*). Modifications can also occur at other positions, resulting in large combinatorial complexity.

tative information about site occupancy and glycan structural diversity are essential.

The best-known forms of protein glycosylation are the N-linked glycans, which are large, branched structures with a conserved core. These glycans are synthesized in a baroque process initiated in the ER and continued in the Golgi. The core oligosaccharide structure is assembled on the cytoplasmic face of the ER, flipped into the lumen of the ER, and transferred co-translationally to asparagine residues of new polypeptides. Monosaccharides are removed from this structure by the action of ER- and Golgi-resident glycosidases. In the Golgi, the glycan is elaborated by the action of a variety of glycosyltransferases, which dictate the degree of branching (tetraantennary complex glycans are possible), as well as the extension and elaboration of the branches. Branches are often extended with repeats of galactose (Gal) β1-4-linked to *N*-acetylactosamine (GlcNAc), forming a disaccharide known as *N*-acetyllactosamine (LacNAc), which can be elaborated with fucose, sulfate, and/or sialic acids (Fig. 3).

Biosynthesis of mucin-type O-linked glycoproteins is initiated by the action of polypeptide *N*-acetylgalactosaminyltransferases (ppGalNAcTs), enzymes that add *N*-acetylgalactosamine (GalNAc) in an α-linkage to serine or threonine residues. This initiating GalNAc is typically modified with an additional sugar or sugars at the 3- and/or 6-position, forming one of the eight core O-linked structures. These core structures are typically extended with repeating LacNAc polymers, and additional branching may occur. The extended glycans are elaborated by additional modifications, including fucosylation, sialylation, and sulfation.

Proteins that are modified with glycosaminoglycan (GAG) chains are known as proteoglycans. GAGs are linear polymers composed of alternating amino sugars (GlcNAc or GalNAc) and uronic acids (glucuronic acid or iduronic acid). The sugar composition and modifications determine the classification of the GAG; common GAGs are heparin, heparan sulfate (HS), chondroitin sulfate (CS), and dermatan sulfate (DS). GAGs are biosynthesized by the stepwise addition of individual monosaccharides to a serine residue. All GAGs share a common tetrasaccharide core, which is then elaborated with the appropriate sugars and modifications to result in the different GAG structures.

Secretory pathway enzymes also synthesize other less common glycoproteins. For example, O-fucose glycans are typically small (four monosaccharides or fewer) glycans that are attached to serine residues. The resulting glycoproteins are known to play important roles in developmentally regulated signaling (15). Glycans initiating with O-linked mannose are also observed and are essential to proper brain, eye, and skeletal muscle function (16). Collagen, a key component of connective tissue, is glycosylated on its hydroxy-lysine residues.

Along with the many forms of secretory pathway glycosylation, proteins can also be glycosylated in the cytoplasm and nucleus. The most well-characterized form of cytoplasmic glycosylation is O-GlcNAc, which consists of a single GlcNAc residue β-linked to a serine or threonine. Modification by O-GlcNAc often occurs at sites that can be alternatively phosphorylated, leading to a reciprocal relationship between the two modifications. Hundreds of proteins have been identified to have the O-GlcNAc modification, but mapping of these glycosylation sites is not yet comprehensive (17).

Glyco-proteomics Workflows: Mass spectrometry workflows for glycan and glycoconjugate characterization are highlighted throughout this book and briefly described below. Workflows include distinct steps that range from nonspecific preparative strategies, glyco-specific enrichment and preparative methods, MS/MS methods, and informatics tools (Fig. 4). Representative experimental designs vary due to the diverse physicochemical properties of the oligosaccharides, peptides, and proteins (*v.i.*), as well as the user application (18). Example applications include:

1. Oligosaccharide population analysis, "glycan profiling."
2. Monosaccharide composition analysis (e.g., neutral, acidic, and amino-sugars, sialic acid content).
3. Glycan sequence, linkage, and antennary profiling (e.g., determination of mannose or fucose content, carbohydrate anomer determination).
4. Bottom-up glycoconjugate site analysis (e.g., N- vs. O-linked site analysis, site occupancy).
5. Glycopeptide sequence analysis.
6. Quantitation (e.g., ITRAQ, SILAC for peptides; QUIBL, IGOT, label-free for glycans).
7. Top-down intact protein glycosylation profiling (e.g., recombinant glycoprotein and antibody glycan profiling).

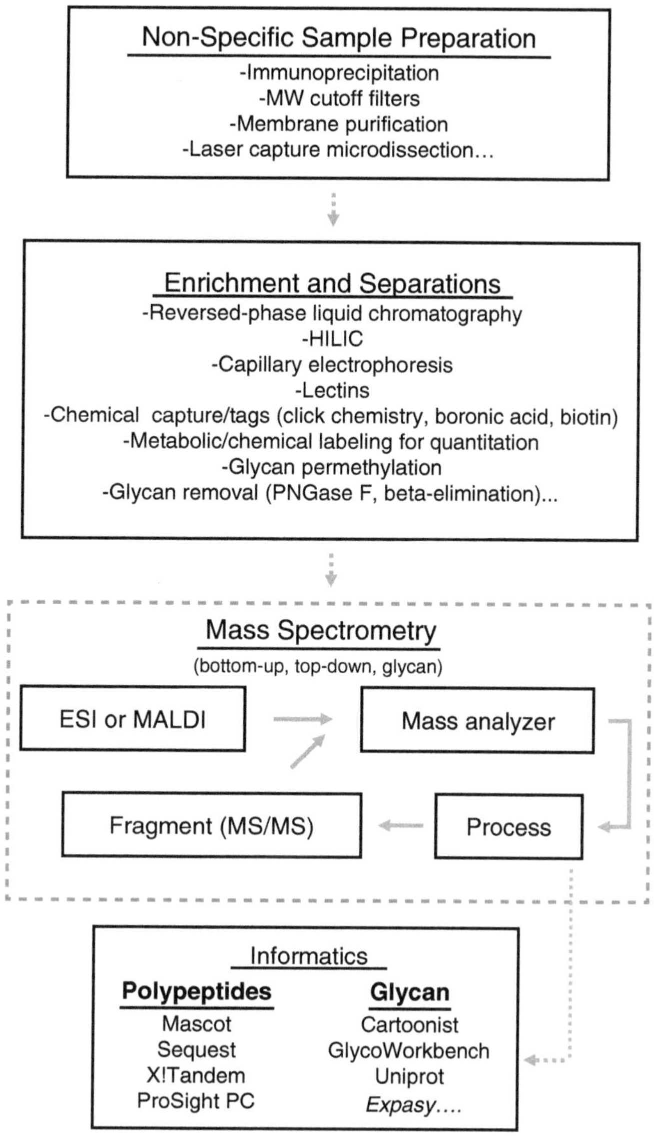

Fig. 4. Steps included in glyco-proteomics workflows (from *top* to *bottom*): (1) nonspecific sample preparation methods, (2) glyco-specific enrichment and separation methods, (3) mass spectrometry, and (4) informatics and data interpretation.

Generally, these workflows are designed for "bottom-up" protocols (19) that begin with glycan removal (e.g., PNGase F or beta-elimination for N- and O-linked glycans, respectively) followed by protease treatment (e.g., trypsin or proteinase K) to cleave proteins into 1–3 kDa peptides. The resulting glycan or polypeptide analyte is then introduced into a mass spectrometer for mass determination. Glycan profiling is a typical starting point and seeks to simultaneously characterize the mass and relative abundance of glycan populations in the sample (20, 21). Similarly, polypeptide analysis allows for a comprehensive evaluation of which glycosylated proteins are present in the sample. In these experiments, gas-phase fragmentation on selected species is commonly used to characterize peptide amino acid sequence and glycan composition and structure (1). Glycan removal from the polypeptide is not a prerequisite in bottom-up methods and many investigators rely upon MS^n methods (*v.i.*) to simultaneously sequence glycan/peptide backbones and identify the modified residue on the polypeptide. Alternatively, "top-down" methods can be used to profile isomeric glycosylation states on intact proteins without prior removal of glycans or protein digestion (22). Below we briefly describe aspects of these workflows in more detail.

Sample enrichment, preparation, and quantitation: Enrichment and labeling strategies used to improve experimental detection limits, dynamic range, and quantitation are critical to most glycoproteomics workflows. Enrichment is particularly important for analysis of complex samples (e.g., human blood) where protein concentration can vary by many orders-of-magnitude (5, 6) and because oligosaccharide populations vary due to spatial and temporal regulation of glycosyltransferases and elaboration of the core glycan by other modifications (e.g., acetylation, sulfatation, fucosylation, sialylation). Enrichment strategies typically start with nonspecific methods to minimize sample complexity (e.g., molecular weight cutoff filters, immunoprecipitation (IP), subcellular organelle enrichment, purification of membrane proteins, and laser capture micro-dissection). However, PTM-specific enrichment methods are perhaps most important because they minimize analyte physicochemical diversity which commonly leads to signal suppression in MS analysis. For example, in glycosylation analysis the hydrophilic and acidic nature of carbohydrates (e.g., sialic acid) tends to suppress ion signal relative to the unmodified hydrophobic peptides. In phospho-proteomics, tools such as immobilized metal affinity chromatography (IMAC), titanium dioxide (TiO_2) chromatography, and immunoprecipitation with phospho-specific antibodies are straightforward enrichment strategies that enable large-scale analysis of thousands of phosphorylation sites from a single sample (23). Glycosylation-specific preparative strategies are less defined because of high chemical heterogeneity of glycans derived from diverse monosaccharide building blocks, anomeric configurations, branching, and elaboration by other chemical moieties (e.g., acetylation and sulfation). Throughout Parts I and II of this book we highlight the most common enrichment and separation strategies available to investigators. These include biological approaches that exploit the diverse sugar recognition specificities of different glycan binding proteins (e.g., lectins) and chemical methods used to specifically capture subsets of glycoconjugate populations (e.g., boronic acid capture on advanced nanoparticles, click chemistry, and oxidative coupling of chemical biotin tags). In addition, our authors provide in-depth analyses of state-of-the-art chromatographic approaches (e.g., hydrophilic interaction chromatography (HILIC), capillary electrophoresis, and nano-reversed-phase liquid chromatography) and advanced instrumental techniques (e.g., ion-mobility MS) that offer complementary separation platforms to improve proteome coverage. When sample amounts are non-limiting, combining these techniques in multidimensional formats provide the greatest overall experimental dynamic range.

In Chapter 13, Orlando provides an excellent overview of metabolic labeling, chemical labeling, and "label-free" strategies used for quantitative glycoproteomics (24). Label-based quantitation methods typically use light and heavy stable-isotopic labels (e.g., $^{16}O/^{18}O$ or $^{14}N/^{15}N$). Differentially labeled samples are mixed and analyzed with MS, and relative quantitation is achieved by comparison of the differentially labeled mass spectral peak intensities within a spectrum. For polypeptides, the labels are introduced either metabolically by labeled amino acids or during sample preparation via labeled chemical tags (e.g., SILAC and ITRAQ, respectively) (25, 26). For oligosaccharides, label-based quantitation is commonly achieved by in vivo incorporation of stable-isotopes via ^{15}N-labeled glutamine in the cell culture or by isotopically labeled permethylation reagents (e.g., IDAWG and QUIBL, respectively (27, 28)). Label-free approaches, commonly used in glycan population analysis, compare the abundances of species in separate experiments that were run under the same experimental conditions (29). Most top-down intact protein methods are label-free in nature, and, as shown by Samuels et al. in Chapter 22, are being exploited by industry to characterize changes in individual protein glycoforms on therapeutic antibodies.

Mass Spectrometry definitions and concepts (30): Mass spectrometry pertains to the study of ionized molecules with a *mass spectrometer*. In order to perform MS, molecules must be in the gas-phase and ionized (e.g., protonated, cationized, or anionized). The mass (often called *molecular weight, m*) is derived from an ionized species in a mass spectrum by its *mass-to-charge* (m/z) ratio. Most modern mass spectrometers have high enough m/z resolution to elucidate the natural variation in the carbon-12/carbon-13 isotopes of the biomolecule (31). These isotopic multiplets are important because they can be assigned at part-per-million (ppm) mass accuracies. An ion's charge (z) is derived with knowledge of m/z difference between adjacent isotopes ($z = 1/(\Delta m/z_{iso1} - \Delta m/z_{iso2})$) (32). Subsequent mass assignment for an isotopic envelope is most commonly reported as the mass of the monoisotopic peak ($^{12}C_{100\%}{}^{13}C_{0\%}$). On low resolution instruments where isotopes are not resolved, the *average mass* is reported after z is derived by deconvolution of related charged species in the spectrum (33). High mass accuracy is important because mass alone can often differentiate competing elemental/chemical compositions of species (e.g., isotopic labels for quantitation) and enable glycan profiling experiments that track carbohydrate population changes between

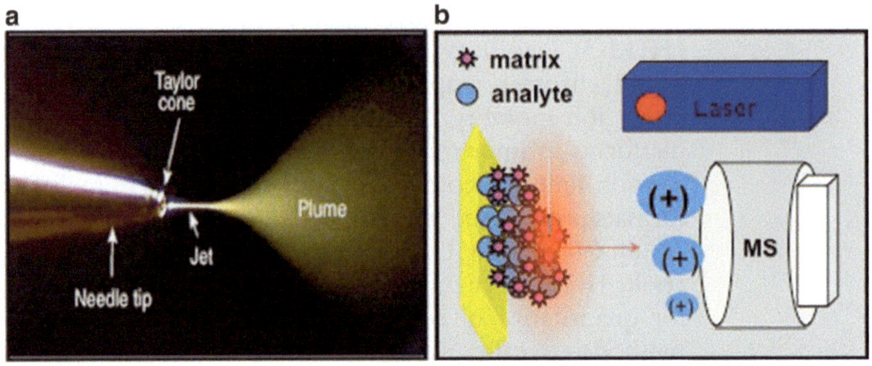

Fig. 5. (**a**) In electrospray ionization, a 1–4 kVdc potential difference between the needle tip and the mass spectrometer inlet (**b**) creates a fine mist—referred to as a "Taylor-cone"—of small highly charged droplets that enter the vacuum system where ions form. (Photo courtesy of New Objective, Inc., ©2000 New Objective, Inc.) (**b**) In matrix-assisted laser desorption ionization (MALDI), analyte mixed with an energy absorbing matrix is irradiated with laser light enabling the desorption of analyte into the gas-phase and where they are ionized in the energetic plume.

samples. In many cases accurate mass eliminates the need for MS/MS steps which significantly add to experiment complexity (34).

Ionization—The routine analysis of polypeptides with MS has been ongoing since the advent of electrospray ionization (ESI) (35) and matrix-assisted laser desorption ionization (MALDI) (36, 37), methods that led to the 2002 Nobel Prize in Chemistry (Fig. 5). These "soft" ionization techniques are significant because they volatilize and ionize analyte under conditions that do not normally break labile amino acid bonds (38). Both ESI and MALDI are suitable for achieving the goal of identification and structural determination of peptides, proteins, and oligosaccharides with molecular weights that range from hundreds to several hundred thousand Daltons (Da, a.k.a. atomic mass units, amu).

In ESI, polypeptides or glycans are suspended in an organic/aqueous solution and aerosolized from an ESI emitter that sits in front of the mass spectrometer inlet. Ions are generated from the fine droplets by mechanisms related to solvent evaporation, droplet fission caused by Coulombic explosion of shrinking charged droplets, and the evaporation of ionized species directly from the droplet surface. To promote the generation of positive (or negative) ions, a small amount (0.1–1%v/v) of acid (or base) is typically added to the sample solution, resulting in a distribution of highly charged (up to $z = 1{,}000$) ions $(M+nH)^{n+}$ that populate the low m/z region of a spectrum (typically m/z 400–3,000). Charge multiplicity is advantageous because virtually any type of mass spectrometer (*v.i.*) can efficiently operate in the low m/z region and MS/MS efficiency often varies with ion charge (39, 40). Glycans are typically permethylated to equalize their chemical properties and improve MS ionization characteristics (41). The power of ESI-MS is best realized when combined "online" with reversed-phase liquid chromatography (denoted LC/MS) which improves dynamic range by reducing sample complexity via chromatographic separation (42). The most sensitive LC/MS implementations use 50–150 μm ID capillary columns with integrated ESI emitters (~5–10 cm in length) at chromatographic flow rates of 50–300 nL/min (termed nano-LC/MS). Experiments with LC/MS commonly take 30–180 min depending on starting sample complexity; however, improved chromatographic performances on ultra-high pressure LC resins (denoted UPLC) (43) have made separation times <30 min more common.

Complementary to ESI, MALDI is a laser desorption/ionization technique that generates gas-phase ions with a solid (or liquid) chemical matrix and a pulsed laser, typically 337 or 355 nm wavelength (38). In MALDI the analyte is mixed with the matrix and co-crystallized on a sample plate. The sample is then irradiated with laser light where energy absorbed by the matrix is transferred to the nonvolatile analyte promoting desorption from the surface. Ions formed in the energetic gas-phase plume are transferred electrostatically into a mass analyzer. For peptides and proteins, MALDI is a soft ionization technique because the laser energy is strongly absorbed by the matrix and not the analyte, preventing thermal degradation and fragmentation of the polypeptide backbone. However, in-source fragmentation of native glycans is still observed in MALDI, necessitating a permethylation step to stabilize the molecule during ionization (27, 28). Common MALDI matrices are α-cyano-4-hydroxycinnamic acid (CHCA), 3,5-dimethoxy-4-hydroxycinnamic acid (sinapinic acid, SA), 2′,4′,6′-trihydroxyacetophenone monohydrate (THAP), and 2,5-dihydroxybenzoic acid (DHB). In contrast to ESI, in MALDI low-charge ions ($z<5$) typically dominate the spectrum. For peptides and proteins, protonated ions are common; however, in glycan characterization cationized species are often present (e.g., $[M+Na]^+$ or $[M+K]^+$). The MALDI-MS method is best used on either simple mixtures or if the chromatographic separations have been performed "off-line" from the mass spectrometer. In the latter case, robotic spotting of LC fractions and matrix onto a MALDI plate must occur prior to analysis.

Mass Spectrometers—Today's state-of-the-art mass spectrometers are hybrid instruments (e.g., Q-TOF, QqQ, tandem TOF-TOF, QqQ-FTMS, IT-FTMS, IT-Orbitrap) that combine two or more of the four common mass analyzers: *quadrupole* (Q), *ion trap* (IT), *time-of-flight* (TOF), and *Fourier transform* (FT), either ion cyclotron resonance (FT-ICR) or Orbitrap (1, 31). While there are numerous types of commercially available instruments, not all mass spectrometers are intended to serve the same function with regards to glycan or glycoconjugate analysis (Table 1). The common features used to differentiate mass analyzer performance include: scan rate (or duty cycle), mass resolution, mass accuracy, and MS/MS capabilities. Shown throughout this book, MALDI-TOF platforms

Table 1
Characteristics of mass analyzers used for glycan and glycoconjugate analysis

Mass analyzer*	Compatible**	Ionization#	Resolution[x]	Spectral duty cycle (s)	Mass accuracy (ppm)[a]	Fragmentation modes
Ion trap	g, pep, LC	ESI, MALDI	1,000	0.02–0.2	100–250	CID[c], ETD, ECD
TOF, TOF/TOF	g, gp, pep	MALDI	15,000	<0.01	5.0–25.0	CID[c], ISD, PSD
Q-TOF	g, gp, pep, LC, IN	ESI, MALDI	40,000	<0.01	5.0–15.0	CAD[b]
Orbitrap	g, gp, pep, LC, IN	ESI	30–60,000	0.01–1.0	2.0–10.0	CID[c], HCD[b], ETD
FT-ICR	g, gp, pep, LC, IN	ESI, MALDI	60–100,000	0.1–1.0	0.5–5.0	CID[c], CAD[b], ECD, IRMPD

*Gas chromatography with mass spectrometry (GCMS), which is not shown, has long been used for monosaccharide composition analysis, antennary profile, and linkage analysis
**Glycan sequencing (g), peptide sequencing (pep), intact proteins (IN), glycan profiling (gp) online LC compatible (LC)
#Previously used with glycosylation analysis
[x]For typical acquisitions
[a]Typical with calibration
[b]In CAD and HCD higher energy collisions are induced by acceleration of ions into a collision chamber
[c]In CID collisions are induced by resonant excitation of ions in a collision chamber

are valuable for rapid glycan composition profiling, peptide and protein mass determination, and MS/MS peptide sequence determination from relatively simple mixtures. State-of-the-art MALDI-TOFs have scan rates that approach 1,000 Hz which enables throughputs of thousands of samples in a few minutes. ESI is commonly used for complex mixture analysis and large-scale LC/MS/MS experiments on ion-trap, Q-TOF and FTMS instruments. Ion trap, Q-TOF, and FTMS are sensitive with high spectral and MS/MS duty cycle, providing efficient analysis for glycan and peptide MS^n sequencing from complex samples. Orbitrap and FT-ICR have commanded significant attention because they provide the highest resolution and mass accuracy available, which greatly aids informatics searches against the large datasets that result from complex mixtures (31).

Tandem Mass Spectrometry (*MS/MS, MS^n*)—Commercial mass spectrometers support a variety of MS/MS approaches, such as collision-induced dissociation (CID) (44, 45), high-energy C-trap dissociation (HCD) (46), infrared multiphoton dissociation (IRMPD) (47), electron capture dissociation (ECD) (48), and electron transfer dissociation (ETD) (49). These techniques are classified as high-energy “threshold” approaches (e.g., CID, HCD, and IRMPD) and low-energy, “nonergodic” (e.g., ECD and ETD) dissociative methods that involve different fragmentation mechanisms and produce distinct types of fragment ions (Fig. 2).

Threshold dissociation occurs by thermal excitation of the polypeptide backbone either through collisions with gas (CID and HCD) or photons from a laser (IRMPD). Fragmentation of polypeptides predominately occurs at amide bonds forming "b" and "y" ions. When applied to glycans, threshold techniques typically result in glycosidic bond cleavages providing valuable information on monosaccharide sequence and branching. Cross-ring fragmentation, which provides information on glycan structure (e.g., high mannose, hybrid, or complex-structures) and antennary information, is also possible. Fragmentation pathways are dependent upon the type of ion formed (e.g., protonated or sodiated), if permethylation was performed, and the charge state (18). For glycoconjugates, activation of vibrational/electronic degrees of freedom of the biomolecule can lead to loss of labile modifications in the gas-phase (e.g., phosphorylation and sugar moieties) (50). Sugar loss prevents localization of the glycan position on the polypeptide. However, such losses can be advantageous because they lead to signature species in the mass spectrum that correspond to common sugar moieties, providing evidence that a biomolecule is glycosylated. Glycan loss also provides an avenue for MS^3 experiments, where the liberated glycan ion is subjected to a second fragmentation.

Dissociation by ECD/ETD is exothermic and is initiated when low energy electrons (<5 eV) cleave the backbone N–Cα bonds forming "c" and "z·" ions. Electrons are supplied in the vacuum system either directly from a cathode (ECD) or indirectly through aromatic radical ions generated in a chemical ionization source (ETD). Fragmentation occurs when the electron is captured by the polypeptide forming a radical cation that undergoes rapid rearrangement and cleaves the backbone before the energy can be dispersed throughout the biomolecule. As a result, secondary processes such as PTM-loss are minimized, which has made ECD/ETD methods of choice when characterizing biomolecules that contain PTMs (51, 52). Like the threshold methods, ECD/ETD is amenable to high-throughput characterization of peptides with online LC/MS acquisition events (39). When performed on glycopeptides, ECD/ETD will commonly fragment just the peptide backbone near the glycosylation site providing precise PTM localization (53). Since ECD/ETD fragmentation complements threshold methods, the approaches are often exploited in parallel to enhance glycan and peptide characterization directly from glycoconjugates (50, 53–55).

Proteomics Informatics: Several recent reviews highlight the specific details of the general software tools and repositories available for peptide and protein characterization from MS/MS datasets (56, 57). To recapitulate, for bottom-up proteomics, Mascot, Sequest, and X!Tandem are commonly used peptide MS/MS search tools (56, 58–62). For top-down, the web-based

ProSightPTM and its commercial analogue ProSightPC are the only widely available protein search engines (63–65); however, others such as BIG-MASCOT (66), and a precursor ion independent top-down algorithm (PIITA) are under development (67). The simplest way to identify a protein from bottom-up methods is to search the list of measured peptide masses against an in silico digest of proteins predicted by the genome. This protein identification method, coined "peptide mass fingerprinting," will reliably identify proteins in simple mixtures. Mass fingerprinting can be extended to complex mixtures if ultra-high mass accuracy is experimentally obtained (2 ppm), in which case peptides are "accurate mass tags" and provide increased selectivity for identification by minimizing the possible candidate matches in large databases (34, 68). Typically, for complex mixtures MS/MS improves the confidence of protein assignment. From MS/MS datasets, the masses of product ions can facilitate de novo sequencing of the original peptide (i.e., sequence tags), which can then be searched against the genome predicted database for matching proteins with the same consensus sequence (69, 70). Alternatively, theoretical MS/MS product ion spectra can be calculated for all of the peptides in the proteome database that match the precursor molecular weight. These theoretical spectra are then correlated to the experimental MS/MS spectra enabling protein identification with a high degree of confidence (71). Because of the size of proteomics datasets, search engines available today provide scoring mechanisms that estimate the probability of the match being random. For the most part, the search engines provide comparable results; however, validation of spectra is still required via decoy methods such as reverse database searches to estimate false discovery rates (72).

Most of the proteomics search engines are easily modified to search for MS datasets. These searches are usually user defined and can include forced modifications, which are applied universally to every instance the amino acid occurs in the database (e.g., alkylation of cysteines). Alternatively, variable modifications allow testing of datasets against both modified and unmodified forms of the peptide at the expense of increased search time. Variable modification searches are useful for identifying unknown or dynamic modification states, but in cases where many variable modifications are probed simultaneously, search specificity drops due to increased probability of spurious matching (56, 57). The complexity of glycan chemistry makes mining MS data with these traditional search engines challenging, necessitating glyco-specific tools to identify and annotate glycosylation MS datasets. In Part V of this book Aoki-Kinoshita (Chapter 17) and Yu et al. (Chapter 18) provide extensive details on the glycosylation specific databases and informatics tools available that are specifically dedicated to the analysis, structural characterization, and prediction of glycoproteins.

References

1. Han X, Aslanian A, Yates JR III (2008) Mass spectrometry for proteomics. Curr Opin Chem Biol 12:483–490
2. Varki A, Cummings RD, Esko JD, Freeze HH, Stanley P, Bertozzi CR, Hart GW, Etzler ME (2010) Essentials of glycobiology, 2nd edn. Cold Spring Harbor, New York
3. Apweiler R, Hermjakob H, Sharon N (1999) On the frequency of protein glycosylation, as deduced from analysis of the swiss-prot data-base. Biochim Biophys Acta 1473:4–8
4. Taniguchi N (2008) Human disease glycomics/proteome initiative (HGPI). Mol Cell Proteomics 7:626–627
5. Pan S et al (2011) Mass spectrometry based glycoproteomics—from a proteomics perspective. Mol Cell Proteomics 10:R110.003251
6. Zaia J (2010) Mass spectrometry and glycomics. OMICS 14:401–418
7. Hart JR et al (2011) Protein expression profiles of C3H 10T1/2 murine fibroblasts and of isogenic cells transformed by the H1047R mutant of phosphoinositide 3-kinase (PI3K). Cell Cycle 10:971–976
8. Sola RJ, Griebenow K (2010) Glycosylation of therapeutic proteins: an effective strategy to optimize efficacy. BioDrugs 24:9–21
9. Varki A et al (2009) Symbol nomenclature for glycan representation. Proteomics 9:5398–5399
10. Roepstorff P, Fohlman J (1984) Proposal for a common nomenclature for sequence ions in mass spectra of peptides. Biomed Mass Spectrom 11:601
11. Domon B, Costello C (1988) A systematic nomenclature for carbohydrate fragmentations in FAB-MS/MS spectra of glycoconjugates. Glycoconj J 5:379–409
12. Messner P (2004) Prokaryotic glycoproteins: unexplored but important. J Bacteriol 186:2517–2519
13. Jarrell KF et al (2010) S-layer glycoproteins and flagellins: reporters of archaeal post-translational modifications. Archaea 2010:612948. doi:10.1155/2010/612948
14. Rudd PM et al (1994) Glycoforms modify the dynamic stability and functional activity of an enzyme. Biochemistry 33:17–22
15. Shao L, Haltiwanger RS (2003) O-fucose modifications of epidermal growth factor-like repeats and thrombospondin type 1 repeats: unusual modifications in unusual places. Cell Mol Life Sci 60:241–250
16. Yoshida-Moriguchi T et al (2010) O-mannosyl phosphorylation of alpha-dystroglycan is required for laminin binding. Science 327:88–92
17. Rexach JE, Clark PM, Hsieh-Wilson LC (2008) Chemical approaches to understanding O-GlcNAc glycosylation in the brain. Nat Chem Biol 4:97–106
18. Morelle W, Michalski JC (2007) Analysis of protein glycosylation by mass spectrometry. Nat Protoc 2:1585–1602
19. McLafferty FW et al (1999) Top down versus bottom up protein characterization by tandem high-resolution mass spectrometry. J Am Chem Soc 121:806–812
20. Shinohara Y et al (2004) Direct N-glycan profiling in the presence of tryptic peptides on MALDI-TOF by controlled ion enhancement and suppression upon glycan-selective derivatization. Anal Chem 76:6989–6997
21. Guillard M et al (2011) Plasma N-glycan profiling by mass spectrometry for congenital disorders of glycosylation type II. Clin Chem 57:593–602
22. Wagner-Rousset E et al (2008) The way forward, enhanced characterization of therapeutic antibody glycosylation: comparison of three level mass spectrometry-based strategies. J Chromatogr B Analyt Technol Biomed Life Sci 872:23–37
23. Schreiber TB et al (2008) Quantitative phosphoproteomics—an emerging key technology in signal-transduction research. Proteomics 8:4416–4432
24. Schulze WX, Usadel B (2010) Quantitation in mass-spectrometry-based proteomics. Annu Rev Plant Biol 61:491–516
25. Blagoev B et al (2003) A proteomics strategy to elucidate functional protein-protein interactions applied to EGF signaling. Nat Biotechnol 21:315–8
26. Ross PL et al (2004) Multiplexed protein quantitation in *Saccharomyces cerevisiae* using amine-reactive isobaric tagging reagents. Mol Cell Proteomics 3:1154–69
27. Atwood JA III et al (2008) Quantitation by isobaric labeling: applications to glycomics. J Proteome Res 7:367–374
28. Orlando R et al (2009) IDAWG: metabolic incorporation of stable isotope labels for quantitative glycomics of cultured cells. J Proteome Res 8:3816–3823
29. Dallas DC et al (2011) N-linked glycan profiling of mature human milk by high-performance microfluidic chip liquid chromatography time-of-flight tandem mass spectrometry. J Agric Food Chem 59:4255–4263
30. Murray KK et al (2005) IUPAC standard definitions of terms relating to mass spectrometry. Abstr Pap Am Chem Soc 229:U123–U123

31. Kellie JF et al (2010) The emerging process of top down mass spectrometry for protein analysis: biomarkers, protein-therapeutics, and achieving high throughput. Mol Biosyst 6:1532–1539
32. Horn DM, Zubarev RA, McLafferty FW (2000) Automated reduction and interpretation of high resolution electrospray mass spectra of large molecules. J Am Soc Mass Spectrom 11:320–332
33. Zhang Z, Marshall AG (1998) A universal algorithm for fast and automated charge state deconvolution of electrospray mass-to-charge ratio spectra. J Am Soc Mass Spectrom 9:225–233
34. Conrads TP et al (2000) Utility of accurate mass tags for proteome-wide protein identification. Anal Chem 72:3349–3354
35. Fenn JB et al (1989) Electrospray ionization for mass-spectrometry of large biomolecules. Science 246:64–71
36. Karas M, Hillenkamp F (1988) Laser desorption ionization of proteins with molecular masses exceeding 10,000 daltons. Anal Chem 60:2299–2301
37. Tanaka K et al (1988) Protein and polymer analyses up to m/z 100 000 by laser ionization time-of-flight mass spectrometry. Rapid Commun Mass Spectrom 2:151–153
38. Zhou M, Veenstra, T (2008) Mass spectrometry: M/z 1983-2008. Biotechniques 44:667–668, 670
39. Swaney DL, McAlister GC, Coon JJ (2008) Decision tree-driven tandem mass spectrometry for shotgun proteomics. Nat Methods 5:959–964
40. Patrie SM et al (2006) Top down mass spectrometry of <60-kDa proteins from *Methanosarcina acetivorans* using quadrupole FRMS with automated octopole collisionally activated dissociation. Mol Cell Proteomics 5:14–25
41. Patrie SM et al (2004) Construction of a hybrid quadrupole/fourier transform ion cyclotron resonance mass spectrometer for versatile ms/ms above 10 kDa. J Am Soc Mass Spectrom 15:1099–1108
42. Niessen WM (1998) Advances in instrumentation in liquid chromatography-mass spectrometry and related liquid-introduction techniques. J Chromatogr A 794:407–435
43. MacNair JE, Patel KD, Jorgenson JW (1999) Ultrahigh-pressure reversed-phase capillary liquid chromatography: isocratic and gradient elution using columns packed with 1.0-micron particles. Anal Chem 71:700–708
44. Wells JM, McLuckey SA (2005) Collision-induced dissociation (CID) of peptides and proteins. Methods Enzymol 402:148–185
45. Bean MF et al (1991) Tandem mass spectrometry of peptides using hybrid and four-sector instruments: a comparative study. Anal Chem 63:1473–1481
46. Olsen JV et al (2007) Higher-energy C-trap dissociation for peptide modification analysis. Nat Methods 4:709–712
47. Little DP et al (1994) Infrared multiphoton dissociation of large multiply charged ions for biomolecule sequencing. Anal Chem 66:2809–2815
48. McLafferty FW et al (2001) Electron capture dissociation of gaseous multiply charged ions by Fourier-transform ion cyclotron resonance. J Am Soc Mass Spectrom 12:245–249
49. Mikesh LM et al (2006) The utility of ETD mass spectrometry in proteomic analysis. Biochim Biophys Acta 1764:1811–1822
50. Hakansson K et al (2003) Combined electron capture and infrared multiphoton dissociation for multistage MS/MS in a Fourier transform ion cyclotron resonance mass spectrometer. Anal Chem 75:3256–3262
51. Wuhrer M et al (2007) Glycoproteomics based on tandem mass spectrometry of glycopeptides. J Chromatogr B Analyt Technol Biomed Life Sci 849:115–128
52. Boersema PJ, Mohammed S, Heck AJ (2009) Phosphopeptide fragmentation and analysis by mass spectrometry. J Mass Spectrom 44:861–878
53. Hakansson K et al (2001) Electron capture dissociation and infrared multiphoton dissociation MS/MS of an N-glycosylated tryptic peptic to yield complementary sequence information. Anal Chem 73:4530–4536
54. Catalina MI et al (2007) Electron transfer dissociation of N-glycopeptides: loss of the entire N-glycosylated asparagine side chain. Rapid Commun Mass Spectrom 21:1053–1061
55. Wu SL et al (2007) On-line LC-MS approach combining collision-induced dissociation (CID), electron-transfer dissociation (ETD), and CID of an isolated charge-reduced species for the trace-level characterization of proteins with post-translational modifications. J Proteome Res 6:4230–4244
56. Deutsch EW, Lam H, Aebersold R (2008) Data analysis and bioinformatics tools for tandem mass spectrometry in proteomics. Physiol Genomics 33:18–25
57. Fenyo D, Beavis RC (2008) Informatics development: challenges and solutions for MALDI mass spectrometry. Mass Spectrom Rev 27:1–19
58. Lisacek F (2006) Web-based MS/MS data analysis. Proteomics 6(suppl 2):22–32

59. Eng JK et al (2008) A fast SEQUEST cross correlation algorithm. J Proteome Res 7:4598–4602
60. Kumar C, Mann M (2009) Bioinformatics analysis of mass spectrometry-based proteomics data sets. FEBS Lett 583:1703–1712
61. Perkins DN et al (1999) Probability-based protein identification by searching sequence databases using mass spectrometry data. Electrophoresis 20:3551–3567
62. Ducret A et al (1998) High throughput protein characterization by automated reverse-phase chromatography/electrospray tandem mass spectrometry. Protein Sci 7:706–719
63. Zamdborg L et al (2007) Prosight PTM 2.0: improved protein identification and characterization for top down mass spectrometry. Nucleic Acids Res 35:W701–W706
64. LeDuc RD et al (2004) Prosight PTM: an integrated environment for protein identification and characterization by top-down mass spectrometry. Nucleic Acids Res 32:W340–W345
65. Taylor GK et al (2003) Web and database software for identification of intact proteins using "top down" mass spectrometry. Anal Chem 75:4081–4086
66. Karabacak NM et al (2009) Sensitive and specific identification of wild type and variant proteins from 8 to 669 kDa using top-down mass spectrometry. Mol Cell Proteomics 8:846–856
67. Tsai YS et al (2009) Precursor ion independent algorithm for top-down shotgun proteomics. J Am Soc Mass Spectrom 20:2154–2166
68. Smith RD et al (2002) An accurate mass tag strategy for quantitative and high-throughput proteome measurements. Proteomics 2:513–523
69. Frank A et al (2005) Peptide sequence tags for fast database search in mass-spectrometry. J Proteome Res 4:1287–1295
70. Sheng QH, Xie T, Ding DF (2000) De novo interpretation of MS/MS spectra and protein identification via database searching. Sheng Wu Hua Xue Yu Sheng Wu Wu Li Xue Bao (Shanghai) 32:595–600
71. Domon B, Aebersold R (2006) Mass spectrometry and protein analysis. Science 312:212–217
72. Peng J et al (2003) Evaluation of multidimensional chromatography coupled with tandem mass spectrometry (LC/LC-MS/MS) for large-scale protein analysis: the yeast proteome. J Proteome Res 2:43–50

Part I

Enrichment and Isolation Methods

Chapter 2

Tandem Lectin Weak Affinity Chromatography for Glycoprotein Enrichment

Zhi Yuan Ma, Yuliya Skorobogatko, and Keith Vosseller

Abstract

In this chapter we describe the application of lectin weak affinity chromatography (LWAC) for the enrichment of peptides modified by O-linked β-*N*-acetylglucosamine (O-GlcNAc). O-GlcNAc is a single carbohydrate moiety post-translational modification of intracellular proteins. The stoichiometry of the modification is low and identification of the sites of O-GlcNAc attachment is challenging. To map O-GlcNAc sites we use the approach where a protein sample of interest is digested with trypsin and subjected to LWAC, which employs weak interaction between lectin wheat germ agglutinin and O-GlcNAc. Obtained sample is enriched with O-GlcNAc-modified peptides, which can be identified by means of mass spectrometry.

Key words: O-linked β-*N*-acetylglucosamine (O-GlcNAc) modification, Lectin weak affinity chromatography, Site mapping, Enrichment, Post-translational modification

1. Introduction

O-linked β-*N*-acetylglucosamine (O-GlcNAc) is a reversible modification of cytosolic and nuclear proteins. A single GlcNAc moiety is attached to serine and threonine residues by O-GlcNAc transferase and removed by O-GlcNAcase (1, 2). O-GlcNAc has recently emerged as a regulatory modification that is important for a variety of cellular processes including protein trafficking and turnover, cell cycle, gene expression, cellular stress response, and signal transduction (3). Defects in O-GlcNAcation have been implicated in diabetes mellitus type II, Alzheimer's disease, and heart failure (3, 4). Significant efforts that are currently being invested into the understanding of function of O-GlcNAc modification create a demand for the mapping of O-GlcNAc sites on proteins.

Jennifer J. Kohler and Steven M. Patrie (eds.), *Mass Spectrometry of Glycoproteins: Methods and Protocols*, Methods in Molecular Biology, vol. 951, DOI 10.1007/978-1-62703-146-2_2, © Springer Science+Business Media, LLC 2013

Sites of O-GlcNAc attachment are mapped on peptides using mass spectrometry. Due to the low stoichiometry of the modification and its lability it is a challenging task. Several approaches that have been developed to overcome those issues focused mostly on the chemical modification of O-GlcNAc to facilitate identification of O-GlcNAc sites during mass spectrometry (MS) and sometimes to enrich for the O-GlcNAc-modified peptides. Examples of these strategies are tagging with [^{3}H] galactose (5, 6) or a biotin tag for high affinity streptavidin chromatography (7, 8), and beta-elimination/Michael addition approaches (9, 10). However the development of novel peptide fragmentation approaches in mass spectrometry (e.g., electron transfer dissociation (ETD)) significantly improved the identification of O-GlcNAc sites and allowed more focused efforts on the enrichment of native O-GlcNAc-modified peptides (11). To enrich for O-GlcNAc-modified peptides, we employed the interaction of O-GlcNAc with the lectin wheat germ agglutinin (WGA).

Lectins are carbohydrate-binding proteins which possess high specificity for their substrates. WGA binds complex carbohydrates that contain O-GlcNAc and sialic acid residues with high affinity (12). The high affinity interaction is achieved due to the presence of four carbohydrate-binding sites on the WGA dimer which presents multiple sites for interactions between WGA and the glycoprotein (13). However, WGA interaction with a single O-GlcNAc is not strong enough to enrich for O-GlcNAc-modified peptides using conventional techniques for affinity chromatography that include washing steps (14, 15). We were able to successfully exploit the WGA/O-GlcNAc interaction by performing isocratic chromatography, where a mixture of peptides is applied at a low flow rate, over a relatively long column packed with WGA coupled to agarose beads in a high salt buffer (16). The majority of peptides elute as a single peak followed by a tail which is strongly enriched for O-GlcNAc-modified peptides. This approach was named lectin weak affinity chromatography (LWAC). LWAC was successfully applied in tandem with mass spectrometry techniques such as collision-induced dissociation (CID) and ETD to map O-GlcNAc sites in subcellular fractions (11, 16, 17). The advantages of the technique are high specificity and absence of false positive identifications due to no prior chemical or enzymatic treatment to facilitate the mapping.

In this chapter we describe how to perform LWAC, starting with a complex mixture of proteins and finishing with a fraction that is highly enriched for O-GlcNAc-modified peptides and is ready for mass spectrometry analyses. We discuss in detail the requirements for the starting material, processing of the sample, packing of the WGA column, and the actual chromatography. Sample preparation prior to LWAC involves tryptic digestion to generate peptides and buffer exchange. Then sample is loaded on the WGA column and isocratically eluted. Collected fractions should be desalted and then they are ready for mass spectrometry. Subsequent mass spectrometry for identifying the O-GlcNAcated residues is discussed in Chapter 15.

2. Materials

2.1. Sample Preparation

Prepare all solutions using deionized water.

Optional:

1. *N*-acetyl-D-glucosamine (Sigma-Aldrich, Inc., St. Louis, MO, #A3286).
2. Streptozotocin, STZ (Sigma-Aldrich, Inc., St. Louis, MO, #S0130).
3. PUGNAc (Sigma-Aldrich, Inc., St. Louis, MO, #A7229).
4. Anti-O-GlcNAc antibody, monoclonal, clone 110.6 (Covance Research Products, Inc., Denver, PA, #MMS-248R).
5. Bovine-alpha crystallin (Sigma-Aldrich, Inc., St. Louis, MO, #L2394).

2.1.1. Trypsin Digestion

1. 100 mM ammonium bicarbonate buffer, pH 8.0. Adjust pH with 1 N NaOH.
2. 0.5 M dithiothreitol (DTT) stock solution in 100 mM ammonium bicarbonate buffer. Store at −20°C.
3. 0.5 M iodoacetamide (IAA) solution in 100 mM ammonium bicarbonate buffer. IAA is photosensitive. Use aluminum foil to protect the solution from light. Prepare fresh IAA solution every time.
4. Sequencing grade modified trypsin (Promega Corporation, Madison, WI, #V511A). Store at −70°C. It may be stored in solution for several weeks at −20°C.

2.1.2. Sample Desalting

1. MacroSpin column (capacity 300 μg) (The Nest Group, Inc., Southborough, MA, #SMM SS18V).
2. 100% acetonitrile (Sigma-Aldrich, Inc., St. Louis, MO, #675415).
3. Washing buffer: 0.1% formic acid in water.
4. Elution buffer: 80% acetonitrile, 25 mM formic acid.

2.2. Long WGA Column Packing

2.2.1. Equilibration of WGA Agarose

1. WGA coupled to agarose, 10 mL (agarose WGA, Vector Laboratories, Inc., Burlingame, CA, #AL-1023).
2. WGA buffer: 25 mM Tris–HCl pH 7.8, 300 mM NaCl, 5 mM $CaCl_2$, 1 mM $MgCl_2$.

3. 40 mL empty glass column for gravity flow to wash WGA agarose slurry.
4. 20 mL empty glass column for chromatography with filter removed. This column will serve as a reservoir for WGA agarose slurry during WGA column packing.

2.2.2. Preparation of the Teflon Tubing for Packing

1. Tubing, teflon PFA, 0.04″ inner diameter, 1/16″ outer diameter, 50′ long, natural color (IDEX Corporation, Northbrook, IL, #1507L). Cut it to produce 39′ long tubing with volume a little less than 10 mL.
2. Two polyetheretherketone (PEEK) unions with fittings (IDEX Corporation, Northbrook, IL, #P-760).
3. Two 0.5 μm frits with PEEK ring (IDEX Corporation, Northbrook, IL, #A-709).

2.3. WGA Isocratic HPLC

1. AKTA Purifier HPLC system (GEHC, Waukesha, WI).
2. 96-well collection plates, 2 mL deep (Greiner Bio-One North America, Inc., Monroe, NC, 2 mL MASTERBLOCK®).

2.4. Sample Clean-Up Using Zip-Tips C18 for MS Analysis

1. ZipTip$_{C18}$, Tip size: P10 (Millipore, Billerica, MA, #ZTC18S096).
2. Reagent A: 0.1% formic acid.
3. Reagent B: 70% acetonitrile, 0.1% formic acid.

2.5. Reuse and Storage of WGA Column

1. WGA column regeneration buffer: 1 M NaCl pH 3, adjust pH with acetic acid.

3. Methods

We were able to successfully enrich for O-GlcNAcated peptides starting with 100–400 μg of complex protein mixture, the amount which is dictated by the properties of the WGA column. Materials we list in the protocol, for example desalting columns, are suited for processing of starting protein amounts in this range. Protein sample should be completely solubilized in your buffer of choice (see Note 1). In our experiments we start with 300 μg of protein solubilized in 100 μL of 6 M Urea Buffer (6 M Urea, 200 mM Tris–HCl, 5 mM EDTA, pH 7.5) (see Note 2).

3.1. Sample Preparation (see Note 2)

3.1.1. Tryptic Digestion

1. *Adjust pH*: Dilute sample six-fold with 100 mM ammonium bicarbonate buffer. Check pH by applying 2 μL of the sample on pH paper. pH should be ≥7.8 (see Note 3).
2. *Reduce cysteines*: Add 0.5 M stock solution of DTT to the sample to final concentration of DTT 10 mM. Vortex. Incubate the sample at 60°C for 60 min.

3. Remove the sample from an incubator and let it equilibrate to room temperature.
4. *Block cysteines by alkylation*: Add 0.5 M stock solution of IAA to the sample to final concentration of IAA 50 mM. Vortex. Incubate sample for 60 min at room temperature in darkness.
5. *Quenching*: Remove the sample from darkness. Add 0.5 M DTT stock solution to the final concentration of DTT 10 mM. Incubate the sample at room temperature for 45 min.
6. *Digestion*: Add trypsin to the sample. Trypsin: sample ratio should be 1:40 (w/w) (see Note 4). Check pH by applying 2 μL of the sample on pH paper. pH should be ≥7.8 (see Note 5). Incubate at 37°C overnight.
7. Sample is ready for desalting.

3.1.2. Sample Desalting (Protocol According to the Manufacturer)

All the centrifugation steps should be performed for 1 min at about 110 × *g*, which equals approximately 800 rpm when using Eppendorf microcentrifuge. We handle digested sample at room temperature.

1. *Acidify the sample*: Add formic acid to final concentration 0.2%.
2. *Assembly*: Remove an end restriction and a cap and place the column in a 2 mL microcentrifuge tube.
3. *Conditioning*: Add 500 μL of 100% acetonitrile on the column. Centrifuge. Discard the flow through.
4. *Washing*: Add 500 μL of washing buffer to the column. Centrifuge. Discard the flow through.
5. Remove the collecting tube and clean any moisture on the exterior of the column with Kimwipe. Place the column in a new 2 mL microcentrifuge tube.
6. *Sample application*: Place the sample (maximum 500 μL) on the column. Centrifuge. Discard the flow through. Apply the rest of the sample on the column. Centrifuge. Discard the flow through.
7. *Washing*: Add 250 μL of washing buffer on the column. Centrifuge. Place the column into new 2 mL microcentrifuge tube.
8. *Elution*: Add 250 μL of elution buffer to the column. Centrifuge. Collect the flow through. Repeat this step once twice.
9. Dry the sample completely in a speed vacuum concentrator (see Note 6 and 7).
10. Dissolve dried sample in 80 μL of WGA buffer (see Note 8).

3.2. Long WGA Column Packing (see Note 9)

3.2.1. Equilibration of WGA Agarose

1. Pour WGA agarose slurry (20 mL) into an empty 40 mL glass column for gravity flow (see Note 9).
2. Wash ten times with 20 mL of WGA buffer under gravity flow (see Note 10).

3. Transfer equilibrated WGA agarose into an empty 20 mL glass column for chromatography (where a filter should be substituted by an end restriction) by rinsing it out with fresh WGA buffer.

3.2.2. Preparation of the Teflon Tubing for Packing

1. Connect the teflon tubing through the adaptors to HPLC.
2. Wash with 70% ethanol: 1 mL/min, 30 mL (see Note 11).
3. Wash with WGA buffer: 1 mL/min, 30 mL.
4. Attach one end to the glass column with WGA slurry.
5. Place a PEEK union with a 0.5 μm frit on the other end of the tubing to permit the flow of WGA buffer and to restrict the passage of WGA agarose particles out of the end of the column.

3.2.3. Packing

1. Connect the glass column with WGA slurry to HPLC.
2. Start WGA buffer flow (0.15 mL/min). The slurry is caused to flow under pressure into the teflon tubing (see Note 12).
3. When packing is done, dismount the teflon tubing from the HPLC and attach the second PEEK union containing 0.5 μm frit to the released tubing end to create a WGA column (see Note 13).

3.3. WGA Isocratic HPLC (See Note 13)

1. After WGA column is attached to HPLC load the sample dissolved in WGA buffer into the 100 μL injection loop (see Note 14).
2. Trigger the injection by a predefined method. Run isocratic WGA buffer at a flow rate of 0.15 mL/min. Monitor pressure and UV (see Note 15 and 16).
3. Collect 1-min fractions using a 96-well collection plate.
4. Pool fractions together, starting from the right shoulder of the major peak, thus creating three enriched for O-GlcNAcated peptides fractions to work with (Fig. 1) (see Note 17).

3.4. Sample Clean-Up Using Zip-Tip C18 for MS Analysis

1. *Acidify sample*: Add formic acid to combined fractions to final concentration 0.2%. Check pH by applying 2 μL of the sample on pH paper. pH should be <4.
2. Rinse Zip-Tip three times by pipetting up and down 10 μL of reagent B.
3. Rinse Zip-Tip ten times by pipetting up and down 10 μL of reagent A.
4. Load sample into tip by repetitive pipetting (ten times) (see Note 18).
5. Rinse Zip-Tip ten times by pipetting up and down 10 μL of reagent A.
6. Elute with 10 μL of reagent B by repetitive pipetting ten times.

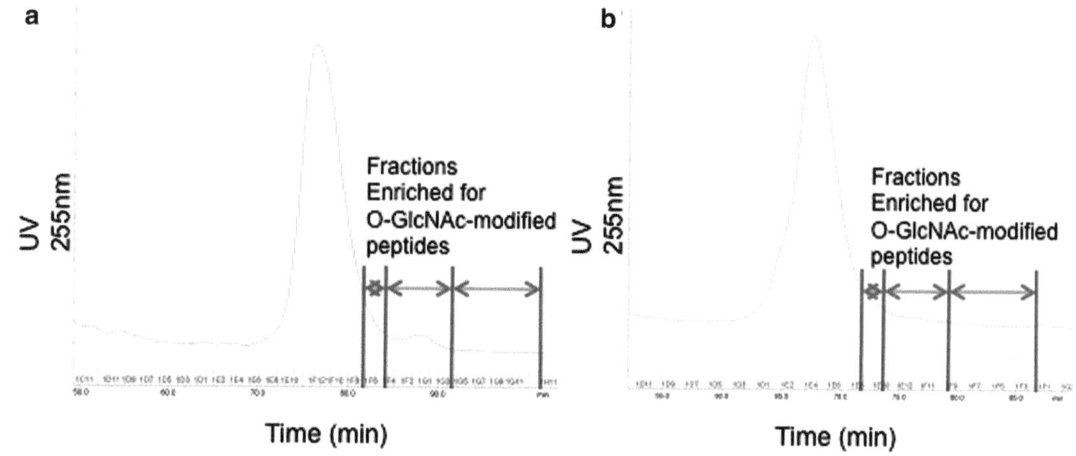

Fig. 1. UV255 trace of successful lectin weak affinity chromatography (LWAC) runs. Peptides, not modified by O-linked β-*N*-acetylglucosamine (O-GlcNAc) elute as a single peak. (**a**) An additional peak roughly corresponding to O-GlcNAc-enriched fraction is observed. (**b**) No additional peak is observed; however, O-GlcNAc-modified peptides still may be identified in indicated fractions.

7. Repeat step 6 using 5 μL of reagent B. Combine eluted fractions.
8. Dry down eluted fractions in a speed vacuum concentrator (see Note 19).

3.5. Reuse and Storage of WGA Column

1. Fit flow stoppers on both ends of the column and store it at 4°C.
2. Reverse if used next time (see Note 15).
3. In cases where chromatographic effectiveness appears to decrease, regenerate the column with WGA column regeneration buffer at flow rate 0.15 mL/min (see Note 20). Then wash the column with 30 mL of WGA buffer at flow rate 0.15 mL/min (see Note 21).

4. Notes

1. In order to maintain enzymatic activity of trypsin, the concentration of detergents (i.e., SDS) and other chaotropic agents (i.e., urea) in the lysis buffer need to be at a compatible concentration. Before trypsinization the sample will be diluted at least six times. In this diluted sample the concentration of urea should be less than 1 M; Triton X-100, NP-40, CHAPS, tween-20 should comprise less than 1%, and SDS less than 0.05%.
2. To obtain your starting material, cellular fractionation is highly recommended. First, extracellular proteins should be removed,

because they carry complex carbohydrates that will contaminate the column. Second, fractionation will reduce the complexity of the sample and will also allow focusing on a particular functional group of proteins. For example, we used cellular fractionation techniques to obtain synaptosomes and postsynaptic density preparations. If you are working with cell culture or rodent animal models, O-GlcNAc levels can be increased by the treatment with *N*-acetyl-D-glucosamine (2–6 h, 7.5 mM in 0.5 mM HEPES, pH 7.5) or O-GlcNAcase inhibitors STZ (1 h, 5 mM) and PUGNAc (6 h, 0.1 mM) (18, 19). In parentheses, the conditions used to treat cultured cells are indicated. To confirm that the protein fraction of interest contains proteins modified by O-GlcNAc, or that the treatment with an O-GlcNAcase inhibitor was successful, sample can be subjected to western blotting with an anti-O-GlcNAc antibody.

3. If pH needs to be further adjusted, further dilute the sample with 100 mM ammonium bicarbonate buffer.
4. Higher protein concentrations in lower sample volume work better for digestion. The ratio of trypsin to sample should be between 1:20 and 1:50 (w/w), i.e., 1 mg of trypsin for every 20–50 mg of protein. We recommend starting with a ratio of 1:40.
5. It is very important to adjust pH ≥7.8 for tryptic digestion.
6. The purpose of this step is to remove acetonitrile and formic acid from the sample. These reagents are more volatile than aqueous components. That is why the sample does not have to be dried completely. Sample volume can be reduced to 10% of the elution volume or sample can be dried completely.
7. Optional: If mass spectrometry is readily available, we recommend checking whether the sample was fully digested by trypsin. After elution make an additional aliquot containing approximately 2 μg of protein, dry it down in speed vacuum concentrator, reconstitute in 0.1% formic acid, and analyze by LC-MS/MS.
8. Make sure that the sample is fully dissolved in WGA buffer. If sample is not fully dissolved, add an additional amount of WGA buffer, do not dilute the sample too much though, because you will be able to load only about 100 μL in a single run. If sample is still not fully dissolved, spin it down and load supernatant on the column.
9. Packing may take up to 4 h, especially when done for the first time. We recommend that the column is prepared ahead of time and stored at 4°C.
10. WGA agarose equilibration may take up to 3 h.
11. Washing with ethanol greatly facilitates column packing.

12. During packing, follow the leading edge of WGA agarose. If agarose is getting stuck in the column, flick the column until the leading edge is freely moving; otherwise, WGA agarose will compress and it will be very difficult to break it apart and force it to move towards the end of the column. It is also possible that agarose slurry will clog the glass column. In this case, stop the flow, and mix the slurry by inverting the glass column up and down several times. Then restart the flow and get back to watching the leading edge of WGA agarose (see Note 13).
13. Optional control: tryptic digest of bovine-alpha crystallin may be used as an optional control for the ability of the prepared column to enrich for O-GlcNAc-modified peptides. Successful run will produce fractions enriched for two peptides, which can be identified by mass spectrometry (16) (Fig. 2.2).
14. If there are bubbles in the sample, spin the sample in a table rotor. Load the sample into the syringe, pull the plunger a little back, move the needle upward, tap the syringe to remove air bubble, and fill the needle with sample by pushing the plunger.
15. During WGA isocratic chromatography, the pressure of the HPLC system may rise slowly due to the compression of agarose resin. The pressure should be never allowed to exceed 5 Mpa (50 bar). One way to work around is to reduce the flow rate to 0.1 mL/min. If multiple sequential runs are being performed, invert the column after each run to release the compression of the preceding run.
16. UV wavelengths 215, 255, or 280 nm can be used. The tail of the major peak will be expected to contain O-GlcNAc-modified peptides. When UV255 or UV215 is used, an additional peak eluting after the major peak sometimes appears. This peak somewhat correlates with the presence of O-GlcNAc-modified peptides, although we find O-GlcNAcated peptides in fractions preceding that peak, and do not observe O-GlcNAcated peptides in the later eluting fractions of this peak.
17. Sometimes a small WGA-enriched O-GlcNAc-modified peak can be observed; sometimes it cannot be seen (Fig. 1). Proceed with the protocol regardless.
18. We usually overlay 1 mL pipette tip on top of 200 μL pipette tip. To do so you need to cut the tip of 1 mL tip with a razor blade. The Zip-Tip is mounted on top of the 200 μL pipette tip. We can handle large volume of the sample by that way.
19. The sample can be stored at −80°C. For mass spectrometry, dissolve it in 0.1% formic acid.
20. The stringent wash buffer is used to remove nonspecifically bound material and bound, if any, complex carbohydrates that may interfere with chromatography.

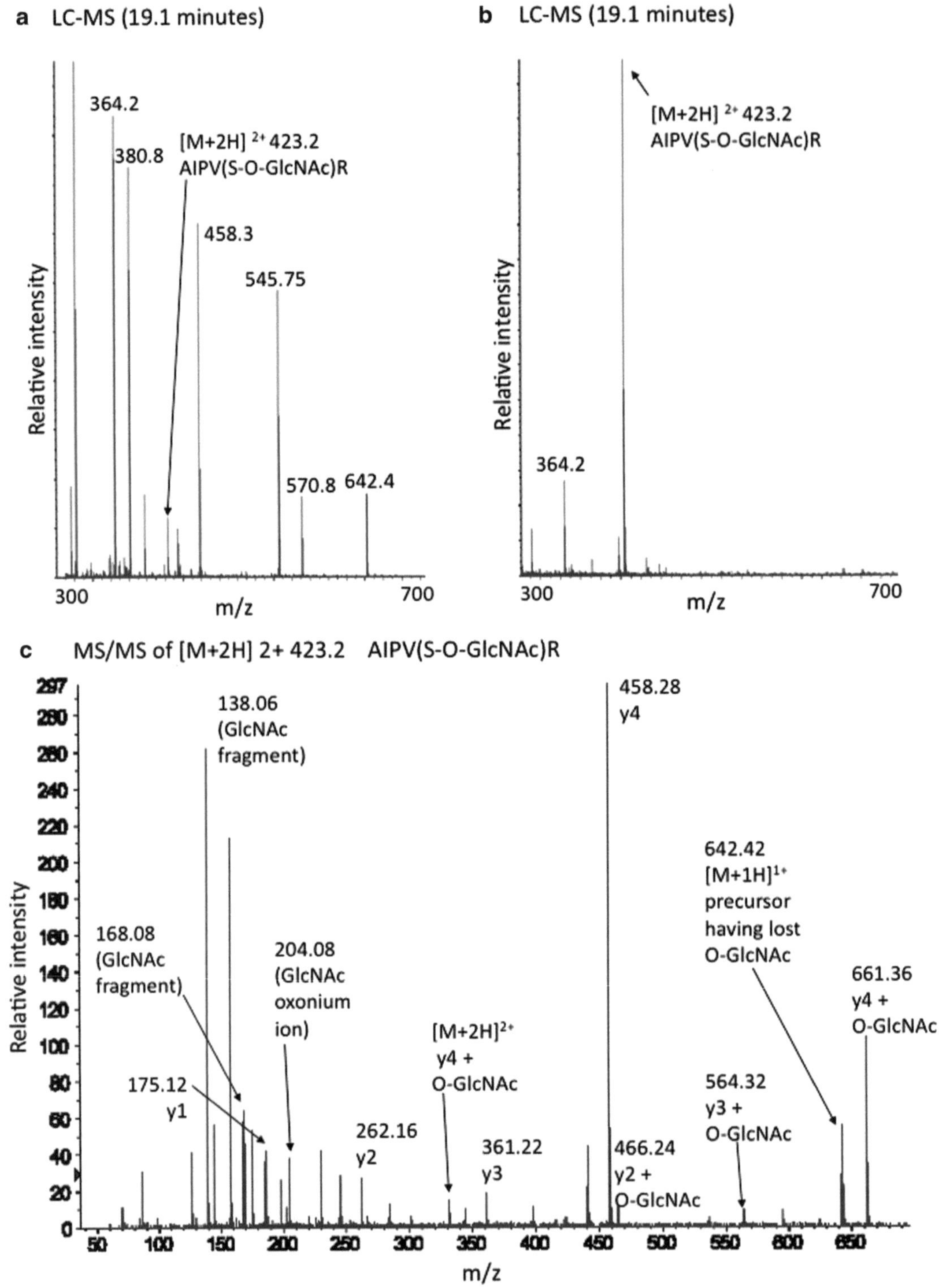

Fig. 2. LWAC enrichment and identification of the O-GlcNAc-modified tryptic peptide AIPV(S-O-GlcNAc)R from bovine alpha-crystallin. LC-MS analysis of the tryptic digest of bovine crystallin before (**a**) and after (**b**) LWAC at 19.1 min. The peptide noted as $[M+2H]^{2+}$ at *m/z* 423.2 is enriched in (**b**, **c**) MS/MS fragmentation spectra of the peptide. (Figures are reproduced with permission from ref. (16)).

21. Regeneration is not recommended as a regular maintenance procedure, because it will lead to further compression of agarose in the column and will decrease the number of available runs. Usually the column can be used up to ten times.

References

1. Hart GW (1997) Dynamic O-linked glycosylation of nuclear and cytoskeletal proteins. Annu Rev Biochem 66:315–335
2. Wells L, Vosseller K, Hart GW (2001) Glycosylation of nucleocytoplasmic proteins: signal transduction and O-GlcNAc. Science 291:2376–2378
3. Hart GW, Housley MP, Slawson C (2007) Cycling of O-linked beta-N-acetylglucosamine on nucleocytoplasmic proteins. Nature 446:1017–1022
4. Watson LJ, Facundo HT, Ngoh GA, Ameen M, Brainard RE, Lemma KM, Long BW, Prabhu SD, Xuan YT, Jones SP (2010) O-linked β-N-acetylglucosamine transferase is indispensable in the failing heart. Proc Natl Acad Sci USA 107:17797–17802
5. Cole RN, Hart GW (1999) Glycosylation sites flank phosphorylation sites on synapsin I: O-linked N-acetylglucosamine residues are localized within domains mediating synapsin I interactions. J Neurochem 73:418–428
6. Greis KD, Gibson W, Hart GW (1994) Site-specific glycosylation of the human cytomegalovirus tegument basic phosphoprotein (UL32) at serine 921 and serine 952. J Virol 68:8339–8349
7. Khidekel N, Ficarro SB, Peters EC, Hsieh-Wilson LC (2004) Exploring the O-GlcNAc proteome: direct identification of O-GlcNAc-modified proteins from the brain. Proc Natl Acad Sci USA 101:13132–13137
8. Nandi A, Sprung R, Barma DK, Zhao Y, Kim SC, Falck JR (2006) Global identification of O-GlcNAc-modified proteins. Anal Chem 78:452–458
9. Vosseller K, Hansen KC, Chalkley RJ, Trinidad JC, Wells L, Hart GW, Burlingame AL (2005) Quantitative analysis of both protein expression and serine/threonine post-translational modifications through stable isotope labeling with dithiothreitol. Proteomics 5:388–398
10. Wells L, Vosseller K, Cole RN, Cronshaw JM, Matunis MJ, Hart GW (2002) Mapping sites of O-GlcNAc modification using affinity tags for serine and threonine post-translational modifications. Mol Cell Proteomics 1:791–804
11. Chalkley RJ, Thalhammer A, Schoepfer R, Burlingame AL (2009) Identification of protein O-GlcNAcylation sites using electron transfer dissociation mass spectrometry on native peptides. Proc Natl Acad Sci USA 106:8894–8899
12. Bhavanandan VP, Katlic AW (1979) The interaction of wheat germ agglutinin with sialoglycoproteins. The role of sialic acid. J Biol Chem 254:4000–4008
13. Gallagher JT, Morris A, Dexter TM (1985) Identification of two binding sites for wheat-germ agglutinin on polylactosamine-type oligosaccharides. Biochem J 231:115–122
14. Neumann D, Kohlbacher O, Lenhof HP, Lehr CM (2002) Lectin-sugar interaction. Calculated versus experimental binding energies. Eur J Biochem 269:1518–1524
15. Ohlson S, Bergstrom M, Leickt L, Zopf D (1998) Weak affinity chromatography of small saccharides with immobilised wheat germ agglutinin and its application to monitoring of carbohydrate transferase activity. Bioseparation 7:101–105
16. Vosseller K, Trinidad JC, Chalkley RJ, Specht CG, Thalhammer A, Lynn AJ, Snedecor JO, Guan S, Medzihradszky KF, Maltby DA, Schoepfer R, Burlingame AL (2006) O-linked N-acetylglucosamine proteomics of postsynaptic density preparations using lectin weak affinity chromatography and mass spectrometry. Mol Cell Proteomics 5:923–934
17. Skorobogatko YV, Deuso J, Adolf-Bergfoyle J, Nowak MG, Gong Y, Lippa CF, Vosseller K (2010) Human Alzheimer's disease synaptic O-GlcNAc site mapping and iTRAQ expression proteomics with ion trap mass spectrometry. Amino Acids 40:765–779
18. Dauphinee SM, Ma M, Too CK (2005) Role of O-linked beta-N-acetylglucosamine modification in the subcellular distribution of alpha4 phosphoprotein and Sp1 in rat lymphoma cells. J Cell Biochem 96:579–588
19. Kwak TK, Kim H, Jung O, Lee SA, Kang M, Kim HJ, Park JM, Kim SH, Lee JW (2010) Glucosamine treatment-mediated O-GlcNAc modification of paxillin depends on adhesion state of rat insulinoma INS-1 cells. J Biol Chem 285:36021–36031

Chapter 3

CSC Technology: Selective Labeling of Glycoproteins by Mild Oxidation to Phenotype Cells

Andreas Hofmann, Damaris Bausch-Fluck, and Bernd Wollscheid

Abstract

Cell surface glycoproteins represent important markers for the phenotyping of healthy and malignantly transformed cells. The mass spectrometry-based cell surface capturing (CSC) technology allows for extensive multiplexed identification and relative quantification of glycoproteins expressed on the cell surface at a given point in time. CSC technology is based on the mild oxidation of glycans from cell surface proteins on living cells. Oxidized glycans are tagged with a bifunctional linker molecule and glycopeptides are subsequently enriched by affinity chromatography. Here, we describe a step-by-step protocol of the CSC technology, which not only enables the identification of cell surface glycoproteins, but also the concurrent determination of protein N-glycosylation sites.

Key words: Cell surface capturing (CSC) technology, Cell surface glycoproteins, Meta-periodate oxidation, Bifunctional linker

1. Introduction

The molecular composition of the plasma membrane determines how a cell can interact with its environment. Cell surface proteins carry out essential cellular functions and show often a cell type-specific expression pattern. Therefore, cell surface proteins are often used for the immunophenotyping of cells. However, immunophenotyping of cells is limited to the number of currently available antibodies and usually only a small number of proteins can concurrently be analyzed with antibody-based technologies, such as flow cytometry or immunohistochemistry. In contrast to antibody-based technologies, the cell surface capturing (CSC) technology allows for discovery-driven identification and quantification of hundreds of expressed cell surface glycoproteins

Jennifer J. Kohler and Steven M. Patrie (eds.), *Mass Spectrometry of Glycoproteins: Methods and Protocols*, Methods in Molecular Biology, vol. 951, DOI 10.1007/978-1-62703-146-2_3, © Springer Science+Business Media, LLC 2013

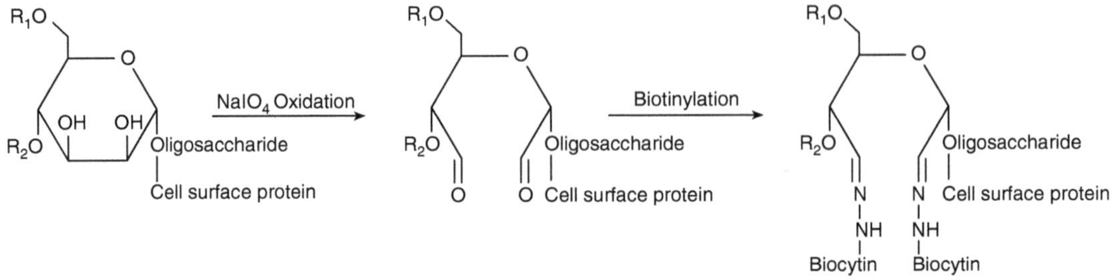

Fig. 1. Schematic overview of the biotinylation of cell surface glycoproteins. Cis-diols of glycans of cell surface proteins are oxidized by $NaIO_4$ to form reactive aldehyde groups. Then, biocytin hydrazide is added to form covalent hydrazone bonds with the aldehyde groups.

by mass spectrometry (MS) in different cell types, including the identification of new or unanticipated cell surface proteins (1).

The CSC technology is based on the mild oxidation of glycans with $NaIO_4$ to generate reactive aldehyde groups from cis-diols (Fig. 1). Then, a bifunctional linker molecule, biocytin hydrazide, is used to form hydrazone bonds with the generated aldehyde groups and to enrich glycopeptides by streptavidin-based affinity chromatography. The covalent hydrazone bonds allow for stringent washing conditions to thoroughly remove unspecific peptides. Finally, N-glycosylated peptides are enzymatically released from streptavidin beads by PNGase F and analyzed by liquid chromatography with MS (LC-MS).

PNGase F catalyzes the hydrolysis of the N-glycosidic bond between the innermost N-acetylglucosamine and the asparagine residue of high mannose, hybrid, and complex oligosaccharides from N-linked glycoproteins. Deglycosylation by PNGase F leads to the deamidation of the NXS/T (X is any amino acid except proline) glycosylation motif, which can be identified by MS through a 0.98 Da mass shift. Thus, the identification of deamidated glycosylation motives allows one to determine the sites of protein N-glycosylation (Fig. 2). The knowledge about N-glycosylation sites can, for example, be used to refine the protein topology prediction of poorly characterized cell surface proteins, since only extracellular domains of cell surface proteins are thought to be glycosylated (2).

The CSC technology is widely applicable to different cell types and model systems. It was, for example, applied to extensively phenotype Drosophila melanogaster cells, mouse myoblasts, and human leukemia cells (3–5). In differentiation and stimulation experiments, CSC analyses enabled to monitor global abundance changes of cell surface glycoproteins. Such analyses provided an extensive overview of induced phenotypic changes in cells in time and space. In-depth knowledge about molecular phenotypes could create new perspectives for improved subclassification of healthy and malignant cells. Furthermore, a detailed molecular analysis of phenotypic changes is a prerequisite for a systems biology understanding of malignantly transformed cells and might also create new perspectives for the development of targeted therapies.

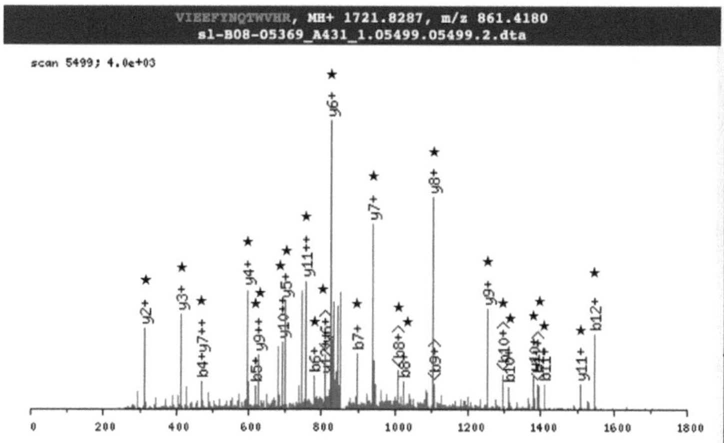

b+	#	AA	#	y+	y2+
100.0762	1	V	13		
213.1603	2	I	12	1622.7603	811.8841
342.2029	3	E	11	1509.6763	755.3420
471.2455	4	E	10	1380.6337	690.8208
618.3139	5	F	9	1251.5911	626.2995
781.3772	6	Y	8	1104.5227	552.7652
896.4042	7	N	7	941.4593	471.2336
1024.4628	8	Q	6	826.4324	413.7201
1125.5104	9	T	5	698.3738	349.6908
1311.5897	10	W	4	597.3261	299.1670
1410.6582	11	V	3	411.2468	206.1273
1547.7171	12	H	2	312.1784	156.5931
	13	R	1	175.1195	88.0637

N(7):+115.03

Fig. 2. MS/MS spectra of the peptide VIEEFYN[115]QTWVHR from the protein SLC2A1. The *left* graph shows the recorded MS/MS spectra 4068, with ions matching the in silica spectra colored and marked with a star. The *right* graph lists all possible theoretical fragments. The shaded fragment ions were found in the measured spectra. The deamidated asparagine was identified with both the b- and the y-ions.

2. Materials

Prepare all solutions with analytical grade reagents and ultrapure water. Store all reagents at room temperature, unless indicated otherwise.

2.1. Cell Surface Protein Biotinylation

1. Labeling buffer: Phosphate buffered saline (PBS) pH 6.5, 0.1% (v/v) fetal calf serum (FCS). Add a tablet for 500 mL PBS (Invitrogen, Carlsbad, CA, USA) to 450 mL water, adjust the pH to 6.5 with 85% (w/v) H_3PO_4, and add 0.5 mL FCS. Mix and make up to 500 mL with water. Store at 4°C.
2. $NaIO_4$ stock solution: 160 mM $NaIO_4$ in labeling buffer. Resuspend 34.22 mg $NaIO_4$ in 0.8 mL labeling buffer and make up to 1 mL with labeling buffer (see Note 1).
3. MACSmix tube rotator (Miltenyi Biotec, Bergisch Gladbach, Germany).
4. Biocytin hydrazide solution: 6.5 mM biocytin hydrazide (Biotium, Hayward, CA, USA) in labeling buffer. Add 25.12 mg biocytin hydrazide to 8 mL labeling buffer, mix, and make up to 10 mL with labeling buffer (see Note 1).

2.2. Membrane Preparation

1. Hypotonic lysis buffer: 10 mM Tris–HCl pH 7.5, 0.5 mM $MgCl_2$. Add 1.21 g Tris base and 47.61 mg $MgCl_2$ to 900 mL water and adjust pH to 7.5 with HCl. Mix and make up to 1 L with water. Store at 4°C.
2. 500 mM MES stock solution pH 6: Add 48.81 g 2-(N-morpholino)ethanesulfonic acid (MES hydrate) to 400 mL water. Adjust pH with NaOH and make up to 500 mL with water.

3. Membrane preparation buffer: 50 mM MES pH 6, 10 mM $MgCl_2$, 450 mM NaCl, 280 mM sucrose. Add 100 mL of the 500 mM MES stock solution pH 6, 0.95 g $MgCl_2$, 26.30 g NaCl, and 95.84 g sucrose to 700 mL water. Mix and make up to 1 L with water. Aliquot in 50 mL tubes and store at –20°C (see Note 2).
4. Membrane wash buffer: 25 mM Na_2CO_3. Add 2.65 g Na_2CO_3 to 900 mL water and make up to 1 L with water.

2.3. Protein Digestion

1. 100 mM NH_4HCO_3 buffer: 100 mM NH_4HCO_3. Add 7.91 g NH_4HCO_3 to 900 mL water and make up to 1 L with water (see Note 3).
2. RapiGest stock solution: 1% (w/v) RapiGest surfactant (Waters, Milford, MA, USA). Add 10 mg RapiGest to 900 μL water. Mix, make up to 1 mL with water, and store at 4°C.
3. VialTweeter (Hielscher, Teltow, Germany) for intense sonication of volumes between 20 μL and 2 mL.
4. Bicinchoninic acid Protein Assay Kit (BCA Protein Assay Kit, Pierce, Rockford, IL, USA).
5. Tris(2-carboxyethyl) phosphine (TCEP) stock solution: 500 mM TCEP. Add 143.33 mg TCEP to 700 μL water, mix, and make up to 1 mL with water. Aliquot and store at –20°C.
6. Iodoacetamide stock solution: 500 mM iodoacetamide. Add 92.48 mg iodoacetamide to 700 μL water, mix, and make up to 1 mL with water. Aliquot and store at –20°C.
7. Sequence grade-modified trypsin (Promega, Madison, WI, USA). Store at –80°C.

2.4. Affinity Chromatography

1. Streptavidin Plus UltraLink Resin (Pierce, Rockford, IL, USA). Store at 4°C.
2. 1 mL Mobicol columns with 35 μm pore size filter (MoBiTec, Goettingen, Germany).
3. Vac-Man laboratory vacuum manifold (Promega, Madison, WI, USA) and vacuum pump.
4. NaCl washing buffer: 5 M NaCl. Add 292.20 g NaCl to 600 mL water, mix, and make up to 1 L with water.
5. Ethylenediaminetetraacetic acid (EDTA) stock solution: 500 mM EDTA. Add 14.61 g EDTA to 80 mL water, adjust pH to 8 with NaOH, and make up to 100 mL with water.
6. Detergent buffer: 137 mM NaCl, 50 mM Tris pH 7.8, 10% glycerol, 0.5 mM EDTA pH 8, 0.1% Triton X-100. Add 6.06 g Tris base to 700 mL water and adjust pH to 7.8 with HCl. Then, add 8.01 g NaCl, 100 mL glycerol, 1 mL EDTA stock solution, and 10 mL Triton X-100 and make up to 1 L with water.

7. $NaHCO_3$ washing buffer: 100 mM $NaHCO_3$ pH 11. Add 8.40 g $NaHCO_3$ to 900 mL water. Adjust pH with NaOH and make up to 1 L with water (see Note 3).
8. Isopropanol solution: 80% (v/v) 2-propanol. Mix 400 mL 2-propanol and 100 mL water.
9. 500,000 units/mL glycerol-free peptide N-deglycosylase F (PNGase F, New England BioLabs, Ipswich, MA, USA). Store at 4°C.

2.5. Desalting of Peptide Solution

1. 10% Formic acid: 10% (v/v) formic acid. Add 1 mL formic acid to 9 mL water.
2. C18 UltraMicro Spin columns (The Nest Group, Southborough, MA, USA) with 0.03–30 μg capacity.
3. 80% Acetonitrile: LC-MS grade water, 80% (v/v) acetonitrile, 0.1% (v/v) formic acid. Add 80 mL acetonitrile and 0.1 mL formic acid to 15 mL LC-MS grade water. Mix and make up to 100 mL with LC-MS grade water (see Note 4).
4. Sample buffer: LC-MS grade water, 5% (v/v) acetonitrile, 0.1% (v/v) formic acid. Add 5 mL acetonitrile and 0.1 mL formic acid to 90 mL LC-MS grade water. Mix and make up to 100 mL with LC-MS grade water (see Note 4).
5. 50% Acetonitrile: LC-MS grade water, 50% (v/v) acetonitrile, 0.1% (v/v) formic acid. Add 50 mL acetonitrile and 0.1 mL formic acid to 45 mL LC-MS grade water. Mix and make up to 100 mL with LC-MS grade water (see Note 4).

3. Methods

3.1. Biotinylation of Cell Surface Proteins

1. Resuspend a cell pellet of approximately 10^8 cells in a 50 mL tube with 40 mL labeling buffer and afterwards pellet cells again by centrifugation. Repeat this washing procedure twice (see Note 5).
2. Resuspend the cells with 49.5 mL labeling buffer and add 500 μL of the $NaIO_4$ stock solution to the cell suspension. The final concentration of the solution is 1.6 mM $NaIO_4$. Incubate the cells at 4°C for 15 min on a MACSmix tube rotator.
3. Pellet the cells by centrifugation and discard the supernatant. Afterwards, wash the cells twice by resuspending the cells in 40 mL labeling buffer and subsequent centrifugation.
4. Resuspend the cell pellet with 10 mL biocytin hydrazide solution and transfer the cell suspension in a 15 mL tube. Incubate the cells at 4°C for 60 min on a MACSmix tube rotator.
5. Transfer the cell suspension in a 50 mL tube and make up to 40 mL with labeling buffer (see Note 6). Pellet the cells by

centrifugation and discard the supernatant. Afterwards, wash the cells twice with 40 mL labeling buffer.

3.2. Cell Lysis and Membrane Preparation

1. Resuspend the cell pellet with 10 mL hypotonic lysis buffer and incubate the cells on ice for 10 min.
2. Transfer the cells in a Dounce homogenizer and homogenize the cells on ice with 40 strokes (see Note 7).
3. Transfer the cell homogenate in a 15 mL tube and centrifuge at 2,000 × *g* for 10 min to remove cell debris and nuclei (see Note 8).
4. Distribute the supernatant equally in two ultracentrifuge tubes and mix in a 1:1 ratio with membrane preparation buffer. Fill up the ultracentrifuge tubes with a 1:1 mix of hypotonic lysis buffer and membrane preparation buffer. Then, centrifuge at 100,000 × *g* for 60 min in order to pellet the cell membranes (see Note 9).
5. Discard the supernatants, combine the membrane pellets in one ultracentrifuge tube, and incubate the membrane pellets on ice in 400 μL membrane wash buffer for 30 min (see Note 10).
6. Fill up the ultracentrifuge tube with hypotonic lysis buffer and centrifuge at 100,000 × *g* for 60 min.

3.3. Protein Digestion

1. Discard the supernatant and transfer the membrane pellet in a 2 mL tube. Add 340 μL of the 100 mM NH_4HCO_3 buffer and 40 μL of the 1% RapiGest stock solution.
2. Indirectly sonicate the tube in a VialTweeter at 4°C until the membrane pellet is completely dissolved (see Note 11).
3. Determine the protein concentration of the solution with the BCA Protein Assay Kit.
4. Add 8 μL of the 500 mM TCEP stock solution to the protein solution and incubate at room temperature for 30 min in order to reduce protein disulfide bonds. Then, add 12 μL of the 500 mM iodoacetamide stock solution to the protein solution in order to alkylate free thiol groups and incubate at room temperature in the dark for 30 min.
5. Add trypsin in a 1:50 protein:protease ratio to the protein solution and incubate at 37°C overnight on a MACSmix tube rotator.

3.4. Affinity Enrichment of Biotinylated Glycopeptides

1. Inactivate trypsin by heating the digestion solution at 95°C for 10 min (see Note 12).
2. Add 350 μL of the streptavidin beads into a Mobicol column and place the Mobicol column on the vacuum manifold (see Note 13). Wash the streptavidin beads with 5 mL of the 100 mM NH_4HCO_3 buffer by cycles of resuspending streptavidin beads

with 400 μL NH_4HCO_3 buffer and subsequently aspirating the liquid.

3. Add the digest solution to the streptavidin beads and rotate the Mobicol column on a MACSmix tube rotator for 1 h at 4°C.
4. Wash the streptavidin beads with 10 mL of the 100 mM NH_4HCO_3 buffer by cycles of resuspending the streptavidin beads with 400 μL NH_4HCO_3 buffer and subsequently aspirating the liquid on the vacuum manifold (see Note 14). Wash in the same manner the beads consecutively with each of the following washing buffers: 10 mL of NaCl washing buffer, 10 mL of detergent buffer, 10 mL of 100 mM NH_4HCO_3 buffer, 10 mL of $NaHCO_3$ washing buffer, and 2 mL of isopropanol solution (see Note 15). Finally, transfer the streptavidin beads in a new Mobicol column and wash the streptavidin beads with 10 mL of the 100 mM NH_4HCO_3 buffer (see Note 16).
5. Resuspend the streptavidin beads with 400 μL NH_4HCO_3 and add 1 μL of the glycerol-free PNGase F. Incubate the suspension at 37°C overnight on a MACSmix tube rotator (see Note 17).
6. Place the Mobicol column in a 2 mL tube and collect released glycopeptides by centrifugation. Afterwards, resuspend the streptavidin beads with 400 μL NH_4HCO_3 and collect the flow through in a 2 mL tube.

3.5. Desalting of the Peptide Sample

1. Combine the two glycopeptide solutions and acidify with 10% formic acid to a pH of 2–3 (see Note 18).
2. Place a C18 column in a 2 mL tube and condition the column by adding 100 μL of 80% acetonitrile into the C18 column. Centrifuge at 100 × *g* for 1 min and afterwards discard the flow-through. Repeat this procedure once with 80% acetonitrile and then flush the column four times with 100 μL sample buffer (see Note 19).
3. Load the glycopeptides on the C18 column and afterwards flush the column three times with 100 μL sample buffer to remove salts.
4. Place the C18 column in a 2 mL tube and elute the glycopeptides by adding 100 μL 50% acetonitrile and centrifugation at 100 × *g* for 1 min. Repeat this elution step once and dry the combined eluates in a SpeedVac concentrator.
5. Resuspend the dried peptides with 25 μL sample buffer and store the peptide sample frozen until analysis by LC-MS.

3.6. LC-MS Analysis and Data Analysis

1. We commonly separate glycopeptides by C18 reversed phase liquid chromatography. Two times 5 μL of peptides are typically separated on an online RP-HPLC column (75 μm inner

diameter and 10 cm length) packed in-house with C18 resin (Magic C18 AQ 3 μm; Michrom Bioresources, Auburn, CA, USA) using a linear gradient from 10% solvent B (water, 0.2% formic acid, and 1% acetonitrile) to 35% solvent B over 39 min at a flow rate of 0.2 μL/min (Eksigent Nano LC system, Eksigent Technologies, Dublin, CA, USA). Comparable chromatographic systems and settings will work equally well.

2. We ionize peptides by nanospray-ESI (Thermo Scientific, Waltham, MA, USA) and acquire MS and MS/MS spectra on a sensitive high mass accuracy mass spectrometer. In our case, this is an LTQ Orbitrap (Thermo Scientific) set to acquire one high resolution MS scan in the Orbitrap followed by three collision-induced dissociation MS/MS scans in the linear ion trap. For a high resolution MS scan, 2×10^6 ions are accumulated over a maximum time of 400 ms and the FWHM resolution is set to 60,000 (at m/z 300). The normalized collision energy is set to 35%. Singly charged ions were excluded from triggering MS/MS scans.
3. Raw data files from the MS instruments are usually converted to mzXML files and searched with Sorcerer-Sequest against the appropriate protein database. Database search criteria include the variable modification of 0.984020 Da for asparagines (representing formerly N-glycosylated asparagines after deamidation through the PNGase F treatment) and static modification of 57.021464 Da for cysteines (representing carbamidomethyl-containing cysteines after alkylation with iodoacetamide). Peptide and protein identifications are statistically validated with the Trans-Proteomic Pipeline TPP (6), containing PeptideProphet (7) and ProteinProphet (8). Identified proteins in different cell types can, for example, be visualized with Cytoscape (9), to obtain an extensive overview of phenotypic differences between different cell types (Fig. 3).

4. Notes

1. Prepare buffers immediately before use.
2. Store the membrane preparation buffer at –20°C, because working under unsterile conditions can lead to contamination of the buffer due to the high sucrose concentration.
3. Prepare bicarbonate buffers freshly because CO_2 escapes over time and the pH value will increase.
4. Always use glassware to transfer formic acid, since formic acid leaches out plasticizer from plasticware.

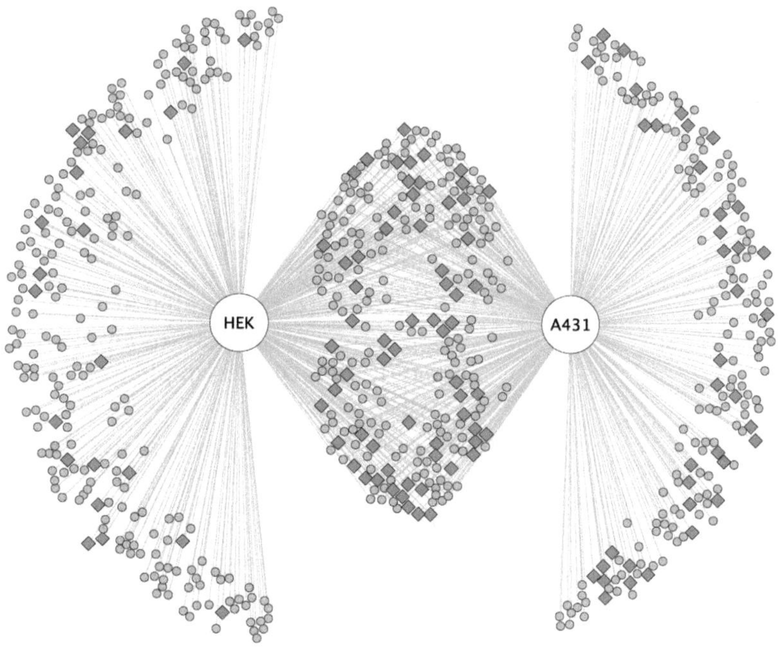

Fig. 3. Cytoscape image of identified cell surface proteins on the two cell lines HEK and A431. Cells are symbolized by large *white* balls, identified cell surface proteins by *yellow* balls or *orange squares* and connected to the cells, on which the protein was found. Proteins represented as *orange squares* are Cluster of Differentiation (CD) annotated molecules. Three protein groups can be distinguished: proteins only identified on HEK cells, proteins only identified on A431 cells, and proteins identified on both cell types.

5. Carry out all centrifugation steps of the protocol at 4°C and put tubes on ice during waiting times. The washing steps help to remove residual cell culture medium and cell debris.
6. When cells are transferred into a different tube, always try to keep the loss of cells as small as possible. For example, after the transfer of the cell suspension in the 50 mL tube, remaining cells in the 15 mL tube can be resuspended with labeling buffer and then added to the cells in the 50 mL tube until the volume of 40 mL is reached.
7. Slowly move the piston of the Dounce homogenizer completely up and down. Reduce the speed if foam is generated or if the homogenizer warms up. The cell lysis can also be carried out by other means. For example, by sonication with the VialTweeter with 20 impulses at half maximum amplitude and half maximum cycle time.
8. The supernatant, containing cell membranes, should be a bit cloudy after the centrifugation step. If the supernatant is completely clear, the homogenization step should be repeated.
9. We usually use a Beckman SW41 swinging bucket rotor. Make sure to fill the ultracentrifuge tubes completely with liquid; otherwise, the tubes could collapse in the buckets during centrifugation.

10. Loosen the membrane pellet from the tube bottom with the pipette tip at each side and try to transfer the membrane pellet as a whole.
11. Depending on the size of the membrane pellet and the cell line it may require long, intense sonication to completely solubilize the membrane pellet. In order to avoid thermal sample degradation, it is recommended to apply cycles of sonication and cooling on ice.
12. Make sure to thoroughly inactivate trypsin. Heat-inactivation of trypsin worked best in our hands, but protease inhibitors or acidification of the sample may also be an option to inactivate trypsin activity.
13. Use filter with 35 μm pore size. Smaller pore size filters tend to clog easily.
14. If you are interested in not biotinylated peptides of the digestion solution, you can collect these peptides by centrifugation of the Mobicol column in a 2 mL tube.
15. Optionally, warm up buffers to 60°C, which may in some cases increase the efficiency to remove unspecific peptides.
16. Streptavidin beads can be easily transferred into a new Mobicol column by resuspending the streptavidin beads with 400 μL of the 100 mM NH_4HCO_3 buffer. The transfer of the streptavidin beads into a new Mobicol column removes peptides, which are unspecifically bound to the Mobicol column.
17. Optionally, incubate the streptavidin beads at 37°C overnight in 50 mM sodium phosphate buffer supplemented with 1% NP-40, which is the manufacturer recommended buffer. However, PNGase F also efficiently releases glycopeptides from streptavidin beads in 100 mM NH_4HCO_3 buffer.
18. The acidification of the glycopeptide solution is necessary for the subsequent desalting step by C18 reversed phase chromatography. Carefully add a few μL of the 10% formic acid and let CO_2 escape by gentle mixing with a pipett tip. Test the pH value of the glycopeptides solution by transferring 1 μL of the solution on a pH indicator paper.
19. Flushing the C18 column with sample buffer can also be carried out on the vacuum manifold by continuously aspirating the liquid.

Acknowledgments

This project was supported by the Krebsliga Zurich, the NCCR Neural Plasticity and Repair, and the InfectX project within the Swiss Initiative in Systems Biology (SystemsX.ch). A.H. was supported in part by the Mueller Fellowship from the Ph.D. Program in Molecular Life Sciences.

References

1. Wollscheid B, Bausch-Fluck D, Henderson C et al (2009) Mass-spectrometric identification and relative quantification of N-linked cell surface glycoproteins. Nat Biotechnol 27(4):378–386
2. Zielinska DF, Gnad F, Wisniewski JR et al (2010) Precision mapping of an in vivo N-glycoproteome reveals rigid topological and sequence constraints. Cell 141(5):897–907
3. Gundry RL, Raginski K, Tarasova Y et al (2009) The mouse C2C12 myoblast cell surface N-linked glycoproteome: identification, glycosite occupancy, and membrane orientation. Mol Cell Proteomics 8(11):2555–2569
4. Hofmann A, Gerrits B, Schmidt A et al (2010) Proteomic cell surface phenotyping of differentiating acute myeloid leukemia cells. Blood 116(13):e26–e34
5. Schiess R, Mueller LN, Schmidt A et al (2009) Analysis of cell surface proteome changes via label-free, quantitative mass spectrometry. Mol Cell Proteomics 8(4):624–638
6. Keller A, Eng J, Zhang N et al (2005) A uniform proteomics MS/MS analysis platform utilizing open XML file formats. Mol Syst Biol 1:2005.0017
7. Keller A, Nesvizhskii AI, Kolker E et al (2002) Empirical statistical model to estimate the accuracy of peptide identifications made by MS/MS and database search. Anal Chem 74(20):5383–5392
8. Nesvizhskii AI, Keller A, Kolker E et al (2003) A statistical model for identifying proteins by tandem mass spectrometry. Anal Chem 75(17):4646–4658
9. Shannon P, Markiel A, Ozier O et al (2003) Cytoscape: a software environment for integrated models of biomolecular interaction networks. Genome Res 13(11):2498–2504

Chapter 4

Use of Boronic Acid Nanoparticles in Glycoprotein Enrichment

Yawei Xu, Lijuan Zhang, and Haojie Lu

Abstract

Glyco-specific enrichment methods for mass spectrometry pretreatment are invaluable for the detection of low abundant glycoproteins or glycopeptides. For example, boronic acid can specifically interact with glycans in nonaqueous or basic aqueous solutions. Here, we describe a glyco-specific enrichment method which uses a boronic acid-functionalized "core-satellite" composite nanoparticle to isolate glycoproteins or glycopeptides from complex biological samples. Furthermore, we also demonstrate detection limit improvements and show how to evaluate the percent recovery from the glycoprotein or glycopeptide enrichment process via SDS-PAGE and $^{16}O/^{18}O$ labeling strategies.

Key words: Glycoprotein, Glycopeptide, Glycosylation, Boronic acid, Enrichment, Recovery

1. Introduction

Glycosylation is one of the most important protein post-translational modifications (PTMs) as it plays the key role in numerous cellular processes (1). Typically, there are two kinds of protein glycosylation: glycans attached to the side-chain nitrogen of asparagine and arginine (N-linked glycosylation), glycans attached to the hydroxy oxygen of serine, threonine, or tyrosine (O-linked glycosylation). Moreover, most glycosylation sites carry various forms of glycans, giving rise to different glycan structures. Therefore, a universal glyco-specific enrichment method is urgently needed for pretreatment of biological samples before mass spectrometric analysis.

Boronic acid could form five- or six-membered cyclic ester with 1,2 or 1,3 diols of saccharides in nonaqueous or basic aqueous solution. Accordingly, it is a good choice to adopt boronic acid as the binding anchor of glyco-specific enrichment methods. Early in 1954, Kuivila et al. noticed that boronic acids can bind with solubilized saccharides and polyols, and postulated the formation

Jennifer J. Kohler and Steven M. Patrie (eds.), *Mass Spectrometry of Glycoproteins: Methods and Protocols*, Methods in Molecular Biology, vol. 951, DOI 10.1007/978-1-62703-146-2_4, © Springer Science+Business Media, LLC 2013

of cyclic esters (2). Subsequently, in 1959, Lorand and Edwards gave the first quantitative analysis of saccharide boronic acid interaction and determined the selectivity of phenylboronic acid towards saccharides (3). In 1995, protein-boronic acid conjugates were used as the matrix to separate low-molecular-mass cis-diols and glycated hemoglobin using affinity chromatography (4).

The introduction of nanotechnology into biological analysis has started a new era of nanoparticle-based enrichment for mass spectrometric analysis (5). Pre-enrichment of glycosylated proteins is no exception and has been facilitated by using various kinds of nanoparticles. In 2005, boronic acid was immobilized onto magnetic beads to isolate glycoproteins since magnetic beads can be easily washed and separated with an external magnetic field (6). In the same year, commercial boronic acid-functionalized magnetic beads were validated, showing good effectiveness in isolating model glycoproteins but with slightly weaker specificity than lectin-based ConA-functionalized magnetic beads (7). Recently, more and more boronic-functionalized matrices, such as agarose, magnetic beads, and mesoporous materials, have been used in enriching glycopeptides/proteins which show remarkable recovery, selectivity, and sensitivity (8–11).

Here, we adopted the boronic acid-functionalized core-satellite composite nanoparticles (12) as the model nanoparticle to show the detailed processes of enriching and isolating glycopeptides/proteins prior to MS analysis (see below). The combination of SiO_2-coated ferrite "core" and numerous "satellites" of gold nanoparticles with lots of boronic acid "anchors" make these composite nanoparticles easy to separate and provide good recovery and high adsorption capacity. Briefly, the synthesis procedure of the composite nanoparticles contains eight steps: First, the magnetic nanoparticle core with narrow size distribution was prepared via a solvothermal reaction (13). Second, the well-washed magnetic nanoparticles were coated with silica through the hydrolysis of tetraethyl orthosilicate (TEOS, 95%, Sinopharm Chemical Reagent, Shanghai, China). Third, the silica-coated magnetic nanoparticles were surface functionalized with thiol groups through hydrolysis of 3-mercaptopropylmethyldimethoxysilane (MPMDMS, 95%, Sigma, St. Louis, MO) in order to conjugate the gold nanoparticle. Fourth, the gold nanoparticles were synthesized through a reduction of $HAuCl_4$ (47.8%, Sinopharm Chemical Reagent, Shanghai, China) based on the previous literature (14). Fifth, the gold nanoparticles were conjugated with the surface-functionalized magnetic nanoparticles through the interaction of thiol groups and gold atoms to form the composite nanoparticles. Sixth, the composite nanoparticles were immersed in 11-mercaptoundecanol (MUD, 97%, Sigma, St. Louis, MO) solution to form a self-assembled monolayer of MUD on the gold surface. Seventh, the composite nanoparticles were immersed into a solution of succinic anhydride and 4-dimethylaminopyridine (DMAP, 99%, Sigma, St. Louis, MO)

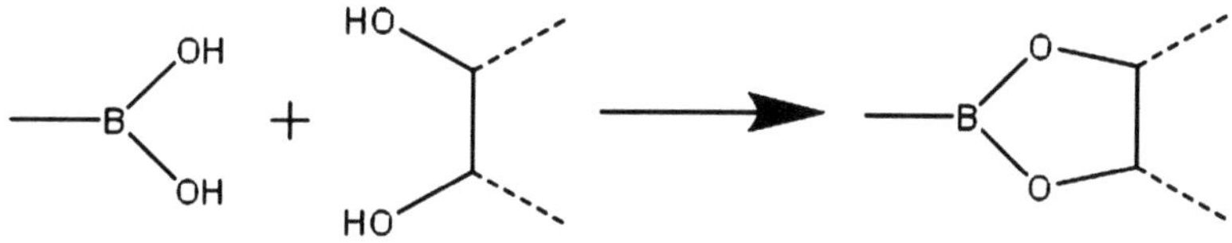

Fig. 1. The structure of the cyclic ester formed between the boronic acid and a glycan.

to convert the gold surface hydroxygroups to carboxy groups. Finally, the surface carboxy groups were activated with 1-(3-dimethylaminopropyl)-3-ethylcarbodiimide hydrochloride (EDC, 95%, GL Biochem, Shanghai, China) and 1-hydroxy-7-azabenzotriazole (HOAt, 95%, GL Biochem, Shanghai, China) to facilitate the grafting of 3-aminophenylboronic acids (APB, 98%, Sigma, St. Louis, MO). As the functional group, the boronic acid from APB could form a cyclic diester with all kinds of glycans which contains a cis-diol group (Fig. 1).

2. Materials

All reagents were of analytical grade or better and used as received without further purification. Deionized water (18.4 MΩ cm) used for all experiments was obtained from a Milli-Q system (Millipore, Bedford, MA, USA). All reagents were prepared and stored under room temperature unless indicated otherwise. All wastes were disposed following waste disposal regulations.

2.1. Glycoprotein Enrichment Procedure

1. Loading buffer: 50 mM ammonium bicarbonate (ABC, 99.5%, Sigma, St. Louis, MO). Add about 100 mL water to a 250-mL volumetric flask. Weigh 988.25 mg ABC and transfer to flask. Mix and make up to 250 mL with water. Store at 4°C.
2. Washing buffer: The same as the loading buffer.
3. Elution buffer: 1% Trifluoroacetic acid (TFA, 99.8%, Merck, Darmstadt, Germany), 50% Acetonitrile (ACN, 99.9%, Merck, Darmstadt, Germany), and 49% H_2O. Add 490 μL H_2O, 500 μL ACN, and 10 μL TFA to an Eppendorf (EP) tube. Mix and store at 4°C.

2.2. Glycoprotein Trypsin Digestion

1. Digestion buffer: 50 mM ABC. Add about 100 mL water to a 250-mL volumetric flask. Weigh 988.25 mg ABC and transfer to flask. Mix and make up to 250 mL with water. Adjust pH to approximately 8.0. Store at 4°C.

2.3. Evaluating the Recovery of Glycoprotein

1. Sodium dodecyl sulfate polyacrylamide gel electrophoresis (SDS-PAGE)-related materials were not listed here on account of the restriction of the length of this chapter. See vendor application sheets for further information.

2.4. Glycopeptide Enrichment Procedure

1. Loading buffer: 50 mM ABC. Add about 100 mL water to a 250-mL volumetric flask. Weigh 988.25 mg ABC and transfer to flask. Mix and make up to 250 mL with water. Store at 4°C.
2. Washing buffer: The same as the loading buffer.
3. Eluting buffer: 1% TFA/50% ACN/49% H_2O. Add 490 μL H_2O, 500 μL ACN and 10 μL TFA to an Eppendorf (EP) tube. Mix and store at 4°C.

2.5. Evaluating the Recovery of Glycopeptides via $^{16}O/^{18}O$ Labeling (see Note 9)

1. Dissolving buffer: 50 mM ABC. Weigh 9.88 mg ABC and transfer to a 1.5-mL Eppendorf tube. Add 1 mL $H_2{}^{18}O$ to the Eppendorf tube. Mix and store at 4°C.

3. Methods

Carry out all procedures at room temperature unless otherwise specified.

3.1. Glycoprotein Enrichment Procedure

1. Boil protein mixture for 5 min to denature the proteins.
2. Dissolve the protein mixture (see Note 1) in 200 μL of loading buffer. Then add 10 μL of 3.3 mg/mL composite nanoparticles (in ethanol) (see Note 2) to form the loading mixture.
3. Shake the loading mixture for 1 h at room temperature to bind the glycoprotein onto the composite nanoparticle.
4. Use a magnet to separate the supernatant from the composite nanoparticles (see Note 3). Add 200 μL of washing buffer and redisperse the composite nanoparticles to form the washing mixture.
5. Shake the washing mixture for 0.5 h at room temperature to release the nonspecifically bound materials from the composite nanoparticle.
6. Use a magnet to separate the supernatant from the composite nanoparticles (see Note 3). Add 200 μL of eluting buffer and redisperse the composite nanoparticle to form the eluting mixture.
7. Shake the eluting mixture for 0.5 h at room temperature to release the glycoprotein from the composite nanoparticle.
8. Use a magnet to separate the supernatant from the composite nanoparticle. Remove the solvent using vacuum centrifugal condenser to collect the glycoproteins and store at 4°C for further analysis.

3.2. Evaluating the Recovery of Glycoprotein

1. Divide the denatured protein mixture (containing 5–20 μg of protein) into two aliquots of equal amount. One of them (denoted as *Sample A*) will be kept untreated until SDS-PAGE analysis.
2. Apply the other aliquot to glycoprotein enrichment procedure (Subheading 3.1) and keep the isolated glycoproteins (denoted as *Sample B*) for SDS-PAGE analysis.
3. Cast the 12% separating gel in a 7.25 cm × 10 cm × 0.7 mm gel cassette. Allow space for the stacking gel and gently overlay with water. Cast the stacking gel and insert a 10-well gel comb immediately without introducing air bubbles (see Note 4).
4. Boil *Sample A and B* for 5 min after adding 10 μL of lysis buffer containing bromophenol blue (BPB) dye. Centrifuge the heated samples at 3,000 × *g* for 30 s to bring down the condensate. Load *Sample A*, *B*, and protein standard onto three adjacent lanes. Electrophorese at 10 mA until the sample has entered the gel and then continue at 20 mA until the dye front has reached the bottom of the gel.
5. Visualize the gel contents with Coomassie Brilliant Blue staining. Scan the stained gel with a gel scanner.
6. The amounts of glycoproteins in *Sample A* and *Sample B* can be detected, quantified, and reported by the software (see Note 5).
7. Divide the amount of glycoproteins in *Sample B* by that of *Sample A* to get the recovery of glycoproteins.
8. To demonstrate the glycoprotein enrichment procedure, a standard protein mixture containing 1 μg of bovine serum albumin (BSA, 96%, Sigma, St. Louis, MO), 1 μg of horseradish peroxidase (HRP, 850 units/mg, Sigma, St. Louis, MO), and 1.6 μg of cytochrome c (95%, Sigma, St. Louis, MO) was adopted as a model mixture. Following the steps in Subheadings 3.1 and 3.2 the result was shown in Fig. 2. According to this figure, only the glycoprotein, HRP, was isolated through method in Subheading 3.1. The calculated recovery of HRP was 71.6%.

3.3. Glycoprotein Trypsin Digestion

1. Dissolve a glycoprotein in digestion buffer, to form sample solution of 1 mg/mL (see Note 6).
2. Boil the sample solution for 5 min to denature the glycoprotein (see Note 7).
3. After the sample cools down to room temperature, add sequence grade trypsin to a final concentration of 1:50 (w/w) of glycoprotein.
4. Keep the sample solution at 37°C overnight with gentle shaking to enable the trypsin digestion.
5. Store the tryptic protein mixture at −20°C for further use.

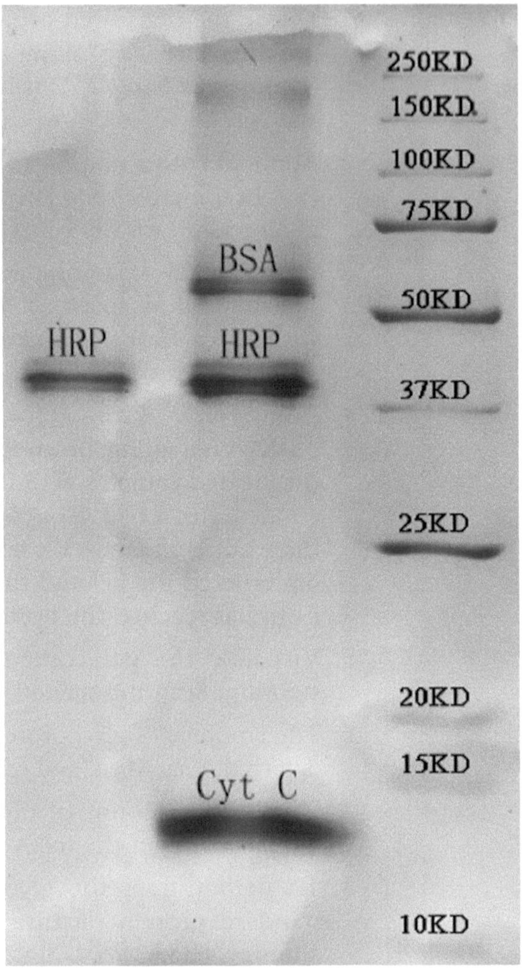

Fig. 2. Sodium dodecyl sulfate polyacrylamide gel electrophoresis (SDS-PAGE) analysis of the eluate after enrichment with the composite nanoparticles (*left*) and the model mixture without enrichment (*middle*).

3.4. Glycopeptide Enrichment Procedure

1. Shake the peptide mixture (see Note 8) in 200 μL of loading buffer. Then add 10 μL of 3.3 mg/mL composite nanoparticles (in ethanol) to form the loading mixture (see Note 2).
2. Shake the loading mixture for 1 h at room temperature to bind the glycoprotein onto the composite nanoparticle.
3. Use a magnet to separate the supernatant from the composite nanoparticles (see Note 3). Add 200 μL of washing buffer and redisperse the composite nanoparticles to form the washing mixture.
4. Shake the washing mixture for 0.5 h at room temperature to release the nonspecifically bound components from the composite nanoparticle.
5. Use a magnet to separate the supernatant from the composite nanoparticles (see Note 3). Add 200 μL of elution buffer and

redisperse the composite nanoparticle to form the eluting mixture.

6. The eluting mixture was shaken for 0.5 h at room temperature to release the glycoprotein from the composite nanoparticle.
7. Use a magnet to separate the supernatant from the composite nanoparticle. Remove the solvent using vacuum centrifugal condenser to collect the glycopeptides and store at 4°C for mass spectrometric analysis.
8. To demonstrate this method, 2 ng/μL tryptic HRP was adopted as a model mixture. Following the steps mentioned in Subheading 3.4 resulted in the spectra shown in Fig. 3. According to Fig. 3c, d, the tryptic HRP before enrichment gave five peaks representing glycopeptides and numerous peaks representing nonglycopeptides. After enrichment, all peaks observed in Fig. 3a, b were assigned to only glycopeptides or their fragments.

3.5. Evaluating the Recovery of Glycopeptides via $^{16}O/^{18}O$ Labeling (See Note 9)

1. Divide a peptide mixture (containing glycopeptides) into two aliquots of equal amount. Denote one as *Sample C* and the other as *Sample D*.
2. Dry *Sample C* using vacuum centrifugal condenser to remove $H_2{}^{16}O$.
3. Dissolve *Sample C* in loading buffer in $H_2{}^{18}O$ (see Note 9) to form the sample solution at the final concentration of 1 mg/mL.
4. Add PNGase F (from *Elizabethkingia meningoseptica*, Sigma, St. Louis, MO), which was dissolved in $H_2{}^{18}O$, to *Sample C* and keep at 37°C for 16 h with gentle shaking to release the glycans from peptides.
5. Boil *Sample C* for 5 min to denature the residual PNGase F (see Note 10).
6. Apply *Sample D* to the glycopeptide enrichment procedure (Subheading 3.4) to isolate glycopeptides. Denote the isolated glycopeptides as *Sample E*.
7. Add PNGase F (dissolved in $H_2{}^{16}O$) to *Sample E* and keep at 37°C for 16 h with gentle shaking to release glycans from peptides.
8. Boil *Sample E* for 5 min to denature the residual PNGase F (see Note 10).
9. Mix *Sample C* and *Sample E* and subject to mass spectrometric analysis.
10. The recovery of glycopeptides enrichment method can be derived from the ratio of the peak area (or intensity) of the signals of *Sample C* (processed with enrichment method but not labeled with ^{18}O) and those of *Sample E* (without enrichment

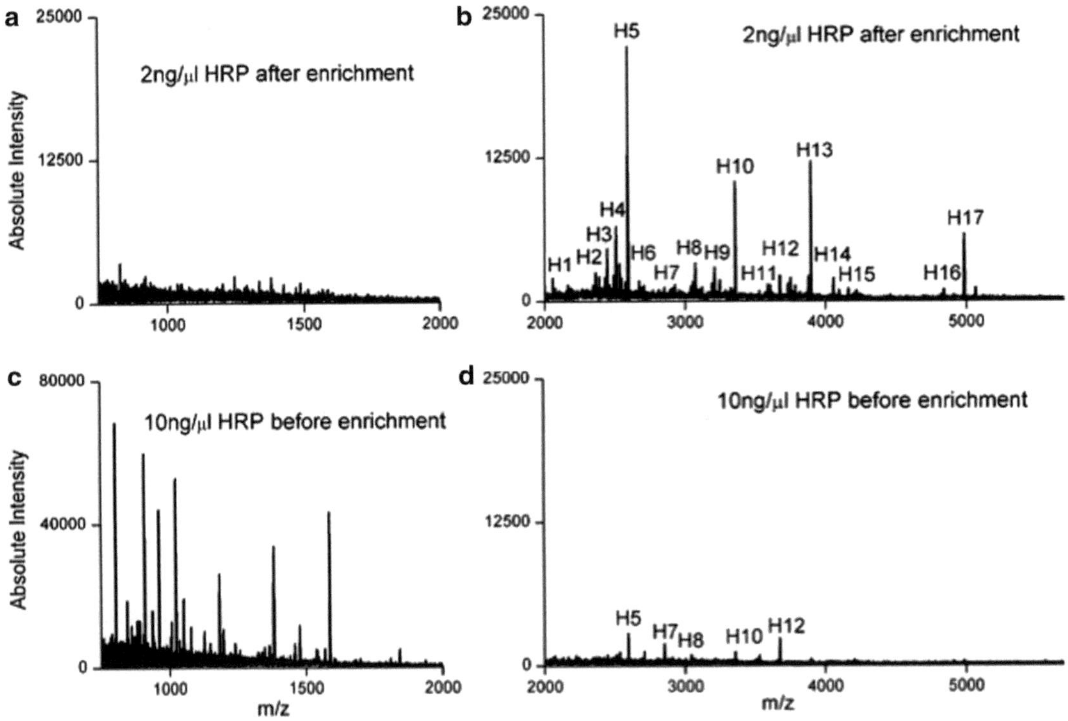

Fig. 3. MALDI mass spectra of 2 ng/μL tryptic horseradish peroxidase (HRP) digestion after enrichment with the composite nanoparticles (**a**, **b**); 10 ng/μL tryptic HRP digestion without enrichment (**c**, **d**). The peaks marked with Arabic numerals represent the glycopeptides or their fragments of tryptic HRP digestion.

process but labeled with ^{18}O). The detailed calculation process is listed in ref. (11, 15).

11. To demonstrate this method, 2 μg of tryptic asialofetuin was adopted as a model mixture. Following the steps mentioned in Subheading 3.5 resulted in the spectrum shown in Fig. 4. The theoretical isotope distribution of the glycopeptide VVHAVEVALATFNAESN#GSYLQLVEISR (N# denotes the N-linked glycosylation site) is 3,017 (61.6%): 3,018 (100%): 3,019 (86.1%): 3,020 (51.9%) (denoted by blue dots in the Fig. 4 inset) was calculated using the software: ChemBioDraw Ultra 11.0. Define "x" the intensity of the ^{16}O-labeled peptide and "y" the intensity of this ^{18}O-labeled peptide. So $((x/y)\times 100\%)$ represents the percent recovery of the glycopeptides. For the example in Fig. 4, based on the areas of the first and the third peak we got a group of equations: $0.62x=13191.8$ and $0.86x+0.62y=34662.2$ with the calculated recovery of 80.6%. Based on the areas of the second and the fourth peak we got another group of equations: $x=21999.7$ and $0.52x+y=35547.2$ with the calculated recovery of 91.2%. The average percent recovery for this glycopeptide was 85.9%.

3.6. Evaluating the Limit of Detection of Glycopeptide

1. Three tryptic glycoproteins (HRP, fetuin, and asialofetuin) were adopted as model samples to evaluate the limit of detection (LOD) of this method. All glycoproteins were tryptic

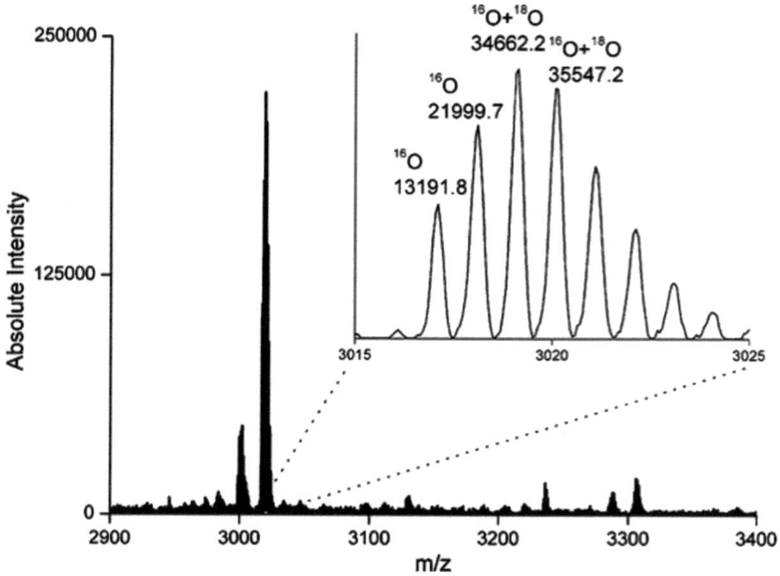

Fig. 4. MALDI mass spectrum of glycopeptide VVHAVEVALATFNAESN#GSYLQLVEISR (N# denotes the N-linked glycosylation site) which consists of partially ^{18}O-labeled N#. The actual isotope distribution of this peptide was enlarged and shown in the inner graph.

digested following the steps in Subheading 3.3. The tryptic glycoproteins were directly sent to mass spectrometric analysis (AXIMA-QIT, SHIMADZU BIOTECH, Japan) to acquire the LOD before enrichment (left column of Table 1). The LOD after enrichment (right column of Table 1) were acquired through the glycopeptide enrichment process in Subheading 3.4 using the same mass spectrometer. As Table 1 shows the LOD of glycopeptides could be improved by 101–200 times through this strategy.

4. Notes

1. Remove protein mixture solvent or concentrate the mixture using vacuum centrifugal condenser. Make sure the final concentration of ABC would not be diluted by the added protein mixture.
2. The volume and the concentration of nanoparticles are based on the adsorption capacity of the nanoparticles used. The data listed here is only appropriate for the composite nanoparticles (12).
3. Make sure sample mixture is collected at the bottom of the Eppendorf tube before applying magnet to prevent the loss of composite nanoparticles.
4. Detailed procedure and the material of SDS-PAGE were not listed here on account of the limit of the length of this chapter. See vendor application notes for further instructions.

Table 1
The limits of detection (LOD) of glycopeptides before and after enrichment

Protein name	LOD (before enrichment) fmol/μL	LOD (after enrichment) fmol/μL
HRP	43.1	0.37
Fetuin	108	0.54
Asialofetuin	54.4	0.54

5. The spot detection and quantification were performed with ImageMaster 2D Platinum 6.0 software (GE Healthcare). The spot volumes (sum of pixel intensities within the spot boundary) of glycoproteins in *Sample A* and *B* were automatically calculated and reported by the software.
6. The pH of the digestion buffer should be maintained at approximately pH 8.0.
7. If the glycoprotein was from the eluate or had been denatured before, this step could be omitted.
8. Normally, peptides were kept in 50 mM ABC which is the same as the loading buffer. Accordingly, the solvent did not need to be removed during the loading process.
9. $H_2{}^{18}O$ was used as the only oxygen atom source during the PNGase F deglycosylation process to substitute one ^{16}O with ^{18}O. In the mass spectrum, the ^{18}O-treated glycopeptides exhibit a positive two Daltons mass shift relative to the ^{16}O peptides. Once mixed with those glycopeptides which were enriched and subsequently deglycosylated with PNGase F in $H_2{}^{16}O$, the percent recovery can be calculated by determining the ratio of the isotopes from an unlabeled ^{16}O peptide to an isotope with summed contributions from ^{16}O- and ^{18}O-labeled isotopes (11). This calculation is based upon knowledge of the percent abundance of the theoretical unlabeled isotopic envelope.
10. Residual PNGase F was denatured to prevent from ^{16}O replacing ^{18}O once mixing ^{18}O-labeled peptide with unlabeled peptide.

Acknowledgments

This work was supported by National Science and Technology Key Project of China (2007CB914100 and 2009CB825607), National Natural Science Foundation of China (31070732 and 21025519), Shanghai Projects (Shuguang, Eastern Scholar and B109).

References

1. Tian Y, Zhou Y, Elliott S, Aebersold R, Zhang H (2007) Solid-phase extraction of N-linked glycopeptides. Nat Protoc 2:334–339
2. Kuivila HG, Keough AH, Soboczenski EJ (1954) Areneboronates from diols and polyols. J Org Chem 19:780–783
3. LorandJP, EdwardsJO(1959) Polyolcomplexes and structure of the benzeneboronateion. J Org Chem 24:769–774
4. Frantzen F, Grimsrud K, Heggli DE, Sundrehagen E (1995) Protein–boronic acid conjugates and their binding to low-molecular-mass cis-diols and glycated hemoglobin. J Chromatogr B 670:37–45
5. Zhang L, Lu H, Yang P (2010) Recent developments of nanoparticle-based enrichment methods for mass spectrometric analysis in proteomics. Sci China Chem 53:695–703
6. Lee JH, Kim Y, Ha MY, Lee EK, Choo J (2005) Immobilization of aminophenylboronic acid on magnetic beads for the direct determination of glycoproteins by matrix assisted laser desorption ionization mass spectrometry. J Am Soc Mass Spectrom 16:1456–1460
7. Sparbier K, Koch S, Kessler I, Wenzel T, Kostrzewa M (2005) Selective isolation of glycoproteins and glycopeptides for MALDI-TOF MS detection supported by magnetic particles. J Biomol Tech 16:407–413
8. Sparbier K, Asperger A, Resemann A, Kessler I, Koch S, Wenzel T, Stein G, Vorwerg L, Suckau D, Kostrzewa M (2007) Analysis of glycoproteins in human serum by means of glycospecific magnetic bead separation and LC-MALDI-TOF/TOF analysis with automated glycopeptidedetection. J Biomol Tech 18:252–258
9. Zhou W, Yao N, Yao G, Deng C, Zhang X, Yang P (2008) Facile synthesis of aminophenylboronic acid-functionalized magnetic nanoparticles for selective separation of glycopeptides and glycoproteins. Chem Commun 43:5577–5579
10. Gontarev S, Shmanai V, Frey SK, Kvach M, Schweigert FJ (2007) Application of phenylboronic acid modified hydrogel affinity chips for high-throughput mass spectrometric analysis of glycated proteins. Rapid Commun Mass Spectrom 21:1–6
11. Xu Y, Wu Z, Zhang L, Lu H, Yang P, Webley PA, Zhao D (2009) Highly specific enrichment of glycopeptides using boronic acid-functionalized mesoporous silica. Anal Chem 81:503–508
12. Zhang L, Xu Y, Yao H, Xie L, Yao J, Lu H, Yang P (2009) Boronic acid functionalized core-satellite composite nanoparticles for advanced enrichment of glycopeptides and glycoproteins. Chem Eur J 15:10158–10166
13. Deng H, Li X, Peng Q, Wang X, Chen J, Li Y (2005) Monodisperse magnetic single-crystal ferrite microspheres. Angew Chem Int Ed 44:2782–2785
14. Grabar KC, Freeman RG, Hommer MB, Natan MJ (1995) Preparation and characterization of Au colloid monolayers. Anal Chem 67:735–743
15. Johnson KL, Muddiman DC (2004) A method for calculating $^{16}O/^{18}O$ peptide ion ratios for the relative quantification of proteomes. J Am Soc Mass Spectrom 15:437–445

Chapter 5

Incorporation of Unnatural Sugars for the Identification of Glycoproteins

Balyn W. Zaro, Howard C. Hang, and Matthew R. Pratt

Abstract

Glycosylation is an abundant post-translational modification that alters the fate and function of its substrate proteins. To aid in understanding the significance of protein glycosylation, identification of target proteins is key. As with all proteomics experiments, mass spectrometry has been established as the desired method for substrate identification. However, these approaches require selective enrichment and purification of modified proteins. Chemical reporters in combination with bioorthogonal reactions have emerged as robust tools for identifying post-translational modifications including glycosylation. We provide here a method for the use of bioorthogonal chemical reporters for isolation and identification of glycosylated proteins. More specifically, this protocol is a representative procedure from our own work using an alkyne-bearing O-GlcNAc chemical reporter (GlcNAlk) and a chemically cleavable azido-azo-biotin probe for the identification of O-GlcNAc-modified proteins.

Key words: Proteomics, Glycosylation, Bioorthogonal chemical reporter, Click chemistry, Azide, Mass spectrometry, O-GlcNAc

1. Introduction

The identification of glycosylated proteins via mass spectrometry (MS) has proven invaluable in elucidating the function of glycosylation events as well as the relationship between mono- and polysaccharides and their substrates. Several technologies have been developed to facilitate this investigation, which typically involves affinity enrichment for target proteins followed by mass spectrometry analysis. Lectins (1) are helpful in isolating and identifying N-linked glycoproteins, but lectins are not generally applicable to other forms of glycosylation. Several antibodies have been raised for the recognition of O-GlcNAc-modified proteins (2); however,

Jennifer J. Kohler and Steven M. Patrie (eds.), *Mass Spectrometry of Glycoproteins: Methods and Protocols*, Methods in Molecular Biology, vol. 951, DOI 10.1007/978-1-62703-146-2_5, © Springer Science+Business Media, LLC 2013

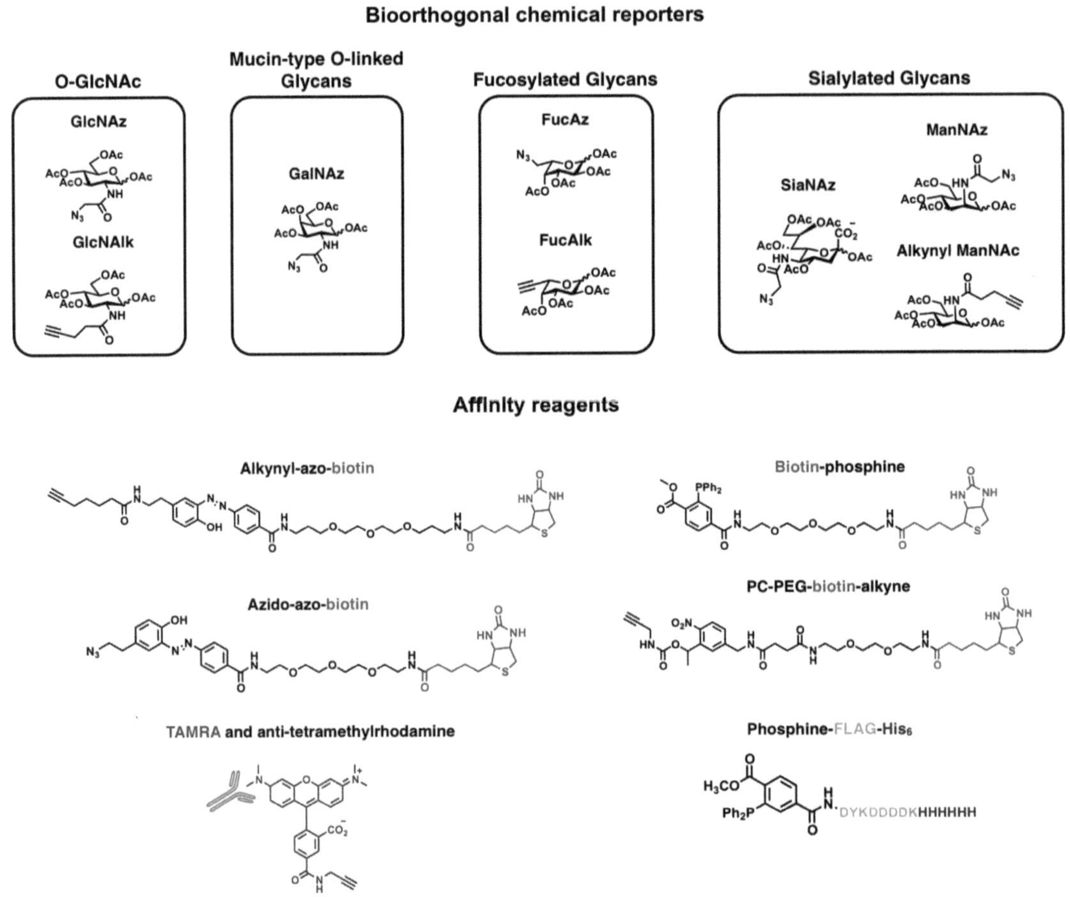

Fig. 1. Bioorthogonal chemical reporters for glycosylation: GlcNAz (21), GlcNAlk (20), GalNAz (4), FucAz (22), FucAlk (23), ManNAz (21), alkynyl ManNAc (23) and SiaNAz (24). Affinity probes: azido-azo-biotin (16), alkynyl-azo-biotin (16), PC-PEG-biotin-alkyne (14), phosphine-FLAG-His_6 (9) and alkynyl tetramethyl-6-carboxyrhodamine (TAMRA) and antibody.

they all display some sequence requirement for the underlying peptide. Chemical reporters are an alternative to these methods. Originally developed in the Bertozzi laboratory, metabolic bioorthogonal chemical reporters deliver unique reactivity to glycosylated proteins (3). There are currently several azide- or alkyne-bearing monosaccharide analogs that can be used as chemical reporters of glycosylation (Fig. 1). This metabolic labeling approach has been effective for the visualization and identification of cell-surface glycoproteins (4–8) and O-GlcNAc-modified proteins (9–12). Post-lysis enzymatic transfer of an azide-modified chemical reporter to O-GlcNAcylated proteins is an alternative delivery method developed by the Hseih-Wilson laboratory (13, 14). A mutant galacytosyltranferase transfers an N-azidoacetylgalactosamine (GalNAz) residue onto O-GlcNAc-modified proteins. Regardless of the method of incorporation, these bioorthogonally functionalized cell lysates can then be subjected

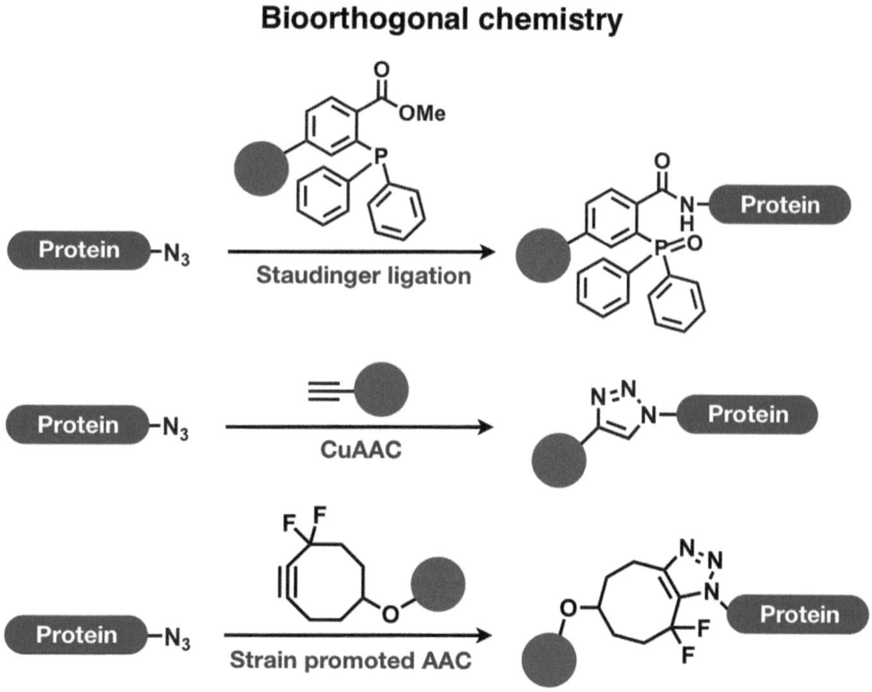

Fig. 2. Bioorthogonal chemistry: Staudinger Ligation, Cu-catalyzed Azide-Alkyne Cycloaddition and Strain Promoted Azide-Alkyne Cycloaddtion.

to Staudinger ligation, Cu-catalyzed Azide-Alkyne Cycloaddition (CuAAC), or Strain-promoted Azide-Alkyne Cycloaddition with a corresponding phosphine-, azide-, or alkyne-modified affinity probe (Fig. 2).

There are several approaches to affinity purification. The Hsieh-Wilson laboratory developed an alkyne-modified fluorescent rhodamine derivative, tetramethyl-6-carboxyrhodamine alkyne (TAMRA), and a corresponding antibody is commercially available that allows for the selective enrichment of rhodamine-modified proteins (13) (Fig. 1). Cleavable azido- and alkynyl-biotin probes facilitate the enrichment and subsequent elution of target proteins for mass spectrometry analysis (15, 16). Notably, the Hart laboratory developed a photo-cleavable probe termed PC-PEG-biotin-alkyne (Fig. 1). Additionally, chemically cleavable biotin reagents that contain an azo moiety that is readily reduced to liberate proteins from streptavidin beads for downstream MS analysis can be used (17–19).

This protocol is representative of the identification of O-GlcNAc-modified proteins (Fig. 3a) but should be readily applicable to other types of glycosylation including mucin-type O-linked glycans and N-linked glycans. Specifically, we

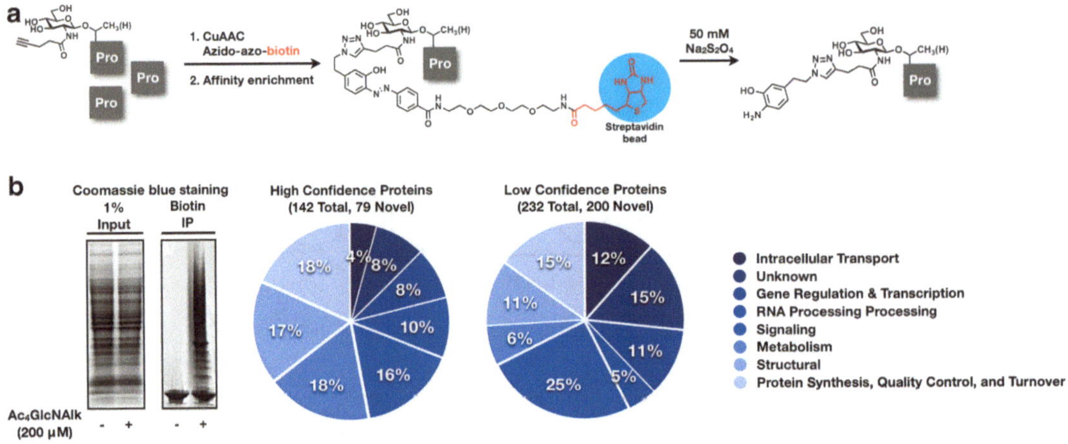

Fig. 3. (**a**) GlcNAlk was metabolically incorporated into NIH-3T3 cells. Following lysis in 4% SDS, soluble proteins were subjected to CuAAC with azido-azo-biotin. Affinity enrichment using streptavidin beads isolated GlcNAlk-modified proteins. The proteins were then liberated from the beads using sodium dithionite (50 mM), separated by SDS-PAGE and subjected to proteolysis. (**b**) Inputs and biotin enriched proteins stained with Coomassie blue. *Large smear at bottom of the gel is streptavidin. LC-MS proteomics were conducted to identify 374 GlcNAlk-modified proteins of high and medium confidence (10).

describe the metabolic incorporation of the alkyne-bearing *N*-acetylglucosamine analog GlcNAlk onto O-GlcNAcylated proteins in NIH-3T3 cells. Following lysis in 4% SDS, soluble proteins were subjected to CuAAC with azido-azo-biotin. Subsequent affinity enrichment with streptavidin beads was used to isolated GlcNAlk-modified proteins. The biotinylated proteins were then liberated from the beads using sodium dithionite (50 mM), separated by SDS-PAGE and subjected to trypsinolysis. Finally, LC-MS was conducted to identify 374 GlcNAlk-modified proteins (Fig. 3b).

2. Materials

All solutions and buffers should be prepared with 18 MΩ H_2O at 25°C. Reagents should be stored at room temperature unless otherwise noted. Dispose of hazardous waste appropriately.

2.1. Materials for Metabolic Incorporation of Chemical Reporters and Preparation of Cell Lysates

1. 1,3,4,6-tetra-O-acetyl-*N*-4-pentynylglucosamine (Ac_4GlcNAlk): 200 mM stock solution in DMSO. Ac_4GlcNAlk was synthesized according to literature procedure (20). To 84.5 mg of Ac_4GlcNAlk add 1 mL DMSO in a microcentrifuge tube. Vortex until homogeneous. Store at −20°C.
2. HyClone Dulbecco's Phosphate Buffered Saline (DPBS) (Thermo Scientific, Rockford, IL, USA): Combine 9.6 g DPBS powder with 1 L H_2O and autoclave.

3. Phenylmethanesulfonylfluoride (PMSF) (Sigma Chemical Company, St. Louis, MO, USA) in H_2O: 250 mM stock solution. Add 43.5 mg PMSF to a microcentrifuge tube and add 1 mL H_2O. Vortex. Store at –20°C.
4. 0.05% SDS Buffer: 0.05% SDS, 10 mM triethanolamine (TEA) pH 7.4, 150 mM NaCl with Complete Mini protease inhibitor cocktail (Roche Biosciences, Indianapolis, IN, USA). Add 25 mg SDS, 746 mg TEA, 4.38 g NaCl to a 1 L graduated cylinder, and add 450 mL H_2O. Mix and adjust pH to 7.4. Add additional H_2O to final volume of 500 mL.
5. Benzonase nuclease (Sigma Chemical Company, St. Louis, MO, USA). Store at –20°C.
6. 4% SDS buffer: 4% SDS, 150 mM NaCl, 50 mM TEA pH 7.4. Combine 40 g SDS, 1.49 g NaCl, 7.46 g TEA, and 950 mL H_2O. Mix and adjust pH to 7.4. Add additional H_2O to final volume of 1 L.
7. Pierce BCA Protein Assay Kit (Thermo Scientific, Rockford, IL, USA).
8. Bovine serum albumin standard (2 mg/mL): 5.52 mg bovine serum albumin (Sigma Chemical Company, St. Louis, MO, USA), 260 μL H_2O, 500 μL 0.05% SDS buffer, 2,000 μL 4% SDS buffer. Aliquot into 500 μL stocks and store at –20°C.
9. 1% Nonidet P-40 (NP-40) buffer: 1% NP-40, 150 mM NaCl, 50 mM TEA pH 7.4, Complete Mini protease inhibitor cocktail (Roche Biosciences, Indianapolis, IN, USA). In a 1 L glass bottle combine 10 g NP-40, 1.49 g NaCl, 7.46 g TEA, and 950 mL H_2O and stir until all reagents go into solution. Adjust pH 7.4. Add additional H_2O to a final volume of 1 L.

2.2. Materials for Click Chemistry, Biotin Enrichment, and Preparation of Samples for LC-MS Analysis

1. Azido-azo-biotin: 10 mM stock solution in DMSO. Azido-azo-biotin was synthesized according to literature procedure (16). To 7.4 mg azido-azo-biotin was added 1 mL DMSO in a microcentrifuge tube. Vortex until solubilized. Pipette out into 100 μL aliquots. Store at –20°C.
2. Tris(2-carboxyethyl)phosphine hydrochloride (TCEP) (Calbiochem, San Diego, CA, USA): 50 mM freshly prepared stock solution in water (see Note 1). Combine 5 mg TCEP and 349 μL H_2O in a microcentrifuge tube. Vortex until TCEP goes into solution. Store on ice during use.
3. Tris[(1-benzyl-1-*H*-1,2,3-triazol-4-yl)methyl]amine (TBTA) (Anaspec, Fremont, CA, USA): 10 mM stock solution in DMSO. Combine 5 mg of TBTA and 942 μL DMSO in a microcentrifuge tube. Vortex until TBTA goes into solution. Divide into

~250 μL aliquots. Store at −20°C solution for short-term or −80°C for 6+ months. Store dry compound at −20°C.

4. $CuSO_4 \cdot 5H_2O$ (Sigma Chemical Company, St. Louis, MO, USA): 50 mM freshly prepared stock solution in water. Combine 5 mg $CuSO_4 \cdot 5H_2O$ with 400 μL H_2O in a microcentrifuge tube. Vortex until solubilized. Store on ice during use.
5. HEPES buffer: 6 M urea, 2 M thiourea, 10 mM HEPES pH 8.0 (see Note 2). Combine 36 mg urea, 15.2 mg thiourea, 238 mg, and 90 mL H_2O. Adjust pH to 8.0 and add H_2O to a final volume of 100 mL.
6. Dithiothreitol (Sigma Chemical Company, St. Louis, MO, USA): 100 mM stock solution in H_2O. Add 15.4 mg dithiothreitol and 1 mL H_2O to a microcentrifuge tube. Vortex until dissolved. Store at −20°C.
7. Iodoacetamide (Sigma Chemical Company, St. Louis, MO, USA): 550 mM freshly prepared stock solution in H_2O. Add 102 mg of iodoacetamide and 1 mL H_2O to a microcentrifuge tube. Vortex until dissolved.
8. Streptavidin agarose resin (Thermo Scientific, Rockford, IL, USA). Store at 4°C.
9. Sodium dithionite solution: freshly prepared 25 mM sodium dithionite, 1% SDS in DPBS (see Note 3). To a 15 mL centrifuge tube add 22 mg sodium dithionite and 5 mL 1% SDS in DPBS. Invert gently until sodium dithionite goes into solution. Sonicate if necessary.
10. YM-10 Centricon 3,000 MWCO filters (Millipore, Billerca, MA, USA).
11. 1× SDS-free loading buffer: 10% glycerol, 0.1% bromophenol blue, 0.7% β-mercaptoethanol. Combine 50 g glycerol, 500 mg bromophenol blue and 500 mL H_2O. Store at room temperature. As needed, add 7 μL β-mercaptoethanol to a 1 mL aliquot of 1× loading buffer. Store the loading buffer containing β-mercaptoethanol at −20°C.
12. Criterion Tris–HCl 4–20% polyacrylamide gel (Bio-Rad, Hercules, CA, USA). Store at 4°C.
13. 50 mM ammonium bicarbonate (ABC) solution in H_2O: In a 250 mL bottle dissolve 395 mg ABC in 100 mL H_2O.
14. Acetonitrile, anhydrous (EMD Chemicals, Gibbstown, NJ, USA).
15. Trifluoroacetic acid (TFA) (Sigma Chemical Company, St. Louis, MO, USA).
16. Trypsin solution: Combine 1 mg trypsin in 15 mL of 50 mM ABC solution in a 15 mL falcon tube. Vortex until solubilized. Aliquot into 1 mL portions. For long-term storage, place at 20°C. Once aliquot is thawed, store at 4°C.

3. Methods

3.1. Metabolic Incorporation of Chemical Reporters and Preparation of Cell Lysates

1. Replace media on cells (twenty 150 mm plates) at 80–85% confluency for 20 mL low-glucose media (see Note 4) containing 200 μM Ac_4GlcNAlk or DMSO vehicle.
2. After 16 h, aspirate off media, and wash cells with 10 mL PBS per plate. Add 2 mL trypsin to each plate and return in incubator for ~3 min or until cells come off the plate.
3. Resuspend cells in 5 mL PBS per plate and combine into two 50 mL centrifuge tubes.
4. Centrifuge at 4°C for 4 min at 3,000 × *g*. Aspirate off supernatant, resuspend both pellets in a total of 30 mL PBS and combine into 1 falcon tube. Centrifuge at 4°C for 4 min at 3,000 × *g*. Aspirate off supernatant. Repeat wash and centrifugation one time.
5. Resuspend washed cell pellet in 200 μL H_2O, 60 μL PMSF, and 500 μL 0.05% SDS buffer. Add 8 μL Benzonase and incubate cells on ice for 30 min.
6. Add 2,000 μL 4% SDS buffer, sonicate the cells in a bath sonicator for 5 min and collect by centrifugation at 20,000 × *g* for 10 min at 15°C. Transfer the soluble fractions to a new 15 mL centrifuge tube.
7. Normalize protein concentration by BCA assay (Pierce, ThermoScientific). Combine 50 parts Reagent A to 1 part Reagent B in 15 mL falcon tube and vortex until green color is homogeneous. Aliquot out 1 mL of working reagent (WR) into a microcentrifuge tube for each sample and an additional four tubes for the standard curve. Pipette 1 μL of soluble lysate into the corresponding centrifuge tube filled with 1 mL WR. For the standard curve add 0, 1, 2, or 4 μL (0, 2, 4, or 8 μg, respectively) of albumin standard to the WR. Place in heat block at 60°C for 30 min. Upon completion, remove all samples from heat block. Transfer the samples to 1 cm plastic cuvettes. With the UV spectrophotometer set to 562 nm, blank the instrument using the standard sample not containing albumin. Take absorbance readings of each sample.

3.2. Click Chemistry, Biotin Enrichment, and Preparation of Samples for LC-MS Analysis

1. In a spreadsheet program, graph the absorbance vs. concentration of the BCA assay albumin standards. Generate a linear best-fit line and determine the concentration of each of the samples using this equation. Dilute the samples with 1% NP-40 buffer to a final concentration of 1 mg/mL (10 mg of total lysate per sample) and transfer to a 50 mL centrifuge tube.
2. Prepare click chemistry cocktail (1,200 μL per 10 mg sample). Combine azido-azo-biotin tag (200 μL, 100 μM, 10 mM

stock solution in DMSO), TCEP (400 μL, 1 mM, 50 mM freshly prepared stock solution in water), tris[(1-benzyl-1-*H*-1,2,3-triazol-4-yl)methyl]amine (TBTA) (200 μL, 100 μM, 10 mM stock solution in DMSO), and $CuSO_4 \cdot 5H_2O$ (400 μL, 1 mM, 50 mM freshly prepared stock solution in water). Vortex gently. Add appropriate amount of click chemistry cocktail to each sample. Vortex gently.

3. Place the samples in the dark and allow the reaction to proceed for 1 h. To quench the reaction and precipitate proteins, add ~10 volumes (12 mL) of ice-cold methanol and place at −80°C overnight.
4. Centrifuge precipitated proteins at 6,000 × *g* for 30 min at 0°C. Wash 3× with 40 mL ice-cold MeOH, taking care to resuspend the pellet each time.
5. Allow the protein pellet to air-dry for 1 h and resuspend in 4 mL of HEPES buffer by bath sonication. Transfer to a new 15 mL centrifuge tube.
6. Incubate captured proteins in freshly made 1 mM dithiothreitol for 40 min to reduce cysteines. Cap cysteines by further incubation with freshly prepared 5.5 mM iodoacetamide for 30 min in the dark.
7. Wash 250 μL streptavidin beads with an equal volume PBS two times and with an equal volume HEPES buffer one time. Resuspend beads in an equal volume HEPES buffer; add beads to captured proteins. Incubate on a rotator for 2 h.
8. Collect beads by centrifugation (2,000 × *g* for 2 min). Wash with HEPES buffer two times, PBS two times, and 1% SDS in PBS two times (10 mL per wash, 2,000 × *g*, 2 min). After the final wash, resuspend beads in 250 μL 1% SDS in PBS and transfer samples to 2 mL dolphin-nosed tubes. Collect beads by centrifugation (2,000 × *g* for 2 min) and carefully pipette away supernatant.
9. Pipette 250 μL of sodium dithionite solution into each sample and incubate for 30 min at room temperature to elute captured proteins. Collect the beads by centrifugation for 2 min at 2,000 × *g* and collect eluent. Repeat elution step with an additional 250 μL of sodium dithionite solution. Combine eluents from both steps.
10. Transfer eluent to a YM-10 Centricon 3,000 MWCO filter and centrifuge at 10,000 × *g* for 30 min. Add an additional 300 μL of PBS into the filter and centrifuge again at 10,000 × *g* for 30 min at room temperature. Transfer the concentrated eluent to a microcentrifuge tube (see Note 5) and dry by SpeedVac overnight (see Note 6).
11. Resuspend the dried pellets in 40 μL 1× SDS-free loading buffer and boil at 98°C for 5 min (see Note 7). Load 36 μL of the

sample onto a Criterion Tris–HCl 4–20% polyacrylamide gel for subsequent in-gel trypsin digestion. Load the remaining sample onto another Criterion Tris–HCl 4–20% polyacrylamide gel for validation of protein candidates by Western blot.

12. Remove each lane of the Criterion Tris–HCl 4–20% gel using a razor blade (see Note 8). Divide each lane evenly into 10 sections. Dice each section into ~0.5 cm squares (see Note 9) and transfer the pieces to a microcentrifuge tube. Add 300 μL of 50 mM ABC and incubate for 15 min. Carefully aspirate away the ABC solution. Repeat 2×.
13. Add 300 μL of a 1:1 solution of 50 mM ABC/acetonitrile and incubate for 30 min. Carefully aspirate away the solution and repeat. Add 300 μL 100% acetonitrile and SpeedVac until dry.
14. Rehydrate gel slices by adding 30 μL trypsin solution and incubate at 37°C in a water bath for 18 h.
15. Add 200 μL 50% acetonitrile in H_2O with 0.1% TFA to elute peptides. Collect eluent and repeat elution. Dry combined eluents by SpeedVac.

3.3. LC-MS Analysis

1. Samples are now ready to be subjected to standard nano-HPLC-MS/MS analysis. Under our conditions, peptides were pressure-loaded onto a 75-μm (inner diameter), 15-cm C18 reverse-phase column, and separated with a gradient running from 95% buffer A (HPLC water with 0.1% (v/v) formic acid) and 5% buffer B (HPLC-grade CH_3CN with 0.1% (v/v) formic acid) to 55% B over 30 min, increased to 95% B over 10 min and held at 95% (v/v) B for 10 min.
2. After one complete MS scan (300–2,000 MW), conduct three data-dependent scans of the *n*th most intense ions with dynamic exclusion enabled. For peptide identification, use SEQUEST version 28 (ThermoFisher Scientific) and search against the appropriate International Protein Index protein sequence database v3.45.v. Compile data using scaffold software (Proteome Software).

4. Notes

1. TCEP degrades over time. Therefore, we recommend storing the reagent at 4°C and purchasing fresh TCEP every 6 months.
2. HEPES buffer should be made fresh for every use. The thiourea can be tricky to get into solution. Do not warm up the solution to solubilize. Sonicate instead.
3. The sodium dithionite solution should be made fresh for each use.

4. Metabolic incorporation conditions should be optimized for the type of glycosylation targeted. We have shown previously that treatment with Ac_4GlcNAlk under low-glucose conditions optimizes the labeling of O-GlcNAc-modified proteins (10).
5. The easiest way to transfer the concentrated eluent is to take a clean microcentrifuge tube, place it over the top of the filter and gently invert the tube right side up. Centrifuge at 500 × *g* for 5 min. Eluent will collect in bottom of new microcentrifuge tube.
6. Set the temperature to 25°C on the SpeedVac and do not allow it to rise above 30°C.
7. Due to the large amount of SDS in the sample, resuspension of the dried pellet can be difficult. The best way to resuspend is to add the 40 μL 1× SDS-free loading buffer and then take a pipette tip, dip it in the loading buffer in the bottom of the tube and then wet the sides of the tube with the buffer to resuspend the sample on the sides. Boil the sample for 5 min at 98°C, take the sample pipette tip, wet the sides again, and then centrifuge at 13,000 × *g* for 1 min. Do not use gel-loading tips to load the sample. Instead use a standard 10–200 μL pipette tip.
8. In order to prevent contamination, use a different razor blade for each lane.
9. Take care in dividing each section into uniform squares as this will improve in-gel trypsin digestion.

Acknowledgments

We thank Leslie Bateman for careful reading of the manuscript. This work was supported by the University of Southern California and the American Cancer Society Grant IRG-58-007-51 (to M.R.P.) and the Ellison Medical Foundation and the National Institutes of Health and General Medical Sciences Grant 1RO1GM087544 O1A2 (to H.C.H.).

References

1. Zielinska DF, Gnad F, Wisniewski JR, Mann M (2010) Precision mapping of an in vivo N-glycoproteome reveals rigid topological and sequence constraints. Cell 141(5):897–907
2. Teo CF et al (2010) Glycopeptide-specific monoclonal antibodies suggest new roles for O-GlcNAc. Nat Chem Biol 6(5):338–343
3. Saxon E, Bertozzi CR (2000) Cell surface engineering by a modified Staudinger reaction. Science 287(5460):2007–2010
4. Hang HC, Yu C, Kato DL, Bertozzi CR (2003) A metabolic labeling approach toward proteomic analysis of mucin-type O-linked glycosylation. Proc Natl Acad Sci USA 100(25):14846–14851
5. Dehnert KW et al (2011) Metabolic labeling of fucosylated glycans in developing zebrafish. ACS Chem Biol 6(6):547–552
6. Dube DH, Prescher JA, Quang CN, Bertozzi CR (2006) Probing mucin-type O-linked

glycosylation in living animals. Proc Natl Acad Sci USA 103(13):4819–4824

7. Zaro BW, Bateman LA, Pratt MR (2011) Robust in-gel fluorescence detection of mucin-type O-linked glycosylation. Bioorg Med Chem Lett 21(17):5062–5066
8. Hanson SR et al (2007) Tailored glycoproteomics and glycan site mapping using saccharide-selective bioorthogonal probes. J Am Chem Soc 129(23):7266–7267
9. Boyce M et al (2011) Metabolic cross-talk allows labeling of O-linked {beta}-N-acetylglucosamine-modified proteins via the N-acetylgalactosamine salvage pathway. Proc Natl Acad Sci USA 108(8):3141–3146
10. Zaro BW, Yang Y-Y, Hang HC, Pratt MR (2011) Chemical reporters for fluorescent detection and identification of O-GlcNAc-modified proteins reveal glycosylation of the ubiquitin ligase NEDD4-1. Proc Natl Acad Sci USA 108(20):8146–8151
11. Sprung R et al (2005) Tagging-via-substrate strategy for probing O-GlcNAc modified proteins. J Proteome Res 4(3):950–957
12. Nandi A et al (2006) Global identification of O-GlcNAc-modified proteins. Anal Chem 78(2):452–458
13. Clark PM et al (2008) Direct in-gel fluorescence detection and cellular imaging of O-GlcNAc-modified proteins. J Am Chem Soc 130(35):11576–11577
14. Wang Z et al (2010) Enrichment and site mapping of O-linked N-acetylglucosamine by a combination of chemical/enzymatic tagging, photochemical cleavage, and electron transfer dissociation mass spectrometry. Mol Cell Proteomics 9(1):153–160
15. Wang Z et al (2010) Extensive crosstalk between O-GlcNAcylation and phosphorylation regulates cytokinesis. Science STKE 3(104):ra2
16. Yang Y-Y, Ascano JM, Hang HC (2010) Bioorthogonal chemical reporters for monitoring protein acetylation. J Am Chem Soc 132(11):3640–3641
17. Fonović M, Verhelst SHL, Sorum MT, Bogyo M (2007) Proteomics evaluation of chemically cleavable activity-based probes. Mol Cell Proteomics 6(10):1761–1770
18. Verhelst SHL, Fonović M, Bogyo M (2007) A mild chemically cleavable linker system for functional proteomic applications. Angew Chem Int Ed 46(8):1284–1286
19. Denny JB, Blobel G (1984) 125I-labeled crosslinking reagent that is hydrophilic, photoactivatable, and cleavable through an azo linkage. Proc Natl Acad Sci USA 81(17): 5286–5290
20. Gurcel C et al (2008) Identification of new O-GlcNAc modified proteins using a click-chemistry-based tagging. Anal Bioanal Chem 390(8):2089–2097
21. Saxon E et al (2002) Investigating cellular metabolism of synthetic azidosugars with the Staudinger ligation. J Am Chem Soc 124(50): 14893–14902
22. Rabuka D, Hubbard SC, Laughlin ST, Argade SP, Bertozzi CR (2006) A chemical reporter strategy to probe glycoprotein fucosylation. J Am Chem Soc 128(37): 12078–12079
23. Hsu T-L et al (2007) Alkynyl sugar analogs for the labeling and visualization of glycoconjugates in cells. Proc Natl Acad Sci USA 104(8):2614–2619
24. Goon S, Schilling B, Tullius MV, Gibson BW, Bertozzi CR (2003) Metabolic incorporation of unnatural sialic acids into Haemophilus ducreyi lipooligosaccharides. Proc Natl Acad Sci USA 100(6):3089–3094

Chapter 6

Characterization of Membrane-Associated Glycoproteins Using Lectin Affinity Chromatography and Mass Spectrometry

Yashu Liu, Jintang He, and David M. Lubman

Abstract

Membrane-associated glycoproteins play critical roles in many biological processes and are often the therapeutic targets for drug discovery. Lectin affinity chromatography is one of the most widely used approaches for enrichment of glycoproteins at the protein level. Here, we describe a strategy for the characterization of membrane glycoproteins including membrane protein extraction, lectin affinity chromatography, protein digestion, and analysis by LC-MS/MS.

Key words: Membrane, Glycoprotein, Lectin affinity chromatography, LC-MS/MS

1. Introduction

Membrane-associated glycoproteins (MAGs) perform important functions in cell signaling, cell–cell interaction and recognition, cell differentiation and growth, and cell movement (1–3). MAGs represent a promising source for disease biomarkers and molecular targets for drug development. Traditional strategies for large-scale characterization of MAGs include two-dimensional polyacrylamide gel electrophoresis (2D-PAGE) coupled MALDI mass spectrometry (4) and liquid chromatography (LC) coupled electrospray ionization tandem mass spectrometry (ESI-MS) (5). The 2D-PAGE based method is limited by low recovery, poor reproducibility, and small dynamic range for detection. LC coupled ESI-MS provides a more powerful approach for the analysis of MAGs.

Lectins are a class of proteins which can specifically bind to the sugar moieties of glycoproteins (6). Lectin affinity chromatography has greatly facilitated the analysis of intact glycoproteins (7–9).

Jennifer J. Kohler and Steven M. Patrie (eds.), *Mass Spectrometry of Glycoproteins: Methods and Protocols*, Methods in Molecular Biology, vol. 951, DOI 10.1007/978-1-62703-146-2_6, © Springer Science+Business Media, LLC 2013

A mixture of proteins containing glycoproteins and non-glycoproteins are incubated with an immobilized lectin, the nonbinding proteins are washed away, and the bound proteins are eluted by competing monosaccharides.

Herein, we describe an effective approach combining membrane-extraction and lectin affinity chromatography for the preparation of MAGs from a human glioblastoma-derived stem-like neurosphere HSR-GBM1 and a traditional human glioblastoma cell line U373 (10). Enriched glycoproteins were then characterized by LC-MS/MS, which enables automated protein identification and quantification.

2. Materials

Prepare all the solutions using ultrapure water (deionized water with a sensitivity of 18 MΩ cm at 25°C) and analytical grade gradients. Prepare all the reagents at room temperature and store them at 4°C (unless indicated otherwise).

2.1. Membrane Protein Extraction Components

1. Membrane protein extraction buffer: 20 mM tris(hydroxymethyl) aminomethane (Tris), pH 7.4, 150 mM NaCl, and 1% (w/v) octyl-β-D-glucopyranoside (see Note 1). Store at 4°C. Add 1% protease inhibitor (Sigma-Aldrich, St. Louis, MO, USA) mixture immediately before use.
2. PBS buffer: phosphate buffered saline (PBS) powder (Sigma-Aldrich, St. Louis, MO, USA) dissolved in distilled water.
3. Dounce glass homogenizer with a tight-fitting pestle (Products, Millville, NJ, USA).
4. Temperature controlled centrifuge (Thermo Fisher Scientific, Waltham, MA, USA).
5. 1.5 mL low-retention eppendorf tubes.
6. Protein Assay Kit (Bio-rad, Hercules, CA, USA) (see Note 2).
7. Cell scraper (Corning Inc., Corning, NY, USA).

2.2. Lectin Affinity Chromatography Components

1. Agarose-bounded peanut agglutinin (PNA) (Vector Laboratories, Burlingame, CA, USA) (see Note 3).
2. Binding buffer: 20 mM Tris–HCl, pH 7.4, 150 mM NaCl, 1 mM $MgCl_2$, 1 mM $CaCl_2$, and 1 mM $MnCl_2$.
3. Elution buffer: 200 mM D-galactose (Sigma-Aldrich, St. Louis, MO, USA) in binding buffer (see Note 4).
4. 2 mL disposable screw endcap spin column (Thermo Fisher Scientific, Waltham, MA, USA).
5. Microcon YM-10 centrifugal filter (Millipore, Billerica, MA, USA).
6. 25 mM ammonium bicarbonate (Sigma-Aldrich, Waltham, MA, USA).

2.3. Trypsin and PNGase F Digestion Components

1. Reducing buffer: 500 mM Tris(2-carboxyethyl)phosphine (TCEP) (Thermo Fisher Scientific, Waltham, MA, USA), store at room temperature (see Note 5).
2. Alkylation buffer: 500 mM iodoacetamide (Sigma-Aldrich, Waltham, MA, USA) solution. Prepare the buffer immediately before use and keep it away from light.
3. TPCK modified sequencing grade porcine trypsin (Promega, Madison, WI, USA).
4. PNGase F (New England Biolabs, Ipswich, MA, USA).
5. Speedvac concentrator system (Thermo Fisher Scientific, Waltham, MA, USA).
6. Formic acid (Sigma-Aldrich, Waltham, MA, USA).

2.4. LC-MS/MS Components

1. Paradigm MG4 micropump system (Michrom Biosciences, Inc., Auburn, CA, USA).
2. Paradigm Platinum Peptide Nanotrap (Michrom Biosciences, Inc., Auburn, CA, USA).
3. Magic C18AQ column (0.1 mm × 150 mm, 5 μm, 200 Å, Michrom Biosciences, Inc., Auburn, CA, USA).
4. Solvent A: 0.3% formic acid in H_2O (HPLC grade) (Sigma-Aldrich, Waltham, MA, USA).
5. Solvent B: 0.3% formic acid in Acetonitrile (HPLC grade) (Sigma-Aldrich, Waltham, MA, USA).
6. LTQ linear ion trap mass spectrometer (Thermo Fisher Scientific, Waltham, MA, USA) (see Note 6).
7. Paradigm Metal Spray Tip (Michrom Biosciences, Inc., Auburn, CA, USA).

3. Methods

The strategy for the characterization of MAGs involves solubilization of membrane proteins, enrichment of glycoproteins, and LC-MS/MS analysis of these proteins. The whole process may take ~1 week from the sample preparation to the data analysis. The work flow of this strategy is shown in Fig. 1.

3.1. Solubilization of Membrane Proteins (1st Day)

Carry out all procedures at 4°C unless otherwise specified.

1. Collect ~10 million cells by scraping and wash the cells twice with PBS, centrifuge at 500 × *g* for 5 min at room temperature. Store the cell pellets in −80°C until use (see Note 7).
2. Suspend the cell pellets in 500 μL of membrane protein extraction buffer, then transfer the suspension to a Dounce glass

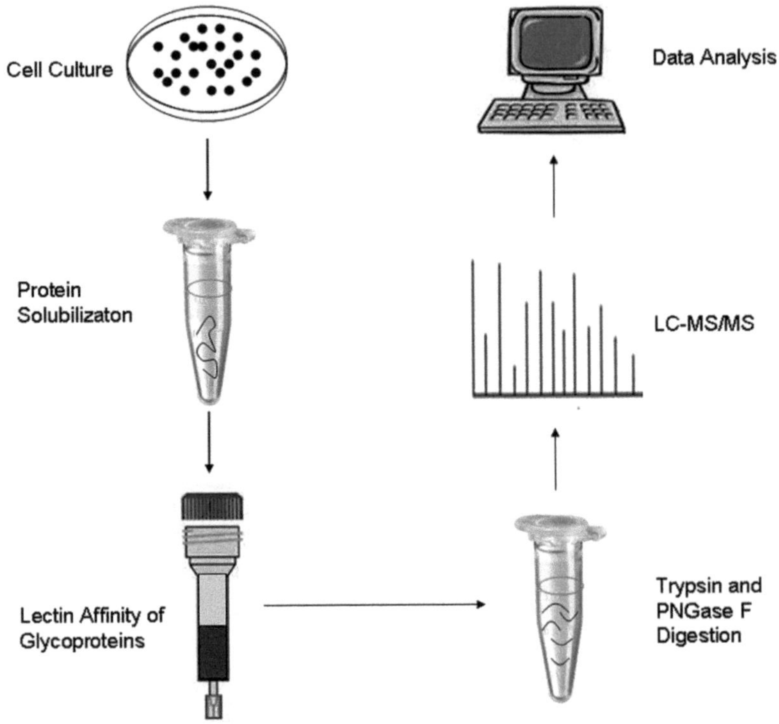

Fig. 1. Workflow of the strategy for the characterization of membrane-associated glycoproteins.

homogenizer and homogenize the cells with 30 strokes using a tight-fitting pestle. Be careful to avoid bubbles.

3. Transfer the cell lysate to a 1.5 mL eppendorf tube and centrifuge at 40,000 × *g* for 30 min at 4°C. Collect the supernatants and measure protein concentration. Do not discard the pellets until confirming that the lysis is successful. Adjust the final protein concentration to 2 mg/mL (see Note 8). The sample should be applied to the enrichment step within the same day of protein extraction. Freeze-thaw cycle may cause the precipitation of membrane proteins.

3.2. Enrichment of Membrane-Associated Glycoproteins by Lectin Affinity (1st Day)

Carry out all procedures at 4°C unless otherwise specified.

1. Pack 1 mL of agarose-bound PNA into a 2 mL Pierce centrifuge column with filters at both ends. Wash the column with 3 mL of binding buffer (see Note 9).
2. Dilute the cell lysate four times with binding buffer, resulting in a protein concentration of ~0.5 mg/mL. Load 1 mL (500 µg) of sample to the column, incubate for 15 min, and collect the flow-through. Reload the flow-through to the lectin column and incubate for another 15 min. Do not discard the recollected flow-through until confirming that the enrichment is successful.

3. Wash the column with 4 mL of binding buffer to remove the nonspecific binding proteins (see Note 10). Do not discard the recollected flow-through until confirming that the enrichment is successful.
4. Add 2 mL of elution buffer to the column, incubate for 10 min and collect the flow-through. Repeat this step and pool the eluted fractions.
5. Concentrate the eluate and change the buffer to 25 mM ammonium bicarbonate with Micron YM-10 centrifugal filter device to a final volume of 200 μL. Determine protein concentration and store the eluate at –80°C until further use (see Note 11).

3.3. Trypsin and PNGase F Digestion (2nd and 3rd Days)

1. Add reducing buffer (500 mM TCEP) to the sample to reach a final concentration of 5 mM and incubate for 30 min at room temperature.
2. Add alkylation buffer (500 mM iodoacetamide) to the sample to a final concentration of 25 mM and incubate in the dark for 20 min.
3. Add ~0.5 μg trypsin to a final ratio between 1:20 and 1:50 and incubate the resulting mixture for 12–16 h at 37°C.
4. Add 1 μL of formic acid to the digest to stop the enzymatic reaction and dry the sample with a speedvac concentrator.
5. Suspend the dried sample with 25 mM ammonium bicarbonate by vortex, then add 0.5 μL of PNGase F (250 U) to the mixture, thoroughly mix and incubate for 16–20 h at 37°C.
6. Add 1 μL formic acid to the digest to stop the enzymatic reaction and dry the sample with a speedvac concentrator. Keep the dried sample at –80°C for further analysis with mass spectrometry.

3.4. LC-MS/MS (4th Day)

1. Dissolve the sample in 0.3% formic acid and thoroughly mix. Centrifuge sample at 30,000 × *g* for 10 min to remove the insoluble material which may be present in the sample and clog the columns.
2. Separate the peptides with the Paradigm MG4 micropump system. A 120 min linear gradient of acetonitrile (solvent B) / water (solvent A) containing 0.3% formic acid is used at a flow rate of 300 nL/min (see Notes 12 and 13). The gradient is as follows:

   0–10% B for 10 min.

   10–30% B for 80 min.

   30–40% B for 10 min.

   40–100% B for 8 min.

   100% B for 1 min.

   100–0% B for 1 min.

   0% B for 10 min.

3. A LTQ linear ion trap mass spectrometer is set up to operate in *positive ion mode* and *data dependent mode*. For each cycle of full mass scan (range of m/z 400–2,000) the five most intense ions in the spectrum were selected for MS/MS analysis. Set the *dynamic exclusion* time at 60s, the ESI spray voltage at 2.5 kV, and the capillary voltage at 30 V. The ion activation was achieved by utilizing helium at a normalized collision energy of 35%. Xcalibur software is used to acquire the data (see Notes 14 and 15).

3.5. Data Analysis (5th Day)

1. SEQUEST algorithm version 27 incorporated in Bioworks software version 3.1 SR1 was used to perform the search of all MS/MS spectra. The search parameters were as follows: (1) *fixed modification*, carbamidomethyl of cysteine; (2) *variable modification*, oxidation of M and asparagine to aspartate; (3) allow two missed cleavages; (4) peptide ion mass tolerance 1.50 Da; (5) fragment ion mass tolerance 0.0 Da; (6) peptide charges +1, +2, and +3 (see Notes 14–17).
2. The SEQUEST results were then filtered using Trans-proteomic Pipeline (TPP) (11) to minimize false positives. TPP is a tool developed by the Aebersold group to generate probabilities of protein identifications based on MS/MS data. In the TPP, the PeptideProphet software uses various SEQUEST scores (*Xcorr*, Δ*Cn*, and *Sp*) to calculate a probability score for each identified peptide. The peptides are then assigned for protein identification using the ProteinProphet software. ProteinProphet is applied to evaluate predictable sensitivity and false positive rates of protein identification (see Note 18). Both the PeptideProphet probability and ProteinProphet probability scores were set to be higher than 0.9 (see Note 19), which resulted in a false discovery rate (FDR) below 1% (12, 13).

A representative MS/MS spectrum of a peptide is shown in Fig. 2. The sequence of the peptide was identified as DIEEGAIVNPGR.

4. Notes

1. The nonionic detergent octyl-β-D-glucopyranoside can improve the solubility of membrane proteins (14). Other substitutes such as NP-40 (15), ASB-14, and Triton X-100 may also be used.
2. The protein concentration is measured by Bradford Protein Assay (16). Other methods such as BCA assay, etc. can also be applied to determine the amount of proteins in the cell lysate.

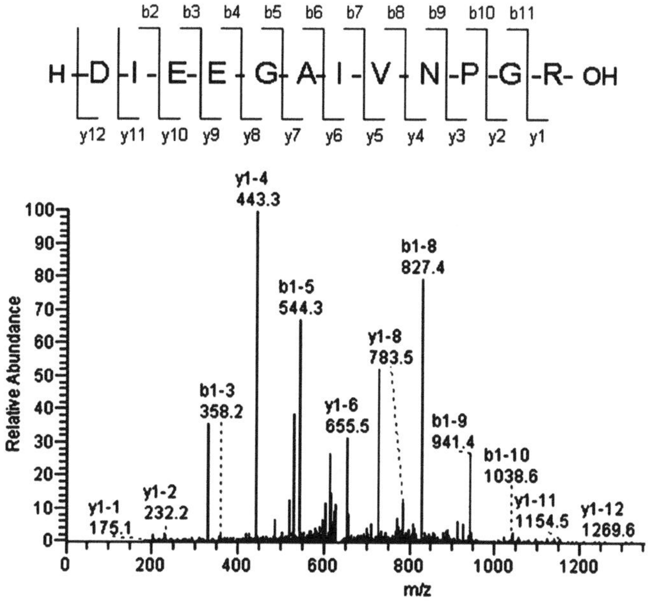

Fig. 2. A representative MS/MS spectrum of a peptide from the glycoprotein receptor-type tyrosine-protein phosphatase zeta. The sequence of the peptide was identified as DIEEGAIVNPGR.

3. PNA is a lectin which has a binding specificity to galactosylated proteins (17). The lectin was chosen because a lectin microarray analysis showed that this lectin could distinguish the cancer stem-like neurosphere HSR-GBM1 from the traditional glioblastoma cell line U373 (10).
4. Elution buffer should be prepared no more than 24 h before use. Different monosaccharides should be chosen for the elution based on the binding preference of the chosen lectins.
5. The working concentration of TCEP is 5 mM. The stock solution should be stored at room temperature and be added to sample to reach a final concentration of 5 mM. Other reducing reagents such as dithiothreitol (DTT) can also be used as a substitute for reducing the disulfide bond.
6. Any mass spectrometer capable of tandem MS with automated data acquisition can be used.
7. Do not use a trypsin-based method to harvest cells because it may cause damage of the membrane-associated proteins.
8. Usually ~1 mg of total protein can be obtained from ten million cells, but the yield may be different for different cell lines.
9. Agarose-bound lectins are usually stored in 50% bead slurry, so 2 mL of the slurry should be added to reach 1 mL of settled resin. The sample volume is usually the same as beads volume. When dealing with multiple samples, it is important to make sure the same volumes of buffer and beads are used for the

enrichment of each sample. Different lectins may have different binding capacity with their preferred glycoproteins; therefore, a preparative test should be performed to determine the loading ratio between lectin and samples.

10. The wash step is applied to remove unbonded proteins from the lectin column. The wash volume can be determined by monitoring the protein concentration of flow-through. For most experiments, it is sufficient to remove the nonspecific bindings by washing 4–6 beads volume.
11. The yield of glycoproteins varies between cell lines and depends on the lectins used. Sometimes the protein concentration after lectin extraction is too low to be measured by a protein concentration kit. Then, SDS-PAGE coupled with a silver staining method can be used to evaluate the protein amount by loading 10–20 μL of sample.
12. The time of the gradient and the solvent gradient can be adjusted to achieve the best performance of peptide separation. For each sample, a test run should be done to determine all the parameters for the separation.
13. Monitor pump pressure to avoid exceeding the limit. Three pumps are used in the Paradigm MG4 micropump system. Pump A is used for solvent A, pump B is used for solvent B, and pump C is used for sample loading. Both the trap column and the separation column should be changed routinely to get the best performance of separation and maintenance of the pumps.
14. The *data dependent mode* means that the mass spectrometer acquires one full MS-scan, followed by 5 MS/MS scans from the top five most intense ions detected in the previous MS scan.
15. *Dynamic exclusion* means that if a peptide is selected and analyzed in MS/MS mode, the ion is excluded for a period of time. Dynamic exclusion is usually employed to increase the number of identified proteins.
16. Other algorithms such as Mascot can also be used for the data searching. There are also other protein databases available for searching, including: Swiss-prot, IPI, NCBI_NR, etc.
17. The parameters used for database searching mostly depend on the type of the mass spectrometer and the type of the sample. Different mass spectrometers have different mass tolerance and resolution.
18. TPP supports different search engines such as SEQUEST, Mascot, etc. The SEQUEST search results are in the form of .out files and TPP analyzes the search results in pepXML format.

19. It is very important to implement the FDR control to reduce false positive identifications. TPP is a straightforward approach for this kind of analysis. Other methods may also be applied.

Acknowledgment

This work was supported by the National Cancer Institute under grant R21 CA124441 (D.M.L.), and the National Institutes of Health Grant No. R01 GM49500 (D.M.L.).

References

1. Gahmberg CG, Tolvanen M (1996) Why mammalian cell surface proteins are glycoproteins. Trends Biochem Sci 21:308–311
2. Ashwell G, Harford J (1982) Carbohydrate-specific receptors of the liver. Annu Rev Biochem 51:531–554
3. Millette CF, Scott BK (1984) Identification of spermatogenic cell plasma membrane glycoproteins by two-dimensional electrophoresis and lectin blotting. J Cell Sci 65:233–248
4. Santoni V, Molloy M, Rabilloud T (2000) Membrane proteins and proteomics: un amour impossible? Electrophoresis 21:1054–1070
5. Wu CC, Yates JR III (2003) The application of mass spectrometry to membrane proteomics. Nat Biotechnol 21:262–267
6. Goldstein IJ, Hayes CE (1978) The lectins: carbohydrate-binding proteins of plants and animals. Adv Carbohydr Chem Biochem 35:127–340
7. Ghosh D, Krokhin O, Antonovici M, Ens W, Standing KG, Beavis RC et al (2004) Lectin affinity as an approach to the proteomic analysis of membrane glycoproteins. J Proteome Res 3:841–850
8. Hart DA (1980) Lectins in biological systems: applications to microbiology. Am J Clin Nutr 33:2416–2425
9. Dai Z, Zhou J, Qiu SJ, Liu YK, Fan J (2009) Lectin-based glycoproteomics to explore and analyze hepatocellular carcinoma-related glycoprotein markers. Electrophoresis 30:2957–2966
10. He J, Liu Y, Xie X, Zhu T, Soules M, DiMeco F et al (2010) Identification of cell surface glycoprotein markers for glioblastoma-derived stem-like cells using a lectin microarray and LC-MS/MS approach. J Proteome Res 9:2565–2572
11. Keller A, Nesvizhskii AI, Kolker E, Aebersold R (2002) Empirical statistical model to estimate the accuracy of peptide identifications made by MS/MS and database search. Anal Chem 74:5383–5392
12. Nesvizhskii AI, Keller A, Kolker E, Aebersold R (2003) A statistical model for identifying proteins by tandem mass spectrometry. Anal Chem 75:4646–4658
13. Schiess R, Mueller LN, Schmidt A, Mueller M, Wollscheid B, Aebersold R (2009) Analysis of cell surface proteome changes via label-free, quantitative mass spectrometry. Mol Cell Proteomics 8:624–638
14. Garner AE, Smith DA, Hooper NM (2008) Visualization of detergent solubilization of membranes: implications for the isolation of rafts. Biophys J 94:1326–1340
15. Wang Y, Ao X, Vuong H, Konanur M, Miller FR, Goodison S et al (2008) Membrane glycoproteins associated with breast tumor cell progression identified by a lectin affinity approach. J Proteome Res 7:4313–4325
16. Bradford MM (1976) A rapid and sensitive method for the quantitation of microgram quantities of protein utilizing the principle of protein-dye binding. Anal Biochem 72:248–254
17. Lotan R, Skutelsky E, Danon D, Sharon N (1975) The purification, composition, and specificity of the anti-T lectin from peanut (Arachis hypogaea). J Biol Chem 250: 8518–8523

Chapter 7

Sialic Acid Capture-and-Release and LC-MS^n Analysis of Glycopeptides

Jonas Nilsson and Göran Larson

Abstract

Extracellular glycoproteins frequently carry terminal sialic acids on their N-linked and/or O-linked glycan structures. In this chapter a sialic acid specific capture-and-release protocol for the enrichment of N- and O-glycopeptides originating from glycoproteins in complex biological samples is described. The enriched glycopeptides are subjected to reversed phase liquid chromatography (LC) interfaced with electrospray ionization and multistage tandem mass spectrometry (MS^n). The glycopeptide precursor ions are fragmented by collision-induced dissociation (CID) for analysis of the glycan parts in the MS^2 spectra. Further fragmentation (i.e., MS^3) of deglycosylated peptide ions results in peptide backbone fragmentation, which is used in protein database searches to identify protein sequences. For O-glycopeptides the use of both CID and electron capture dissociation (ECD) fragmentation of the peptide backbone with intact glycans still attached are used to pinpoint the glycosylation sites of glycopeptides containing several Ser/Thr residues. The step-by-step protocols for fragmentation analyses of O- and N-glycopeptides enriched from human cerebrospinal fluid are described.

Key words: Sialic acid, Periodate oxidation, Glycopeptide, CID, ECD

1. Introduction

Glycosylation represents the most complex but also the most common post-translational modification of proteins (1, 2). Specific glycosyltransferases are responsible for the initiation and stepwise addition of each monosaccharide to build up the final glycans. The glycans are usually either O-glycosidically linked to Ser and Thr residues (O-glycans), or N-glycosidically linked to Asn residues (N-glycans) of the Asn-X-Ser/Thr/Cys, X ≠ Pro consensus sequence. The repertoire of glycoproteins in a proteome differs depending upon which glycosyltransferases, sugar-nucleotides, acceptors (proteins), and

Jennifer J. Kohler and Steven M. Patrie (eds.), *Mass Spectrometry of Glycoproteins: Methods and Protocols*, Methods in Molecular Biology, vol. 951, DOI 10.1007/978-1-62703-146-2_7, © Springer Science+Business Media, LLC 2013

glycosidases that are available, which add extra dimensions of complexity to any biological system. A common theme for N- and O-glycans is that they frequently are terminated with sialic acids, of which *N*-acetyl-5-neuraminic acid is the typical one for human glycoproteins (3). Sialic acid is a critical constituent of the sialyl Lewis x epitope that mediates selectin binding in cellular homing mechanisms (4), and is also an active part of host cellular receptors for adhesion and infection of a range of pathogenic human viruses (5). It is thus important to qualitatively and quantitatively characterize protein glycosylation in order to understand physiological processes where glycans are involved. In the field of glycomics the glycans are typically detached from their proteins, and analyzed with liquid chromatography (LC) and/or MS (6–9). The important issue regarding which proteins in a preparation that are glycosylated and specifically at what sites can however not be addressed by this approach. To characterize the site-specific glycosylation profiles of proteins it would be highly advantageous to study glycopeptides where it is possible to extract both glycosylation site and glycan structure information from the same molecule (10–12). Glycopeptides may be obtained by the *in-solution* or *in-gel* (13) protease digestions of glycoproteins. A complicating matter is however that the digestion will result in a complex mixture of peptides and glycopeptides, of which many will co-elute during the LC, and through ion suppression hamper the MS analysis of individual glycopeptides. Naturally, the situation gets even more difficult if one wants to study glycopeptides originating from proteomic samples such as plasma or cerebrospinal fluid (CSF), which contain complex mixtures of glycoproteins and unglycosylated proteins. Unglycosylated albumin is for instance the dominating protein in plasma and in CSF with a concentration of 35–50 mg/mL (14) and 0.1–0.3 mg/mL, respectively. For the efficient analysis of glycopeptides it is thus important to develop methods aimed at enriching glycopeptides from the vast majority of unglycosylated peptides in protease digested biological samples. The hydrophilic character of glycopeptides has been exploited for their enrichments in hydrophilic interaction chromatography (HILIC) (15–19). A second strategy has been to use glycan-binding lectins for glycopeptide purification purposes (20–25). Thirdly, affinity binding using titanium dioxide and graphite matrices has been used for the enrichment of sialylated glycopeptides (26–28). A common theme for these three approaches is that they are based on noncovalent interactions and harsh washing conditions can often not be used to remove nonspecific binding originating from various unglycosylated peptides, which interfere in the enrichment and LC-MS^n steps. For glycopeptide capture-techniques based on covalent binding on the other hand it is amendable to utilize harsh washing of the solid phase in order to remove most nonspecific binding and accomplish relatively pure glycopeptide samples.

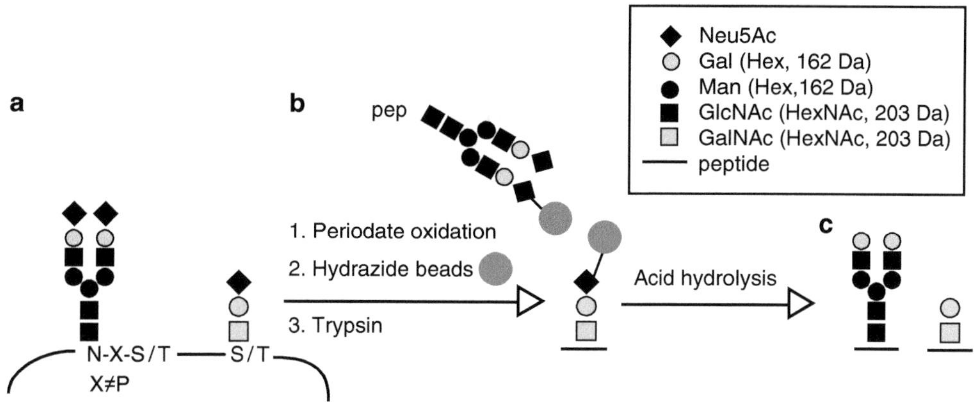

Fig. 1. Capture-and-release protocol to enrich sialylated glycopeptides originating from glycoproteins in complex biological samples. (**a**) Both N-glycans (left structure) and O-glycans (right structure) frequently carry terminal sialic acids (Neu5Ac). (**b**) The sample is periodate oxidized, hydrazide beads are added and the captured glycoproteins are trypsin digested. (**C**) The desialylated N- and O-glycopeptides are released by formic acid hydrolysis. The depiction of monosaccharides is according to the consortium of functional glycomics (CFG, a full list is available at http://glycomics.scripps.edu/CFGnomenclature.pdf). Neu5Ac, *N*-acetyl-5-neuraminic acid (sialic acid); Gal, galactose; Man, mannose; GlcNAc, *N*-acetylglucosamine; GalNAc, *N*-acetylgalactosamine; Hex, hexose; HexNAc, *N*-acetylhexosamine.

We have developed a method for the covalent capture-and-release of tryptic N- and O-glycopeptides from biological samples (29). In short, sialic acid-terminated glycoproteins are selectively periodate oxidized to carry an aldehyde functional group on their glycerol chain (Fig. 1a) which are then covalently captured onto hydrazide beads via a hydrazone linkage (Fig.1b). The hydrazide beads are commercially available from Bio-Rad. The captured glycoproteins are then trypsin digested while remaining on the beads, tryptic peptides are removed by extensive washing and the tryptic glycopeptides are released from the beads by acid hydrolysis of the pH sensitive sialic acid glycosidic bond (Fig. 1c). We thus specifically enrich for sialic acid containing glycopeptides over two selective steps, and the sialic acids are consequently lost from the glycopeptides as a part of the enrichment protocol.

For the analysis of enriched glycopeptides we have used nano-reversed phase liquid chromatography electrospray ionization (ESI) interfaced with multistage tandem mass spectrometry (nLC-ESI-MS^n) on a linear ion trap quadrupole-Fourier transform-ion cyclotron resonance (LTQ-FT-ICR) mass spectrometer. The molecular weights of precursor ions are measured in the ICR cell. By selection and collision-induced dissociation (CID) of glycopeptide precursors in the LTQ we observe glycosidic fragmentation patterns in the MS^2 spectrum. Through further CID of deglycosylated ions, to get the second-generation fragment ions (MS^3), the peptide backbones are fragmented, which is used in Mascot database searches to identify the peptides (and thus also the glycoproteins). We characterized 36 N-linked and 44 O-linked glycosylation sites on glycoproteins from human CSF (29), and submitted the protein glycosylation data to the UniprotKB/

Swissprot database (see http://www.uniprot.org/uniprot/?query=citation:19838169 for a list of the entries). We recently immunopurified α-dystroglycan from a human skeletal muscle sample, did *in-gel* trypsin digestion and identified O-glycopeptides and peptides by LC-MS^n analysis (30). In parallel, we used the sialic acid capture-and-release methodology on immunopurified α-dystroglycan samples, which provided us with complementary information of α-dystroglycan glycosylation, and demonstrated the high sensitivity and specificity of the protocol (30).

A method for periodate oxidation followed by covalent capture of glycoproteins onto hydrazide beads for glycoproteomics purposes was originally developed in the Aebersold lab (31–34). They used peptide-*N*-glycosidase F (PNGase F) to liberate previously N-glycosylated peptides from the beads. Thus, N-glycosylation sites could be mapped but no information about N-glycan structures or the analysis of O-glycopeptides was possible to obtain. Methods for periodate oxidation, covalent capture, and release of N-glycosylated peptides for MS analysis have also been developed independently in the Nishimura lab (35, 36).

2. Materials

2.1. CSF Samples and Model Glycoproteins

1. The CSF samples are collected in the clinic by lumbar puncture for the diagnosis of suspected infection of the central nervous system. We use the remaining parts of such CSF samples in our studies, and aliquotes are picked from those that were found to be normal based on the assayed white cell count and blood-brain barrier function. The CSF samples are de-identified (removal of all patient information), before they arrive in our lab (see Note 1).
2. Transferrin purified from human serum (Sigma-Aldrich, St. Louis, USA) and bovine fetuin from fetal calf serum (Sigma-Aldrich, St. Louis, USA).
3. Phosphate buffered saline (PBS, 10 mM phosphate, 100 mM NaCl, pH 7.2).

2.2. Periodate Oxidation and Capture onto Hydrazide Beads

1. 0.04 M periodic acid, in water, freshly prepared.
2. 50% Glycerol (glycerol: water, 1:1 *by volume*).
3. Capture buffer (100 mM sodium acetate, 100 mM NaCl, pH 4.5).
4. Sephadex PD-10 columns (GE healthcare, Uppsala, Sweden).
5. Hydrazide beads (Bio-Rad, Hercules, USA).

2.3. On-Bead Alkylation, Trypsin Cleavage, and Release of Glycopeptides

1. 50 mM ammonium bicarbonate buffer (NH_4HCO_3, pH 8), freshly prepared (see Note 2).

2. 10 mM dithiothreitol (DTT) in ammonium bicarbonate buffer, freshly prepared.
3. 50 mM iodoacetamide (IAA), in ammonium bicarbonate buffer, freshly prepared.
4. 0.05% Tween-20 in PBS.
5. 8 M urea in ammonium bicarbonate buffer.
6. 0.1 μg/μL trypsin in resuspension buffer (sequencing grade porcine trypsin, Promega, Madison, USA).
7. Prelubricated microcentrifuge tubes (Costar, Cambridge, USA).
8. 50% Acetonitrile (acetonitrile:water, 50:50 *by volume*).
9. 20% Acetonitrile (acetonitrile:water, 20:80 *by volume*).
10. 0.1 M formic acid, freshly prepared.

3. Methods

3.1. Sialic Acid Capture-and-Release of Glycopeptides

3.1.1. Mild Periodate Oxidation of Sialylated Glycoproteins

1. CSF samples (10 mL) were centrifuged at 1,800 × *g* for 10 min to remove insolubles, portioned in 1 mL-fractions, and stored at –80°C before analysis.
2. Human transferrin (0.1 mg in 100 μL PBS), bovine fetuin (0.1 mg in 100 μL PBS), or CSF samples (1.0 mL) in 1.5 mL-microcentrifuge tubes are placed on an ice/water cooling bath for 10 min.
3. 0.04 M periodic acid is also placed on the ice/water-cooling bath for 10 min.
4. Add 50 μL of the cold 0.04 M periodic acid to the cold 1 mL protein samples and 5 μL 0.04 M periodic acid to 0.1 mL protein samples so the final periodic acid concentration is 2 mM (1.9 mM).
5. Briefly vortex the tubes and place them on the ice/water bath for 10 min. Vortex the tubes briefly twice more during the 10 min incubation period.
6. Add 5 μL of 50% glycerol to the tubes and vortex in order to quench further oxidation of glycoproteins.
7. Allow the reaction mixtures to reach room temperature.
8. Equilibrate Sephadex PD-10 columns, one for each glycoprotein sample, with at least 30 mL capture buffer (see Note 3).
9. Dilute oxidized transferrin and/or fetuin samples to 500 μL with coupling buffer (see Note 4).
10. Cap the PD-10 column in the bottom. Add one glycoprotein sample to each PD-10 column and remove the cap.

11. When the sample solutions have absorbed into the column material add 2 mL capture buffer for 0.5 mL samples and 1.5 mL capture buffer for 1 mL samples. These 2.5 mL fractions constitute the void volume of the column and are discarded.
12. Place a 10-mL tube under each column. Add 3.5 mL capture buffer to the columns and collect the samples.

3.1.2. Capture of Periodate Oxidized Glycoproteins onto Hydrazide Beads

1. Wash the hydrazide beads (50 μL per sample) twice with 0.5 mL capture buffer by the use of a tabletop centrifuge and add a suspension of beads corresponding to 50 μL to each sample (see Note 5).
2. Wrap the tubes with aluminum foil to exclude light and agitate the tubes at room temperature (~23°C) for 12–16 h.
3. Let the tubes stand in an upright position for 10 min and then gently aspirate most of the buffer, or centrifuge briefly.
4. Transfer the suspension of hydrazide beads to 1.5 mL-microcentrifuge tubes.
5. Wash the beads with two portions of 0.5 mL capture buffer.
6. Wash the beads with three portions of 0.5 mL ammonium bicarbonate buffer, and then carefully discard as much as possible of the buffer without removing any beads.

3.1.3. Trypsin Digestion of Captured Glycoproteins

1. Add 0.5 mL of 10 mM DTT in ammonium bicarbonate buffer to each tube, vortex and place at 37°C for 30 min.
2. Wash the beads with a 0.5 mL portion of ammonium bicarbonate buffer and remove as much as possible of the buffer.
3. Add 0.5 mL of 50 mM IAA in ammonium bicarbonate buffer to each tube and vortex at room temperature for 30 min.
4. Wash the beads with three portions of 0.5 mL ammonium bicarbonate buffer.
5. Add 0.5 mL of 8 M Urea in ammonium bicarbonate buffer and vortex for 15 min.
6. Wash the beads with three portions of 0.5 mL ammonium bicarbonate buffer.
7. Wash the beads with three portions of 0.5 mL 0.05% Tween-20 in PBS (see Note 6).
8. Wash the beads with three portions of 0.5 mL ammonium bicarbonate buffer.
9. Transfer the beads to prelubricated microcentrifuge tubes and remove as much as possible of the buffer.
10. Add 100 μL of ammonium bicarbonate buffer to each tube and add 10 μL (1 μg) trypsin to each tube and vortex.

11. Incubate at 37°C for 10–16 h.
12. Remove the supernatant to a new prelubricated microcentrifuge tube and wash the beads with 200 μL of 50% acetonitrile in water that are then pooled with the supernatant and lyophilized (see Note 7).

3.1.4. Release of Glycopeptides from the Hydrazide Beads

1. Wash the beads twice with 0.5 mL of 50% acetonitrile.
2. Wash the beads twice with 0.5 mL of 20% acetonitrile.
3. Wash the beads twice with 0.5 mL water.
4. Add 0.5 mL of 0.1 M formic acid to each tube and place tubes in a heating block or in an oven at 80°C for 1 h.
5. Transfer the supernatant to a new prelubricated microcentrifuge tube.
6. Wash the beads with 0.5 mL of 50% acetonitrile and pool this with the supernatant from the previous step.
7. Lyophilize and store at −20°C before LC-MS^n analysis (see Note 8).

3.2. Tandem Mass Spectrometry

A complete step-by-step protocol for performing the LC-MS^n runs is not provided here. In principal, any LC-MS setups aimed at shotgun proteomics with the ability to do multistage CID may be used. We regularly use an LTQ-FT-ICR hybrid instrument (Thermo Scientific, USA) but e.g., Orbitrap (Thermo Scientific) and ion-trap instrumentations capable of multistage CID also works well. The following steps should be guidelines for the MS analysis of glycopeptides enriched via the capture-and-release protocol.

1. Dissolve the glycopeptide fraction in a sufficient volume of 0.1% formic acid to match the loop-volume of the LC-system. We typically dissolve the samples in 20 μL, centrifuge and inject 8 μL.
2. Use a C18 reversed-phase column with a 50 min gradient from 0 to 50% acetonitrile in 0.1% formic acid at 200–300 nL/min.
3. The full MS are acquired in FT-ICR mode with a resolution of 100,000 after an accumulation threshold of 500,000 in the linear ion trap.
4. For CID fragmentation the normalized collision energy is set to 30% for MS^2 and MS^3 events. For ECD fragmentation an arbitrary energy settings of 4 and 5 in consecutive fragmentation events are used.
5. Use profile mode in the collection of MS^2 and MS^3 spectra in order to assess charge states of fragment ions.

6. Use a repeat count of two for each precursor *m/z* and then exclude it for one min to exclude multiple fragmentation events of the same precursors.
7. For each MS^2 spectrum perform data dependent MS^3 on five of the most intense peaks.
8. Run standard LC-MS/MS on the supernatant fraction from Subheading 3.1.3, step 12 to assay which glycoproteins that had been captured.

3.3. Data Analysis of Glycopeptide Fragment Spectra

3.3.1. Finding the Glycopeptides

1. Manually inspect the MS^2 and MS^3 spectra and check for the presence of oxonium ions (37) at *m/z* 204 $[HexNAc]^+$, 366 $[HexHexNAc]^+$, 512 $[dHexHexHexNAc]^+$, and 528 $[Hex_2HexNAc]^+$, (see Note 9). Typical MS^2 and MS^3 spectra of glycopeptides are shown in Figs. 3, 5, and 7. Filter the ion chromatogram to show only ions at *m/z* 366 in the collection of MS^2 spectra in order to quickly find glycopeptide spectra (Fig. 2).
2. In addition, check for delta *m/z* values of 162, 81, 54, and 40.5 for pairs of singly, doubly, triply, and quadruply protonated ions, respectively, corresponding to the neutral loss of Hex; and delta *m/z* values of 203, 101.5, and 67.7 for pairs of singly, doubly, and triply protonated ions, corresponding to the neutral loss of HexNAc, respectively. A proton is also often expelled during the loss of monosaccharide units from glycopeptide ions giving rise to charge-reduced fragment ions.

3.3.2. Glycan Fragment Analysis of O-Glycopeptides

An example of a full MS scan, MS^2, and MS^3 spectra of an O-glycopeptide, which was enriched from human CSF, is shown in Figs. 3, 4, and 5. The peptide sequence is ESKPQAGTARP QDVNR, corresponding to residues 119–134 of insulin-like

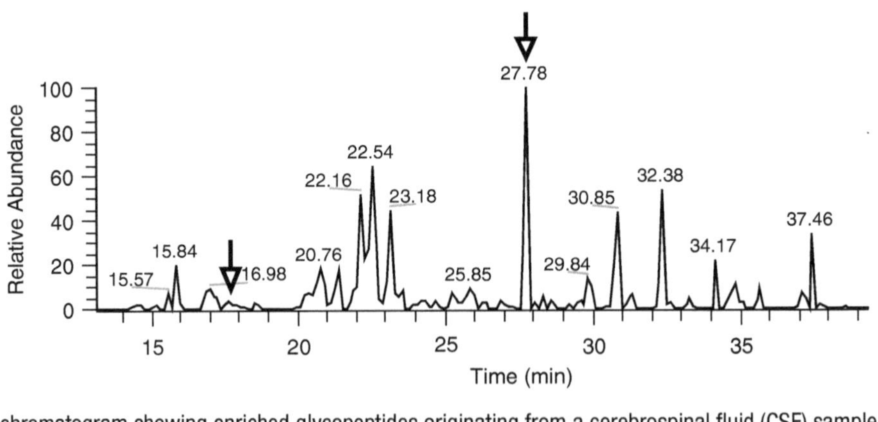

Fig. 2. Ion chromatogram showing enriched glycopeptides originating from a cerebrospinal fluid (CSF) sample. The chromatogram is extracted to show ion intensity at *m/z* 366 for the collected MS^2 spectra as a function of retention time. The *arrows* show the position of *m/z* 366 ions for the O-glycopeptide (~17 min) described in Subheadings 3.3.1, 3.3.2, 3.3.3, 3.3.4 and 3.3.5 and the N-glycopeptide (~28 min) described in Subheading 3.3.6.

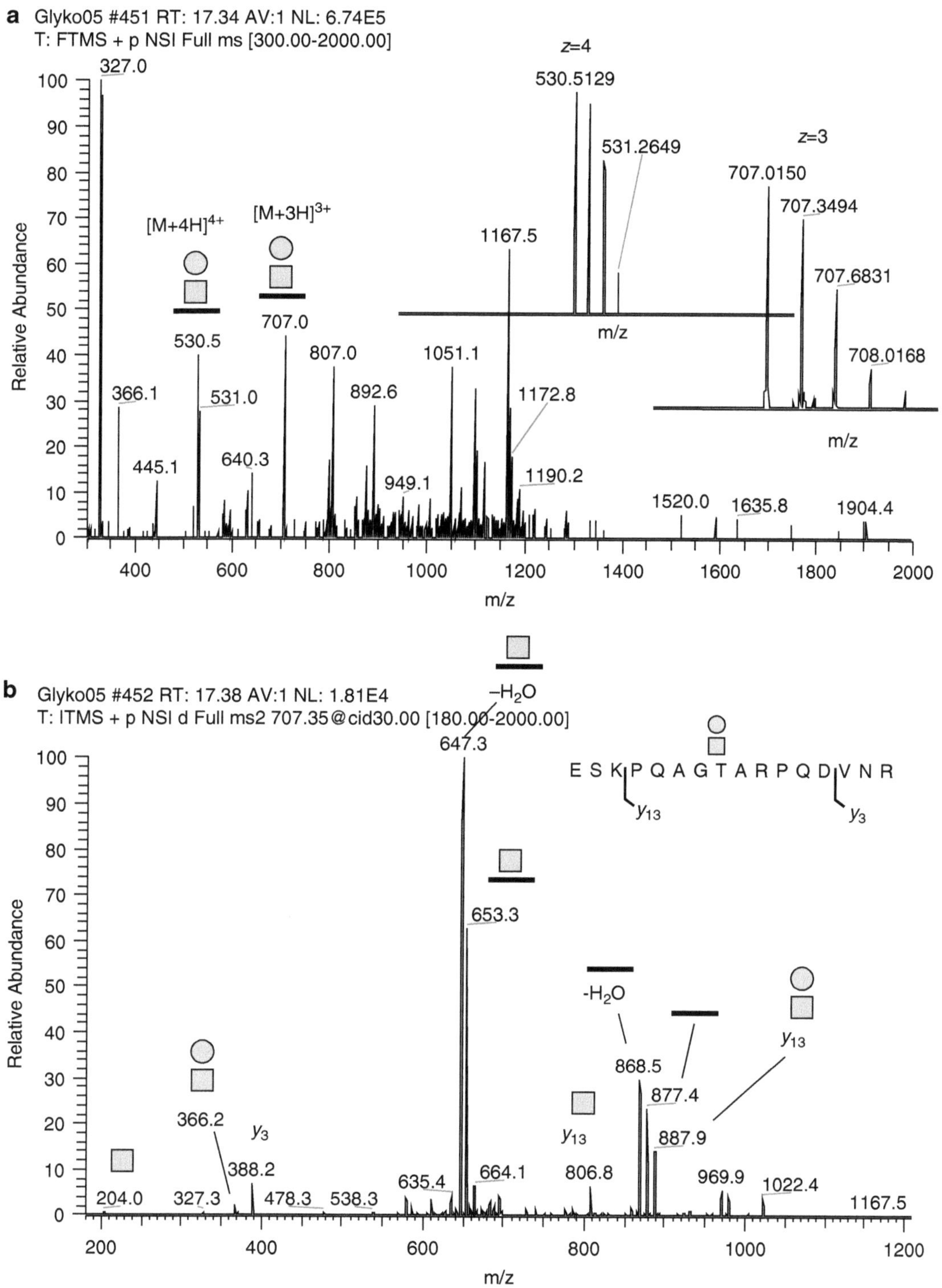

Fig. 3. Collision-induced dissociation (CID) MS^2 of an O-glycopeptide from human CSF. (**a**) Full MS scan at 17.34 min showing the $[M+3H]^{3+}$ and $[M+4H]^{4+}$ precursors. Expansions are shown in the inserts. (**b**) CID MS^2 spectrum of the $[M+3H]^{3+}$ precursor at *m/z* 707.

growth factor-binding protein 6 (IBP6_HUMAN in the UniprotKB/Swissprot database). This tryptic peptide contains a Lys-Pro and an Arg-Pro pair within the sequence that cannot be cleaved by trypsin, which results in multiply protonated $[M+4H]^{4+}$ and $[M+3H]^{3+}$ precursor ions at *m/z* 530.5 and *m/z* 707.0, respectively (Fig. 3a) and thus in relatively low *m/z* values. The specific O-glycosylation sites of this protein have been described before (38), including Thr-126 that is the glycosylation site of the glycopeptide presented here. The O-glycans of IBP6 were demonstrated to protect the protein from degradation by blocking the access of proteases to proteolytic sites, which prolonged its interaction with insulin-like growth factor II (39).

1. Paste full-scan MS spectra together with the consecutive MS^2 and MS^3 fragmentation spectra containing information regarding date, file names, elution times, *m/z* values, and charge states into Adobe Illustrator or Microsoft Powerpoint documents (or similar) for documentation purposes. These spectra can also serve in the process of making publication and presentation figures.
2. CID of the $[M+3H]^{3+}$ precursor at *m/z* 707.0 (Fig. 3a) into the MS^2 spectrum (Fig. 3b) results in glycan fragmentation. The *m/z* 204 and 366 oxonium ions are typically observed. Manually inspect the MS^2 spectrum to characterize the glycan structure and identify the peak corresponding to the unglycosylated peptide ion. The neutral loss of Hex is readily observed (*m/z* 653.3) and a peak corresponding to the loss of HexHexNAc and a proton is also observed (*m/z* 877.4). No further glycan fragmentation of the precursor ion is observed and *m/z* 877.4 thus represents the intact peptide ion. The characterization of additional fragment peaks (for instance *m/z* 806.8, which is [*y*13+HexNAc]) is discussed in Subheading 3.3.4. The glycan is thus composed of HexHexNAc-*O*-Ser/Thr in accordance with the core 1 O-glycan structure (Galβ3GalNAc-*O*-Ser/Thr), which must have been sialylated on the Gal and/or the GalNAc (Neu5Acα3Galβ3[Neu5Acα6]GalNAc-*O*-Ser/Thr) in the native protein.

3.3.3. Identification of the Peptide Sequence of O-Glycopeptides

For CID fragmentation of peptide ions mainly *b*- and *y*-type peptide backbone fragment ions are formed, whereas for ECD of peptides and glycopeptides *c*- and *z*-ions are mainly formed (Fig. 4a, see also Subheading 3.3.5). CID-MS^3 of the CID-MS^2 generated peptide ion at *m/z* 877.4 (Fig. 3b) results in fragmentation into the *b*- and *y*-series of ions (Fig. 4b). The MS^3 spectrum is used in protein database searches to identify the peptide, and thus also the protein that it belongs to. Note that the $[M+4H]^{4+}$ precursor at *m/z* 530.5 (Fig. 3a) and the MS^2 generated peptide ion at *m/z*

a

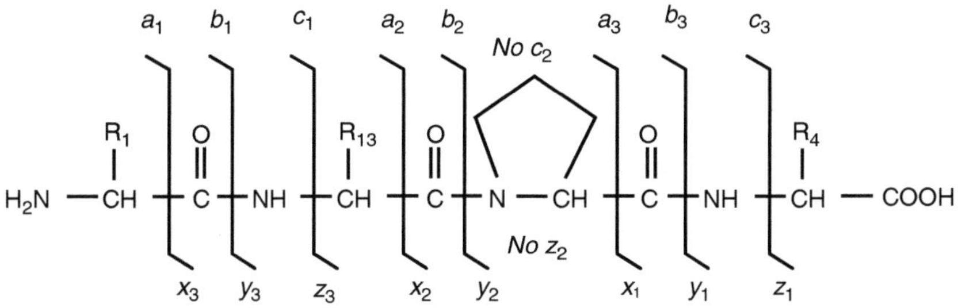

b

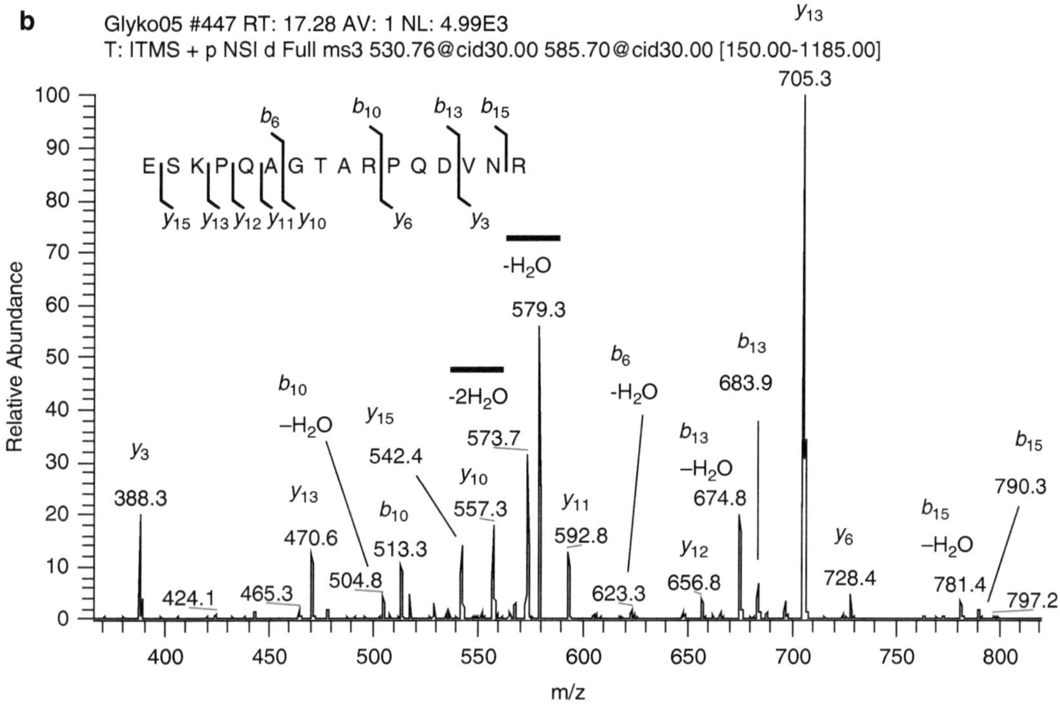

c

```
COM=071129_Glyko_05; RT:17.28 min; Mass: 530.5129 4H+
Glycan:hexhexnac
BEGIN IONS
CHARGE=2+
PEPMASS=877.4518
309.010376       532.093262
326.984253       676.929504
364.647095       343.281830
388.122864       6088.138184
389.114136       1287.201782
406.074219       248.089752
.
.
797.204651       309.190216
1022.315369      731.699280
1023.292603      573.210815
1024.369995      208.718292
END IONS
```

Fig. 4. CID fragmentation of an O-glycopeptide from human CSF *continued*. (**a**) The different N-terminal ions (*a*, *b*, and *c*) and C-terminal ions (*x*, *y* and *z*) that can form during peptide fragmentation. A Pro residue is included in the sequence demonstrating the absence of *c2* and *z2* ions for this peptide. (**b**) CID MS[3] of the peptide ion. Selected *b*- and *y*-series of fragment ions are annotated in the spectrum and in relation to the sequence (**c**) The Mascot search file (truncated), which was generated from the MS[3] spectrum.

585.7 (spectrum not shown) were used for generating the MS^3 spectrum in Fig. 4b.

1. Save the MS^3 spectrum of the peptide ion in the same folder as the Readw application (http://tools.proteomecenter.org/wiki/index.php?title=Software:ReAdW).
2. Open the command prompt and navigate to the Readw folder. Use Readw to convert the profile spectrum into a centroided peak list, in the open file format mzxml, by typing: "readw filename.raw c."
3. Open the generated file with the Mmass application (40, 41), available from http://www.mmass.org and export it to Mascot general format for use with the Mascot search engine at http://www.matrixscience.com or an in-house Mascot server (Fig. 4c) (see Note 10).
4. Calculate the accurate molecular weight of the peptide ion by taking the *m/z* value of the monoisotopic peak from the Full MS (Fig. 3a), multiply with the charge state, subtract the mass of the glycan and two protons and type in the doubly protonated *m/z* value:

$$((530.5129 \times 4) - 2 \times 1.0073 - 365.1322) / 2 = 877.4518$$

5. For ICR and Orbitrap measured parent ion masses a mass tolerance of 5 or 10 ppm can be used.
6. Set the fixed modification to Cys carbamidomethyl and variable modifications to Met oxidation; ammonia-loss for N-terminal Cys carbamidomethyl (42); N-terminal Asn to pyroglutamic acid conversion; and Asn to Asp conversion, which is prone to form for the Asn-Gly sequence (43).
7. Use a fragment mass tolerance of 0.6 Da and set the instrument type to *ESI-TRAP*.
8. Search human sequences of the Swissprot database (if a human sample was used) but if there is no hit change it to NCBI to account for possible sequence variation.
9. Use *Trypsin* as the enzyme and set *max missed cleavages* to 2, but if there is no hit change the enzyme to *semitrypsin* to account for tryptic peptides that for instance include an N-terminal signal peptide cleavage site. Lastly, change enzyme to *none* to account for non-trypsin cleavages at both ends (see Note 11).
10. Add a commentary to the .mgf file for documentation purposes (COM=text), such as in Fig. 4c, which will appear in the peptide summary report.

The Mascot search of the .mgf file (Fig. 4c) gave the ESKPQAGTARPQDVNR peptide from IBP6_HUMAN as a unique match. The Mascot score was 26 where a score of 22 indicates identity or extensive homology at $p<0.05$. The "expect value" was 0.025 and the experimental value of the monoisotopic molecular weight was 1752.8890 Da which was +1.35 ppm off from the theoretical value of 1752.8867 Da.

3.3.4. Pinpointing the O-Glycosylation Site

Occasionally CID induced peptide fragmentation takes place in the presence of a (partially) intact glycan, which sometimes can be used to pinpoint the glycosylation site in the presence of several Ser/Thr residues. This is particularly common for Pro containing peptides where prominent *y*-ions (and sometimes the corresponding *b*-ions), resulting from favorable fragmentation at the N-terminal side of Pro, are observed (44, 45) (Fig. 4a); see for instance the peaks corresponding to *y*13, [*y*13+HexNAc] and [*y*13+HexHexNAc] in Figs. 3b, 4b and 5a.

1. Inspect the MS^2 (Fig. 3b) and MS^3 spectra of the [peptide+HexNAc] ion (Fig. 5a) to find peptide fragments with an intact HexNAc or HexHexNAc glycan.
2. Use the MS-product tool at the protein prospector homepage (http://prospector.ucsf.edu) to list the possible glycosylated peptide fragments (Fig. 5b). The mass of each glycan is typed in parenthesis after the proposed glycan attachment site (alternatively the terms *hexnac* and *hexnachex* may be used). Choose to show the *b*- and *y*-ion series of peptide backbone fragments.

The prominent presence of [*y*13+HexNAc] at *m/z* 807.3 (*m/z* 806.8 in Fig. 3b) and *m/z* 538.6 in Fig. 5a; and [*y*13+HexHexNAc] at *m/z* 887.9 in Fig. 3b are used to unequivocally assign the glycosylation site to Thr-126 as opposed to Ser-120.

3.3.5. Identification of O-Glycosylation Site with ECD Fragmentation

For glycopeptide precursors with relatively low *m/z* values, and $[M+3H]^{3+}$ or $[M+4H]^{4+}$ charge states, it is possible to pinpoint glycosylation sites within O-glycopeptides by the analysis of ECD fragmentation spectra. ECD typically leads to peptide fragmentation into c- and z-type ions (Fig. 4a) (46), while glycans are left intact (47). As an alternative, electron transfer dissociation (ETD) fragmentation using e.g., an LTQ-orbitrap instrument can be used to attain similar fragmentation patterns to determine O-glycosylation sites (48, 49).

1. Run LC-MS with ECD fragmentation and find glycopeptide fragmentation spectra by the knowledge of precursor *m/z* values from previously analyzed CID fragmentation spectra.
2. Use the found ECD fragmentation spectrum (Fig. 6a) together with the list of possible *c*- and *z*-ions from the MS-product tool (see Subheading 3.3.4, step 2) and assign the glycosylation site (Fig. 6b) (see Note 12). Sometimes hydrogen transfer takes place and *c+1* and *z+1* ions may appear. The mass accuracy is very good for these ICR-measured fragment ions, and the *m/z* values are often correct to the second decimal place. Since no visible glycan fragmentation appears in the ECD spectra the

a

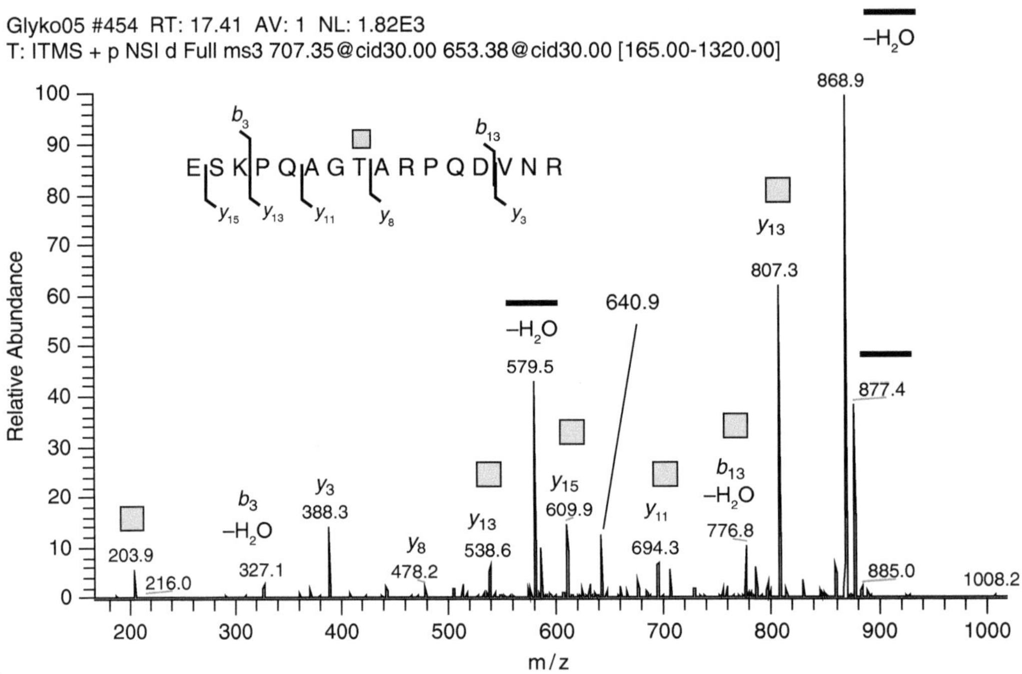

b

ESKPQAGT(203.08)ARPQDVNR

User AA Formula 1: **C2 H3 N1 O1**

MH^{+1}(av)	MH^{+1}(mono)	MH^{+2}(av)	MH^{+2}(mono)	MH^{+3}(av)	MH^{+3}(mono)
1957.9336	1956.9740	979.4705	978.9906	653.3162	652.9962

[–] **Main Sequence Ions**

b	b^{+2}				y	y^{+2}
---	---	1	**E**	16	---	---
217.0819	109.0446	2	**S**	15	1827.9314	914.4693
345.1769	173.0921	3	**K**	14	1740.8994	870.9533
442.2296	221.6185	4	**P**	13	1612.8044	806.9058
570.2882	285.6477	5	**Q**	12	1515.7516	758.3795
641.3253	321.1663	6	**A**	11	1387.6931	694.3502
698.3468	349.6770	7	**G**	10	1316.6559	658.8316
1002.4745	501.7409	8	**T(203.08)**	9	1259.6345	630.3209
1073.5116	537.2594	9	**A**	8	955.5068	478.2570
1229.6127	615.3100	10	**R**	7	884.4697	442.7385
1326.6654	663.8364	11	**P**	6	728.3686	364.6879
1454.7240	727.8657	12	**Q**	5	631.3158	316.1615
1569.7510	785.3791	13	**D**	4	503.2572	252.1323
1668.8194	834.9133	14	**V**	3	388.2303	194.6188
1782.8623	891.9348	15	**N**	2	289.1619	145.0846
---	---	16	**R**	1	175.1190	88.0631

Fig. 5. CID fragmentation of an O-glycopeptide from human CSF *continued*. (**a**) MS^3 of the [peptide+HexNAc] ion (*m/z* 653 in Fig. 3b). (**b**) The list of *b*- and *y*-ions, including a HexNAc (203.08 Da) on the Thr residue.

a

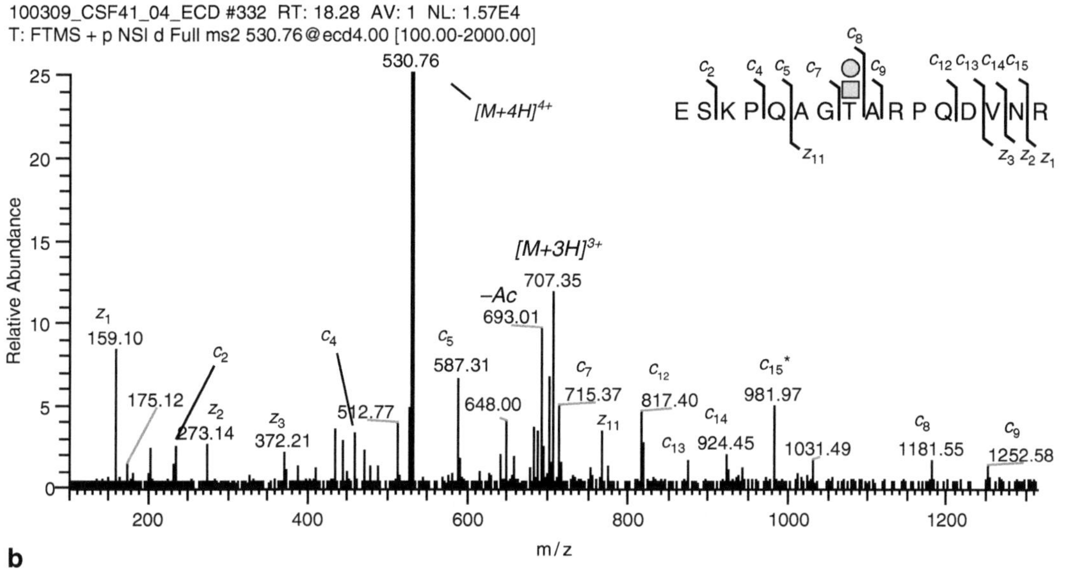

b

ESKPQAGT(hexnachex)ARPQDVNR

User AA Formula 1: **C2 H3 N1 O1**

Elemental Compositon: C85 H144 N27 O36

MH^{+1}(av)	MH^{+1}(mono)	MH^{+2}(av)	MH^{+2}(mono)	MH^{+3}(av)	MH^{+3}(mono)
2119.9858	2119.0262	1060.4966	1060.0167	707.3336	707.0136

[–] **Main Sequence Ions**

c	c^{+2}				z	z^{+2}
---	---	1	**E**	16	---	---
234.1084	117.5579	2	**S**	15	1973.9649	987.4861
---	---	3	**K**	14	1886.9328	943.9701
459.2562	230.1317	4	**P**	13	---	---
587.3148	294.1610	5	**Q**	12	1661.7851	831.3962
658.3519	329.6796	6	**A**	11	1533.7265	767.3669
715.3733	358.1903	7	**G**	10	1462.6894	731.8483
1181.5532	591.2802	8	**T(hexnachex)**	9	1405.6680	703.3376
1252.5903	626.7988	9	**A**	8	939.4881	470.2477
---	---	10	**R**	7	868.4510	434.7291
1505.7442	753.3757	11	**P**	6	---	---
1633.8028	817.4050	12	**Q**	5	615.2971	308.1522
1748.8297	874.9185	13	**D**	4	487.2385	244.1229
1847.8981	924.4527	14	**V**	3	372.2116	186.6094
1961.9411	981.4742	15	**N**	2	273.1432	137.0752
---	---	16	**R**	1	159.1002	80.0538

Fig. 6. Electron capture dissociation (ECD) fragmentation of an O-glycopeptide from human CSF. (**a**) The precursor at *m/z* 530 is the same as for Fig. 3a. The $[M+4H]^{4+}$ precursor at *m/z* 530 and the charge-reduced $[M+3H]^{3+}$ precursor at *m/z* 707 are visible in the spectrum. –Ac, Loss of an acetyl group from the charge-reduced precursor. (**b**) The list of possible c- and z-ions.

absence of a glycan on a series of fragment ions strongly suggests its presence on the opposing side of the peptide. Again, the glycosylation site is pinpointed to Thr-126 as opposed to Ser-120.

3.3.6. Analysis of N-Glycopeptides

The N-glycoproteins present in plasma/serum (for instance transferrin and fetuin) typically carry sialic acid terminated complex bi-, tri-, and tetraantennary glycan structures (Fig. 1), which may also be fucosylated on one of the antennas and/or on the innermost GlcNAc (1). The MS^2 spectra of desialylated N-glycopeptides are characterized by the presence of oxonium ions at *m/z* 366 and 528, and also *m/z* 512 if an antenna of the N-glycan contains a fucose residue (dHex). The base peak in the CID MS^2 spectrum is a fragment that typically has lost 365 Da (HexHexNAc) and a proton from the precursor (Fig. 7b). In general, only glycan fragmentation is visible in the MS^2 and MS^3 spectra, except in the MS^3 of the [peptide+HexNAc] ion, where the peptide backbone fragmentation is used to identify the peptide sequence via Mascot database searching.

1. Identify MS^2 and MS^3 spectra that contain peaks at *m/z* 366 and 528 and show a glycan fragmentation pattern (Fig. 7b) in support of an N-glycan structure.
2. Identify a candidate for the [peptide+HexNAc] ion (*m/z* 804.33 in Fig. 7b) and determine the mass loss compared to the precursor. In Fig. 7b the difference is ~1,420 Da in support of $Hex_5HexNAc_3$, which strongly suggests that the glycan is a complex biantennary structure with $Hex_5HexNAc_4$ composition.
3. Save the MS^3 spectrum where the [peptide+HexNAc] ion has been fragmented (spectrum not shown).
4. Convert the spectrum with the Readw application and generate a Mascot search file, as in Subheading 3.3.3, steps 1–3.
5. Calculate an accurate precursor mass for the Mascot search file by subtracting the mass of the glycan minus one HexNAc ($Hex_5HexNAc_3$ = 1419.5022 Da) and type in the *m/z* value for the doubly protonated precursor. For the example shown in Fig. 7a.

   $$((1009.7631 \times 3) - 1.0073 - 1419.5022) / 2 = 804.3899$$

6. Use a variable amino acid modification of Asn in the Mascot configuration that allows the addition of HexNAc (203.0794 Da) and neutral loss of 203.0794 and 0 Da, which are in line with the observed CID fragmentation of [peptide+HexNAc] ions.
7. Setup further search terms such as in Subheading 3.3.3, steps 6 and 7.
8. Set the mass accuracy of the precursor to 20 ppm (see Note 13).
9. Include the term "SEQ=b-NX[STC]" after the BEGIN IONS line in the Mascot search file to only search for peptide sequences containing the N-glycosylation consensus sequence.
10. Perform the Mascot search as in Subheading 3.3.3, steps 8–10.

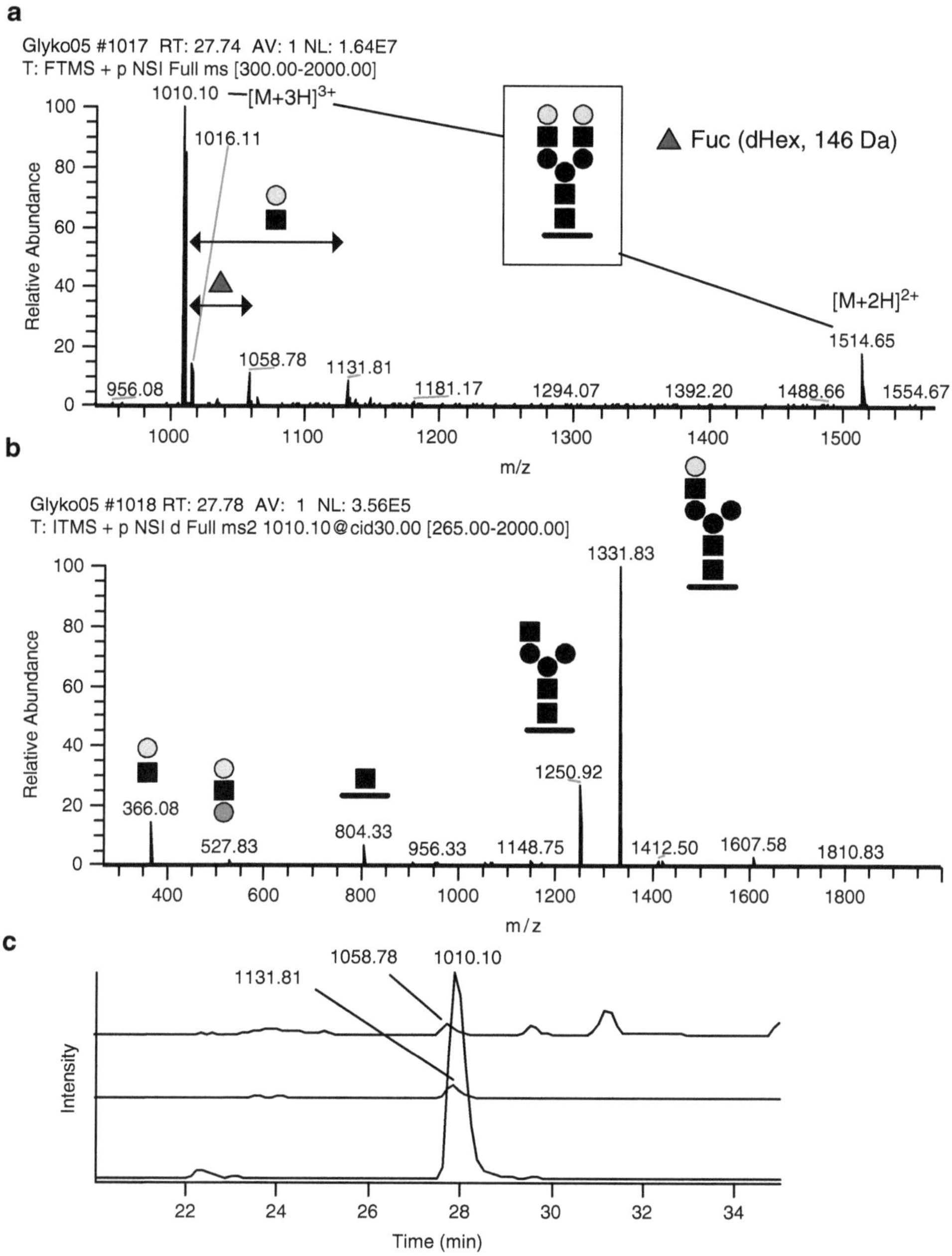

Fig. 7. CID fragmentation of an N-glycopeptide from human CSF. (**a**) The full MS scan at 27.74 min shows the simultaneous elution of glycopeptide glycoforms with the same peptide sequence. (**b**) CID MS^2 spectrum of the $[M+3H]^{3+}$ precursor at *m/z* 1010 in (**a**), which has a complex biantennary structure. (**c**) Extracted ion chromatograms for the three detected glycopeptide glycoforms shown in (**a**)

11. If there is no hit change the N-glycosylation constraint to "SEQ=c-N[KR]" so that only peptides containing an Asn followed by Lys or Arg, which are on the C-terminal due to trypsin cleavage, are searched.
12. Identify further glycopeptides sharing the same peptide sequence, which typically are co-eluting in the ion chromatogram.

The presence of peaks that are 146 Da (dHex) and 365 Da (HexHexNAc) heavier than that of the complex biantennary glycopeptide in Fig. 7a strongly indicate the simultaneous presence of fucosylated biantennary and triantennary structures, respectively.

13. Extract the ion chromatograms for the *m/z* values in the full MS scans, corresponding to the isotopic envelopes of the individual glycopeptide precursors, and integrate using the Xcalibur program (Fig. 7c), (see Note 14).

The Mascot search of the [peptide+HexNAc] MS^3 spectrum identified SWPAVGNCSSALR, where Cys is modified with carbamidomethyl, and Asn is modified with HexNAc as a unique match. This sequence corresponds to the 181–193 region of human hemopexin (HEMO_HUMAN), and includes the well-established Asn-187 glycosylation site (50). The integration of the extracted ion chromatograms (Fig. 7c) gave 88, 6, and 6% for the bi-, tri-, and fucosylated biantennary glycopeptides, respectively.

4. Notes

1. Ensure that you follow all ethical and legal regulations before using any clinical samples in your experiments.
2. Ammonium bicarbonate is dissolved in deionized water, and the pH should be close to 8. Make no attempt to adjust the pH, and thus introduce insoluble salt, since the purpose of using ammonium bicarbonate is to remove it as NH_3 and CO_2 during the lyophilization of samples.
3. Alternatively other desalting columns using G-25 material such as the miditrap or minitrap columns can be used if smaller elution volumes and thus smaller capture volumes are desired.
4. The main reason for using model glycoproteins are for experimental control purposes, and should thus be used under the same conditions as biological samples. This is the reason why model protein samples were diluted to 500 μL.
5. Use a standard micropipette to transfer hydrazide bead suspensions. The hydrazide beads are suspended in isopropanol so ensure that the container is capped properly and stored in a refrigerator. If an excessive amount of isopropanol has evaporated a sufficient amount of fresh isopropanol can be added to the container.
6. Stop point, the beads may be stored in 0.5 mL Tween 20 in PBS at 4°C for a few days.
7. This *supernatant* fraction contains tryptic peptides (and glycopeptides), which were cleaved off from the captured glycopro-

teins. An LC-MS/MS shotgun proteomics analysis of this fraction provides information regarding which glycoproteins that were captured. Stop point, the beads may be stored in 0.5 mL 20% acetonitrile at 4°C for a few days.

8. Store glycopeptide samples lyophilized at −20 or −80°C. When they have been dissolved in 0.1% formic acid the MS analysis should be done within a few days, and spare samples should be re-lyophilized. Substantial formylation (addition of CO to hydroxyl groups) of the glycan parts have been observed after prolonged storage of dissolved samples at 4°C, and also at −20°C.
9. The absolute identity of individual monosaccharides and their anomeric linkages cannot be addressed in the CID fragmentation of glycopeptides. However, a glycan with $Hex_5HexNAc_4$ composition and a glycosidic fragmentation profile compatible with a complex biantennary structure also most likely is indeed a complex biantennary structure. Individual HexHexNAc oxonium ions at *m/z* 366 originating from N-glycopeptides can however either be GalGlcNAc or ManGlcNAc. The same argumentation is valid for O-glycosylation where HexHexNAc-*O*-Ser/Thr is compatible with core 1 Galβ3GalNAc-*O*-Ser/Thr, but could in principle also be core 8 (Galα3GalNAc-*O*-Ser/Thr) for some of the glycopeptides.
10. A Mascot search file is a plain-text file in the format shown in Fig. 4b. The *m/z* values of peptide fragment ions are listed with their intensities. An individual query starts with a *BEGIN IONS* entry and ends with *END IONS*.
11. The formic acid treatment at 80°C results in selective hydrolysis of the Asp-Pro peptide bond (51). This can in some cases be advantageous due to the degradation of large tryptic glycopeptides into smaller-sized peptides more suitable for the LC-MS^n analysis.
12. The fragmentation yield in ECD is low and it is necessary to expand the spectrum to observe the fragment peaks at the 1–10% relative abundance level compared to the precursor. The *c*- and *z*-type ions result from peptide backbone fragmentation of the N-Cα bond. This excludes formation of *c*-ions ending with the residue before Pro, and *z*-ions starting with Pro, due to the ring structure of Pro that includes the N-Cα bond (Fig. 4a).
13. Even though the mass accuracy of an FT-ICR instrument is excellent it is somewhat poorer at higher *m/z* values, which are often the case for N-glycopeptides. This is the reason why 20 ppm was used for N-glycopeptides.
14. The relative amounts of each glycopeptide glycoform are estimated based on the assumption that different glycoforms are enriched and ionized equally well.

Acknowledgment

We thank Adnan Halim, Ammi Grahn, Ulla Rüetschi, Gunnar Brinkmalm, Henrik Zetterberg, and Kaj Blennow for individual contributions to the glycoproteomic project. Funding from the Swedish Medical Research Council (8266) and governmental ALF grants to the Sahlgrenska University Hospital are gratefully acknowledged.

References

1. Varki A et al (2009) Essentials of glycobiology. Cold Spring Harbor Laboratory Press, New York
2. Ohtsubo K, Marth JD (2006) Glycosylation in cellular mechanisms of health and disease. Cell 126:855–867
3. Cohen M, Varki A (2010) The sialome—far more than the sum of its parts. OMICS 14:455–464
4. Magnani JL (2004) The discovery, biology, and drug development of sialyl Lea and sialyl Lex. Arch Biochem Biophys 426:122–131
5. Olofsson S, Bergstrom T (2005) Glycoconjugate glycans as viral receptors. Ann Med 37:154–172
6. Orlando R (2010) Quantitative glycomics. Methods Mol Biol 600:31–49
7. Vanderschaeghe D, Festjens N, Delanghe J, Callewaert N (2010) Glycome profiling using modern glycomics technology: technical aspects and applications. Biol Chem 391:149–161
8. Zaia J (2008) Mass spectrometry and the emerging field of glycomics. Chem Biol 15:881–892
9. Karlsson H et al (2009) High-throughput and high-sensitivity nano-LC/MS and MS/MS for O-glycan profiling. Methods Mol Biol 534:117–131
10. Carr SA, Huddleston MJ, Bean MF (1993) Selective identification and differentiation of N- and O-linked oligosaccharides in glycoproteins by liquid chromatography-mass spectrometry. Protein Sci 2:183–196
11. Demelbauer UM, Zehl M, Plematl A, Allmaier G, Rizzi A (2004) Determination of glycopeptide structures by multistage mass spectrometry with low-energy collision-induced dissociation: comparison of electrospray ionization quadrupole ion trap and matrix-assisted laser desorption/ionization quadrupole ion trap reflectron time-of-flight approaches. Rapid Commun Mass Spectrom 18:1575–1582
12. Kawasaki N, Itoh S, Yamaguchi T (2009) LC/MSn for glycoprotein analysis: N-linked glycosylation analysis and peptide sequencing of glycopeptides. Methods Mol Biol 534:239–248
13. Shevchenko A, Tomas H, Havlis J, Olsen JV, Mann M (2006) In-gel digestion for mass spectrometric characterization of proteins and proteomes. Nat Protoc 1:2856–2860
14. Anderson NL, Anderson NG (2002) The human plasma proteome: history, character, and diagnostic prospects. Mol Cell Proteomics 1:845–867
15. Zauner G, Koeleman CAM, Deelder AM, Wuhrer M (2010) Protein glycosylation analysis by HILIC-LC-MS of Proteinase K-generated N- and O-glycopeptides. J Sep Sci 33:903–910
16. Scott NE et al (2011) Simultaneous glycan-peptide characterization using hydrophilic interaction chromatography and parallel fragmentation by CID, HCD and ETD-MS applied to the N-linked glycoproteome of Campylobacter jejuni. Mol Cell Proteomics 10(2):M000031–MCP201
17. Mysling S, Palmisano G, Højrup P, Thaysen-Andersen M (2010) Utilizing ion-pairing hydrophilic interaction chromatography solid phase extraction for efficient glycopeptide enrichment in glycoproteomics. Anal Chem 82:5598–5609
18. Hägglund P et al (2007) An enzymatic deglycosylation scheme enabling identification of core fucosylated N-glycans and O-glycosylation site mapping of human plasma proteins. J Proteome Res 6:3021–3031
19. Wada Y, Tajiri M, Yoshida S (2004) Hydrophilic affinity isolation and MALDI multiple-stage tandem mass spectrometry of glycopeptides for glycoproteomics. Anal Chem 76:6560–6565

20. Kubota K et al (2008) Analysis of glycopeptides using lectin affinity chromatography with MALDI-TOF mass spectrometry. Anal Chem 80(10):3693–3698
21. Darula Z, Medzihradszky KF (2009) Affinity enrichment and characterization of mucin core-1 type glycopeptides from bovine serum. Mol Cell Proteomics 8:2515–2526
22. Kaji H et al (2003) Lectin affinity capture, isotope-coded tagging and mass spectrometry to identify N-linked glycoproteins. Nat Biotechnol 21:667–672
23. Hashii N et al (2009) Identification of glycoproteins carrying a target glycan-motif by liquid chromatography/multiple-stage mass spectrometry: identification of Lewis x-conjugated glycoproteins in mouse kidney. J Proteome Res 8:3415–3429
24. Zhao J, Simeone DM, Heidt D, Anderson MA, Lubman DM (2006) Comparative serum glycoproteomics using lectin selected sialic acid glycoproteins with mass spectrometric analysis: application to pancreatic cancer serum. J Proteome Res 5:1792–1802
25. Bunkenborg J, Pilch BJ, Podtelejnikov AV, Wiśniewski JR (2004) Screening for N-glycosylated proteins by liquid chromatography mass spectrometry. Proteomics 4:454–465
26. Larsen MR, Jensen SS, Jakobsen LA, Heegaard NHH (2007) Exploring the sialiome using titanium dioxide chromatography and mass spectrometry. Mol Cell Proteomics 6:1778–1787
27. Larsen MR, Højrup P, Roepstorff P (2005) Characterization of gel-separated glycoproteins using two-step proteolytic digestion combined with sequential microcolumns and mass spectrometry. Mol Cell Proteomics 4:107–119
28. Selby DS, Larsen MR, Calvano CD, Jensen ON (2008) Identification and characterization of N-glycosylated proteins using proteomics. Methods Mol Biol 484:263–276
29. Nilsson J et al (2009) Enrichment of glycopeptides for glycan structure and attachment site identification. Nat Methods 6:809–811
30. Nilsson J, Nilsson J, Larson G, Grahn A (2010) Characterization of site-specific O-glycan structures within the mucin-like domain of -dystroglycan from human skeletal muscle. Glycobiology 20:1160–1169
31. Zhang H, X-j L, Martin DB, Aebersold R (2003) Identification and quantification of N-linked glycoproteins using hydrazide chemistry, stable isotope labeling and mass spectrometry. Nat Biotechnol 21:660–666
32. Zhang H et al (2005) High throughput quantitative analysis of serum proteins using glycopeptide capture and liquid chromatography mass spectrometry. Mol Cell Proteomics 4:144–155
33. Tian Y, Zhou Y, Elliott S, Aebersold R, Zhang H (2007) Solid-phase extraction of N-linked glycopeptides. Nat Protoc 2:334–339
34. Zhang H, Aebersold R (2006) Isolation of glycoproteins and identification of their N-linked glycosylation sites. Methods Mol Biol 328:177–185
35. Kurogochi M et al (2007) Reverse glycoblotting allows rapid-enrichment glycoproteomics of biopharmaceuticals and disease-related biomarkers. Angew Chem Int Ed Engl 46: 8808–8813
36. Kurogochi M et al (2010) Sialic acid-focused quantitative mouse serum glycoproteomics by multiple reaction monitoring assay. Mol Cell Proteomics 9(1):2354–2368
37. Huddleston MJ, Bean MF, Carr SA (1993) Collisional fragmentation of glycopeptides by electrospray ionization LC/MS and LC/MS/MS: methods for selective detection of glycopeptides in protein digests. Anal Chem 65:877–884
38. Neumann GM, Marinaro JA, Bach LA (1998) Identification of O-glycosylation sites and partial characterization of carbohydrate structure and disulfide linkages of human insulin-like growth factor binding protein 6. Biochemistry 37:6572–6585
39. Marinaro JA, Neumann GM, Russo VC, Leeding KS, Bach LA (2000) O-glycosylation of insulin-like growth factor (IGF) binding protein-6 maintains high IGF-II binding affinity by decreasing binding to glycosaminoglycans and susceptibility to proteolysis. Eur J Biochem 267:5378–5386
40. Strohalm M, Hassman M, Košata B, Kodíček M (2008) mMass data miner: an open source alternative for mass spectrometric data analysis. Rapid Commun Mass Spectrom 22:905–908
41. Strohalm M, Kavan D, Novák P, Volný M, Havlícek V (2010) mMass 3: a cross-platform software environment for precise analysis of mass spectrometric data. Anal Chem 82:4648–4651
42. Krokhin OV, Ens W, Standing KG (2003) Characterizing degradation products of peptides containing N-terminal Cys residues by (off-line high-performance liquid chromatography)/matrix-assisted laser desorption/ionization quadrupole time-of-flight measurements. Rapid Commun Mass Spectrom 17:2528–2534

43. Krokhin OV, Antonovici M, Ens W, Wilkins JA, Standing KG (2006) Deamidation of -Asn-Gly- sequences during sample preparation for proteomics: consequences for MALDI and HPLC-MALDI analysis. Anal Chem 78: 6645–6650
44. Breci LA, Tabb DL, Yates JR, Wysocki VH (2003) Cleavage N-terminal to proline: analysis of a database of peptide tandem mass spectra. Anal Chem 75:1963–1971
45. Loo JA, Edmonds CG, Smith RD (1993) Tandem mass spectrometry of very large molecules. 2. Dissociation of multiply charged proline-containing proteins from electrospray ionization. Anal Chem 65:425–438
46. Roepstorff P, Fohlman J (1984) Proposal for a common nomenclature for sequence ions in mass spectra of peptides. Biomed Mass Spectrom 11:601
47. Mirgorodskaya E, Roepstorff P, Zubarev RA (1999) Localization of O-glycosylation sites in peptides by electron capture dissociation in a Fourier transform mass spectrometer. Anal Chem 71:4431–4436
48. Alley WR, Mechref Y, Novotny MV (2009) Characterization of glycopeptides by combining collision-induced dissociation and electron-transfer dissociation mass spectrometry data. Rapid Commun Mass Spectrom 23:161–170
49. Perdivara I et al (2009) Elucidation of O-glycosylation structures of the beta-amyloid precursor protein by liquid chromatography-mass spectrometry using electron transfer dissociation and collision induced dissociation. J Proteome Res 8:631–642
50. Takahashi N, Takahashi Y, Putnam FW (1984) Structure of human hemopexin: O-glycosyl and N-glycosyl sites and unusual clustering of tryptophan residues. Proc Natl Acad Sci USA 81:2021–2025
51. Marcus F (1985) Preferential cleavage at aspartyl-prolyl peptide bonds in dilute acid. Int J Pept Protein Res 25:542–546

Part II

Sample Preparation

Chapter 8

In-Solution Digestion of Glycoproteins for Glycopeptide-Based Mass Analysis

Eden P. Go, Kathryn R. Rebecchi, and Heather Desaire

Abstract

Glycopeptides are generated from the enzymatic digestion of glycoproteins with a specific or nonspecific protease. Whether this enzymatic conversion of glycoproteins into glycopeptides and peptides is done in-solution or in-gel, an efficient digestion protocol is one of the key components of a successful outcome in a mass spectrometry-based experimental workflow. This chapter outlines an optimized in-solution digestion protocol to prepare samples for glycopeptide-based mass analysis.

Key words: Glycopeptide, Mass spectrometry, In-solution digestion, Protein glycosylation

1. Introduction

The ability to analyze protein glycosylation has improved considerably due to recent technological advances in mass spectrometry (MS), separation methodologies, and affinity/enrichment schemes (1–11). While these developments have undoubtedly broadened the scope of experiments that could be performed, the sample preparation step is often the bottleneck of a successful outcome. Thus, it is important to implement efficient and effective protocols to prepare samples for MS-based analysis of protein glycosylation.

For most experiments, MS-based analysis of protein glycosylation necessitates an integrated workflow wherein an array of methods including glycoprotein/glycopeptide-based enrichment/affinity schemes, deglycosylation steps, online/offline chromatographic separation, and a combination of mass spectrometric platforms are employed (2, 12, 13). The combination of methods that are implemented will largely depend on the information that is sought,

Jennifer J. Kohler and Steven M. Patrie (eds.), *Mass Spectrometry of Glycoproteins: Methods and Protocols*, Methods in Molecular Biology, vol. 951, DOI 10.1007/978-1-62703-146-2_8, © Springer Science+Business Media, LLC 2013

the type of sample being analyzed, and the available instrumentation and expertise. To date, workflows centered around glycopeptide analysis are becoming mainstays in defining the glycosylation profile and the extent of glycosylation at a specific glycosylation site as well as in assessing changes in glycosylation due to changes in cellular processes. This workflow typically entails proteolytic digestion of glycoproteins with a specific or nonspecific protease to generate a glycopeptide/peptide mixture followed by a separation and/or enrichment step prior to mass analysis. One key component in this workflow that is often underestimated is the proteolytic digestion protocol used to cleave the glycoprotein into a mixture of peptides and glycopeptides. Glycoproteins can be digested in-solution or within a gel. In either protocol, subtle changes can greatly affect the outcome of any MS-based experiment. While standard protocols that are ubiquitously found online or in literature are being adopted in many workflows, these protocols are most often not optimal. In this chapter, an optimized in-solution digestion procedure is outlined. This protocol is used routinely in our laboratory in offline and online ESI LC/MS, MALDI MS, and direct infusion MS ESI analysis of glycoproteins obtained from commercial sources and those that are made recombinantly (14–18). A hydrophilic enrichment glycopeptide extraction procedure described initially by Wada and coworkers (19, 20) is outlined for the analysis of glycopeptides using direct infusion ESI or MALDI MS. We adapted this glycopeptide extraction procedure for the qualitative and quantitative analysis of glycoproteins (16, 18).

2. Materials

All buffers and stock solutions are prepared from high purity grade (>99%) reagents. All reagents were obtained from Sigma Aldrich, St. Louis MO unless otherwise indicated. Dissolve all reagents using deionized water with resistivity of 18.2 MΩ or greater (Millipore Direct-Q3 Water Purification System, Billerica, MA) and use only low binding/low retention siliconized microcentrifuge tubes. Reducing and alkylating reagents must be prepared fresh for every experiment. Discard remaining stock solution of the reducing and alkylating reagent after use, since they degrade once dissolved. All waste must be discarded according to proper waste disposal procedures.

2.1. Buffer and Stock Solutions

1. Digestion Buffer (see Note 1): 100 mM Tris(hydroxymethyl) aminomethane (Tris, Part Nos. T3253, T1503)/3 mM ethylenediaminetetraacetic acid (EDTA, Part No. 431788), pH 8.5. Weigh about 0.04198 g of Tris–HCl, 0.08887 Tris base, and 0.00877 g EDTA then transfer to a 10 mL conical

centrifuge tube. Add 10 mL of deionized water and mix. Verify the pH of the buffer by pipetting 30 μL of the buffer onto pH paper.

2. Reducing Reagent: 100 mM dithiothreitol (DTT, Part No. 43817) (see Note 2). In a microcentrifuge tube, dissolve 1.54 mg DTT in 100 μL of digestion buffer. Prepare fresh solution every digestion.
3. Alkylating Reagent: 500 mM iodoacetamide (IAM, Part No. I1149): In a microcentrifuge tube, dissolve 9.25 mg IAM in 100 μL of digestion buffer. Prepare the solution just prior to use.
4. Protease (see Note 3): Dissolve the protease (sequencing grade) in deionized water at a concentration of 1 μg/μL prior to digestion. Remaining protease stock must be aliquoted and frozen immediately at −20°C for short-term storage or at −80°C for long storage (see Note 4).

2.2. Hydrophilic Affinity Glycopeptide Extraction (see Note 5)

1. Sepharose® CL-4B beads (Part No. CL4B200).
2. Binding Solution: Prepare solution containing 5:1:1 (v/v) of 1-butanol (Fisher Scientific, Pittsburgh, PA. Part No., A383):ethanol (Fisher Scientific, Pittsburgh, PA, Part No., AC39769):deionized H_2O.
3. Elution Solution: Prepare solution containing 1:1 (v/v) solution of ethanol:H_2O.

3. Methods (see Note 6)

3.1. Protein Digestion

1. Glycoprotein sample. Samples from cell lysates, tissue extracts, biofluids, or recombinant expression must be purified or, in the case of serum samples, depleted of high abundant proteins prior to digestion. Samples that are used as standards and are obtained from commercial sources should be at least 95% pure. Glycoproteins from any source should be free of glycerol, phosphates, and detergents. Salts must be kept to a minimum (<100 mM). If the sample is lyophilized, dissolve ~100 μg of the glycoprotein in <25 μL of digestion buffer, to a final concentration of ≥4 mg/mL in a microcentrifuge tube. If the glycoprotein is in solution, determine the protein concentration by BCA assay or absorbance (21–23). Ideal glycoprotein concentration should be at least 4 mg/mL. Glycoprotein samples in solution often have a pH that lies in the physiological pH region. Make sure that the pH of the sample is in the optimal pH range of the protease that will be used before digestion. This can be done by spotting a small amount of the protein solution onto pH paper.

2. Sample denaturation. Denature the sample by heating in a dry bath at 100°C for 10 min (see Note 7). Cool the sample to room temperature then immediately add urea (Part No. U5378) to a final concentration of 6 M (see Note 8). The amount of urea will depend on the total volume of the sample. For example, if the total volume of the glycoprotein sample is 100 μL, weigh about 0.03604 g urea then add directly to the sample and dissolve.
3. Reduction (see Note 9). Add 100 mM DTT solution to the sample to give a final concentration of 10 mM. Mix the sample and spin it down quickly to ensure that no liquid adheres to the side of the microcentrifuge tube. Incubate for 1 h at room temperature.
4. Alkylation (see Note 10). Add 500 mM IAM solution to the sample to give a final concentration of 25 mM. Vortex and spin it down quickly to ensure that no liquid adheres to the side of the microcentrifuge tube. Incubate for 1 h at room temperature in the dark (see Note 10).
5. Quench excess IAM (see Note 11). Neutralize excess IAM by adding 100 mM DTT solution to the sample to a final concentration of 30 mM. Mix the sample and spin it down quickly to ensure that no liquid adheres to the side of the microcentrifuge tube. Incubate for 30 min at room temperature.
6. Digestion of samples (see Note 12). Add enough digestion buffer to dilute the urea to a final concentration of <1 M (see Note 13). Add protease to the sample at a protein:enzyme ratio of 30:1 to 50:1 (w/w). Mix the sample and spin it down quickly to ensure that no liquid adheres to the side of the microcentrifuge tube. Incubate for 18 h at 37°C.
7. The next day. Add additional protease to the sample at a protein:enzyme ratio of 60:1 to 100:1 (w/w) to ensure complete digestion. Incubate for an additional 4 h at 37°C.
8. Stop the digestion by acidifying the sample mixture with glacial acetic acid (Part No. 338826). Add 1 μL of acid for every 100 μL of sample. The digest can be analyzed directly by liquid chromatography mass spectrometry or subjected to hydrophilic affinity glycopeptide extraction step (see below) for direct infusion or MALDI MS experiments. Samples can be stored at −20°C until analysis (see Note 14).

3.2. Hydrophilic Affinity Glycopeptide Extraction for Direct Infusion ESI or MALDI MS Experiments

1. Extract the glycopeptides in the digest mixture by adding 25 μL Sepharose® CL-4B beads and 0.8–1.0 mL of binding solution to the digested glycoprotein sample (see Note 15). Shake the sample:Sepharose® CL-4B mixture gently for 45 min in a rotator. Centrifuge at 2,200 × *g* and discard the supernatant.

2. Add another 1 mL of binding solution to the sample:Sepharose® CL-4B mixture then shake gently in a rotator for 5–10 min, centrifuge at 2,200 × *g* and discard the supernatant.
3. Repeat step 2 3× to ensure that peptides are washed off.
4. Extract the glycopeptides by adding 1 mL of elution solution and shake the sample gently for 45 min. Centrifuge at 2,200 × *g* and transfer supernatant to clean microcentrifuge tube. Repeat the extraction two more times and transfer supernatant to a clean microcentrifuge tube. Pool all extracted samples. Evaporate the glycopeptide extract to dryness in a SpeedVac. Reconstitute in a solvent appropriate for mass spectrometry.

4. Notes

1. Other mass spectrometry compatible buffers such as ammonium bicarbonate (NH_4HCO_3, Part No. 09830) can be used as digestion buffer. When using NH_4HCO_3, prepare 100 mM digestion buffer by dissolving 79.056 mg NH_4HCO_3 in 10 mL deionized water and adjust the pH to 8.5.
2. DTT can be substituted with tris(2-carboxyethyl)phosphine hydrochloride (TCEP). TCEP (Part No. 75259) has wider pH range (pH 2–11) compared to DTT (pH 7–9) (24). It is more stable than DTT because it is resistant to oxidation. When using TCEP, prepare 50 mM stock solution TCEP by dissolving 1.43 mg of TCEP in 100 μL of digestion buffer. Add TCEP to the sample to a final concentration of 5 mM. The addition of TCEP will make the sample acidic. Make sure the sample pH is basic before alkylation. Neutralize excess IAM with DTT. One disadvantage of using TCEP is that it can catalyze peptide backbone cleavage, especially adjacent to cysteines (25).
3. The choice of protease depends on the glycoprotein that will be digested. A wide variety of proteases with broad specificity are currently available for proteolytic digestion (Table 1). Among these proteases, sequencing grade trypsin (Promega, Madison, WI, Part No. V5111) is the most commonly used, due to its high cleavage specificity, stability in the presence of organic solvents, an optimum working pH range of 7–9, ability to digest insoluble proteins, costs, and availability (26–28). Trypsin cleaves at the C-terminus of arginine (R) and lysine (K) residues except when K/R is followed by proline. It generates peptides with optimal lengths and relatively high ionization efficiency. However, potential missed cleavage could occasionally occur when utilized glycosylation sites are in close proximity to the tryptic cleavage sites. In some cases, where peptides are too

Table 1
Commonly used protease for in-solution digestion of glycoproteins

Protease	Specificity
Trypsin	C-terminus of arginine (R) and lysine (K) residues except when K/R is followed by proline (P)
Chymotrypsin	C-terminus of phenylalanine (F), leucine, (L), and tryptophan (W), tyrosine (Y), except when these residues are followed by proline (P)
Endoproteinase Lys-C	C-terminus of lysine (K) except when K is followed by proline (P)
Endoproteinase Glu-C	C-terminus of glutamic (E) and aspartic (D) acid depending on the buffer used for digestion. When phosphate buffer is used, both D and E are cleaved. In ammonium bicarbonate or ammonium acetate (pH 4), only E are cleaved
Endoproteinase Asp-N	N-terminus of aspartic (D) and cysteic acid residues
Proteinase K	Broad specificity with preference to C-terminus of aromatic or aliphatic amino acid residues especially alanine (A). Proteinase K-digested glycoprotein generated glycopeptides with 2–8 amino acid residues
Pronase	Broad specificity. Pronase cleaves the peptide back bone except for regions where glycosylation is present. Pronase-digested glycoproteins yield glycopeptides with 4–8 amino acids

long (>15 residues), using a combination of two proteases or more is necessary. A combination of enzymes such as endoproteinase Glu-C and trypsin can reduce the mass of the glycopeptide thereby facilitating detection and analysis. Also, in some cases, a combination of enzymes is necessary if two glycosylation sites appear on a single tryptic peptide and their glycoforms must be individually identified. It should be noted that shortening the length of the peptide will also change the relative ionization efficiency of the glycopeptide.

4. Most proteases will be activated immediately and undergo autolysis once they are dissolved. Do not leave the protease stock at room temperature or at 4°C for later use.
5. This affinity/enrichment method is based on the hydrophilic interaction between the glycan and agarose gel matrix. The interaction does not depend on the type of glycan but is dependent on the size of glycans. The glycopeptide enrichment protocol is straightforward and does not require a desalting step. The resulting glycopeptide pool can be analyzed by direct infusion ESI or MALDI MS.
6. In Table 2 the editors have included some other helpful suggestions for improved protein digestions that were highlighted in other portions of the book.

Table 2
Helpful hints from other chapters for improved protein digestions

Hint	Chapters
Always use fresh reagents	All
Prepare buffers/solutions immediately before use	All
Use 1D gel or MALDI-TOF to confirm digestion completion	11, 19
Suggested concentration limits of common detergents or denaturants: SDS, 0.05%; OG (octyl β-D-glucopyranoside), 0.1%; NP-40, 0.1%; Triton X-100, 0.1%; Tween 20, 0.1%; CHAPS, 0.2%; urea, <1 M	2, 19
Lysylendopeptidase with trypsin enhances cleavage of Lys-Xxx bonds	16
Rapid microwave digestion	16
Proteinase K for nonspecific proteolysis	10
High concentration/small volumes enhance digestion efficiency	2
Proteases immobilized to beads can be advantageous	10

7. If the samples tend to aggregate upon heating, skip the thermal denaturation step.
8. Urea must be added to the sample on the day the sample is digested to avoid carbamylation.
9. DTT and IAM are the most commonly used reducing and alkylating reagents for the modification of cysteine residues. The amount of DTT and IAM required for complete reaction will depend on the number of disulfide bonds and the size of the protein. Proteins should be reduced with 2–10-fold molar excess DTT over the total number of disulfides and alkylated with 1–5-fold molar excess IAM over the number of sulfhydryl groups. Reduction with DTT and alkylation must be done at basic pH otherwise the reaction will be inefficient. If the sample does not contain any disulfide bonds, skip the reduction and alkylation step and proceed to step 6.
10. In general, iodoalkylating reagents are light-sensitive. In the presence of light, available iodide ions in the reaction mixture are converted to iodine. This conversion tends to make the sample pH acidic during the course of reaction resulting to an inefficient alkylation.
11. Excess alkylating reagent must be neutralized using 5–10-fold molar excess DTT to prevent overalkylation. To avoid overalkylation without the addition of DTT, chloroacetamide can be used as an alternative alkylating reagent (29). Choloroacetamide that has the same reactivity as IAM but has high selectivity to alkylating cysteine residues.

12. Excess DTT and urea can be removed by centrifugal filtration using molecular weight cut-off filters (MWCO, Millipore, Bellirica, MA, Part No. UFC501024) with appropriate MWCO. Care must be exercised to prevent or minimize sample losses during this step. It is recommended to dilute the reduced and alkylated glycoprotein samples to twice its original volume before passing the samples to the MWCO filter. Add 20 μL buffer to the inverted sample reservoir when recovering the protein. This step is useful in concentrating sample, buffer exchange, as well as cleaning up samples containing salts, detergent, excess denaturants, reducing/alkylating reagents, and low molecular weight contaminants that could affect efficient digestion (30, 31).
13. Effort should be made to ensure that the final concentration of urea is <1 M for efficient digestion. However, if the final dilution will result in a very dilute sample, either reduce the initial urea concentration to 5 M or dilute the sample to a final concentration to no more than 2 M final urea concentration. It is preferable to keep the sample concentration as high as possible.
14. To ensure reproducibility, samples should not be subjected to more than three freeze and thaw cycles. Aliquot the digested sample if necessary.
15. Typically, 200 μg of the glycoprotein digest is added to a microcentrifuge tube containing 25 μL packed volume of Sepharose CL-4B. To reduce nonspecific interference from non-glycosylated peptides, addition of divalent salt such as calcium chloride to the binding solution to a final concentration of 1 mM is recommended.

Acknowledgements

This work was supported by NIH grant R01RR026061 and an NSF Career Award (NSF grant No. 0645120) to HD.

References

1. Hirabayashi J (2004) Lectin-based structural glycomics: glycoproteomics and glycan profiling. Glycoconj J 21:35–40
2. Budnik BA, Lee RS, Steen JAJ (2006) Global methods for protein glycosylation analysis by mass spectrometry. Biochim Biophys Acta 1764:1870–1880
3. Morelle W, Canis K, Chirat F, Faid V, Michalski JC (2006) The use of mass spectrometry for the proteomic analysis of glycosylation. Proteomics 6:3993–4015
4. Morelle W, Michalski JC (2007) Analysis of protein glycosylation by mass spectrometry. Nat Protoc 2:1585–1602
5. Calvano CD, Zambonin CG, Jensen ON (2008) Assessment of lectin and HILIC based enrichment protocols for characterization of serum glycoproteins by mass spectrometry. J Proteomics 71:304–317
6. Zaia J (2008) Mass spectrometry and the emerging field of glycomics. Chem Biol 15:881–892

7. Ito S, Hayama K, Hirabayashi J (2009) Enrichment strategies for glycopeptides. In: Packer NH, Karlsson NG (eds) Glycomics: methods and protocols. Humana Press, New York, NY
8. North SJ, Hitchen PG, Haslam SM, Dell A (2009) Mass spectrometry in the analysis of N-linked and O-linked glycans. Curr Opin Struct Biol 19:498–506
9. Yates JR, Ruse CI, Nakorchevsky A (2009) Proteomics by mass spectrometry: approaches, advances, and applications. Annu Rev Biomed Eng 11:49–79
10. Nicoli R, Martel S, Rudaz S, Wolfender JL, Veuthey JL, Carrupt PA, Guillarme D (2010) Advances in LC platforms for drug discovery. Exp Opin Drug Discov 5:475–489
11. Nilsson T, Mann M, Aebersold R, Yates JR III, Bairoch A, Bergeron JJ (2010) Mass spectrometry in high-throughput proteomics: ready for the big time. Nat Methods 7:681–685
12. Geyer H, Geyer R (2006) Strategies for analysis of glycoprotein glycosylation. Biochim Biophys Acta 1764:1853–1869
13. Krishnamoorthy L, Mahal LK (2009) Glycomic analysis: an array of technologies. ACS Chem Biol 4:715–732
14. Go EP, Irungu J, Zhang Y, Dalpathado DS, Liao HX, Sutherland LL, Alam SM, Haynes BF, Desaire H (2008) Glycosylation site-specific analysis of HIV envelope proteins (JR-FL and CON-S) reveals major differences in glycosylation site occupancy, glycoform profiles, and antigenic epitopes' accessibility. J Proteome Res 7:1660–1674
15. Irungu J, Go EP, Zhang Y, Dalpathado DS, Liao HX, Haynes BF, Desaire H (2008) Comparison of HPLC/ESI-FTICR MS versus MALDI-TOF/TOF MS for glycopeptide analysis of a highly glycosylated HIV envelope glycoprotein. J Am Soc Mass Spectrom 19:1209–1220
16. Zhang Y, Go EP, Desaire H (2008) Maximizing coverage of glycosylation heterogeneity in MALDI-MS analysis of glycoproteins with up to 27 glycosylation sites. Anal Chem 80:3144–3158
17. Go EP, Chang Q, Liao HX, Sutherland LL, Alam SM, Haynes BF, Desaire H (2009) Glycosylation site-specific analysis of clade C HIV-1 envelope proteins. J Proteome Res 8:4231–4242
18. Rebecchi KR, Wenke JL, Go EP, Desaire H (2009) Label-free quantitation: a new glycoproteomics approach. J Am Soc Mass Spectrom 20:1048–1059
19. Wada Y, Tajiri M, Yoshida S (2004) Hydrophilic affinity isolation and MALDI multiple-stage tandem mass spectrometry of glycopeptides for glycoproteomics. Anal Chem 76:6560–6565
20. Tajiri M, Yoshida S, Wada Y (2005) Differential analysis of site-specific glycans on plasma and cellular fibronectins: application of a hydrophilic affinity method for glycopeptide enrichment. Glycobiology 15:1332–1340
21. Walker J (2002) The bicinchoninic acid (BCA) assay for protein quantitation. In: Walker JM (ed) The protein protocols handbook, 2nd edn. Humana Press, Totowa, NJ
22. Aitken A, Learmonth M (1996) Protein determination by UV absorption. In: Walker JM (ed) The protein protocols handbook, 2nd edn. Humana Press, Totowa, NJ
23. Noble JE, Bailey MJA (2009) Quantitation of protein. In: Burgess RR, Deutscher MP (eds) Methods in enzymology, 2nd edn. Academic, London, UK
24. Han JC, Han GY (1994) A procedure for quantitative determination of tris(2-carboxyethyl)phosphine, an odorless reducing agent more stable and effective than dithiothreitol. Anal Biochem 220:5–10
25. Liu P, O'Mara BW, Warrack BM, Wu W, Huang Y, Zhang Y, Zhao R, Lin M, Ackerman MS, Hocknell PK, Chen G, Tao L, Rieble S, Wang J, Wang-Iverson DB, Tymiak AA, Grace MJ, Russell RJ (2010) A tris (2-carboxyethyl) phosphine (TCEP) related cleavage on cysteine-containing proteins. J Am Soc Mass Spectrom 21:837–844
26. Olsen JV, Ong SE, Mann M (2004) Trypsin cleaves exclusively C-terminal to arginine and lysine residues. Mol Cell Proteomics 3:608–614
27. Russell WK, Park ZY, Russell DH (2001) Proteolysis in mixed organic-aqueous solvent systems: applications for peptide mass mapping using mass spectrometry. Anal Chem 73:2682–2685
28. Strader MB, Tabb DL, Hervey WJ, Pan CL, Hurst GB (2006) Efficient and specific trypsin digestion of microgram to nanogram quantities of proteins in organic-aqueous solvent systems. Anal Chem 78:125–134
29. Nielsen ML, Vermeulen M, Bonaldi T, Cox J, Moroder L, Mann M (2008) Iodoacetamide-induced artifact mimics ubiquitination in mass spectrometry. Nat Methods 5:459–460
30. Manza LL, Stamer SL, Ham AJL, Codreanu SG, Liebler DC (2005) Sample preparation and digestion for proteomic analyses using spin filters. Proteomics 5:1742–1745
31. Wisniewski JR, Zougman A, Nagaraj N, Mann M (2009) Universal sample preparation method for proteome analysis. Nat Methods 6:359–362

Chapter 9

Nano-HPLC-MS of Glycopeptides Obtained After Nonspecific Proteolysis

Gerhild Zauner, Carolien A.M. Koeleman, André M. Deelder, and Manfred Wuhrer

Abstract

Liquid chromatography-tandem stage mass spectrometry of glycopeptides is a powerful tool for the site-specific glycosylation analysis of glycoproteins. Using fetuin as a model substance, we describe a protocol for glycopeptide dissection using nonspecific proteolysis by proteinase K. Proteolysis is achieved using dissolved or immobilized enzyme. For glycopeptide separation three different nanoHPLC separation principles are compared, namely hydrophilic interaction liquid chromatography (HILIC), C18-reverse phase (RP), and graphitized carbon HPLC. Chromatographically resolved glycopeptides are analyzed by nano-electrospray ionization multistage mass spectrometry for identification of the glycan as well as the peptide moiety. Using this approach, site-specific information on protein glycosylation is obtained.

Key words: Glycopeptides, Mass spectrometry, LC-MS, HILIC, Reverse phase, Graphitized carbon

1. Introduction

Protein glycosylation plays an important role in processes such as cell differentiation, cell adhesion, host–pathogen interaction, immune responses, and cancer metastasis (1). Methods to elucidate the molecular structures and attachment sites of protein-linked glycans are crucial in many aspects as for example to target pharmacologically interesting protein–carbohydrate interactions (2) or to unravel host–pathogen interactions (3).

The most common approach for obtaining information on the site-specific glycosylation of a protein involves its proteolytic cleavage and the subsequent analysis of the resulting glycopeptides.

Jennifer J. Kohler and Steven M. Patrie (eds.), *Mass Spectrometry of Glycoproteins: Methods and Protocols*, Methods in Molecular Biology, vol. 951, DOI 10.1007/978-1-62703-146-2_9, © Springer Science+Business Media, LLC 2013

Trypsin is the protease most widely applied for this purpose, and recently the pros and cons of this enzyme have been described (4). A major drawback which may be encountered when using trypsin for glycoprotein digestion is the lack of cleavage sites in the neighborhood of the glycosylation sites. Moreover, trypsin may fail to act efficiently on highly post-translational modified proteins, due to steric hindrance by the glycan moieties (5). This may result in glycopeptides which are too big and carry too many heterogeneous modifications to be efficiently analyzed (6). Moreover, in many mass spectrometric analyses the large non-glycosylated peptides that exist in tryptic digests can suppress the signals of glycopeptides, which often show poorer ionization efficiencies (7).

Here we employ a nonspecific protease, proteinase K, which overcomes part of the limitations of trypsin). Proteinase K cleavage of the model glycoprotein fetuin resulted in glycopeptide preparations with small peptide moieties (from one to eight amino acids). Noteworthy, non-glycosylated peptide stretches are usually cleaved to the level of single amino acids and therefore do not interfere with the mass spectrometric analysis of the proteinase K-generated glycopeptides (4).

Another aspect of the analysis is obviously the choice of an efficient chromatographic separation method for the glycopeptide mixture after digestion, especially with the prospect of identifying glycopeptides from samples of greater complexity (5). Here, three different types of separation principles, namely hydrophilic interaction liquid chromatography (HILIC), C18-reverse phase (RP), and graphitized carbon HPLC, which all have been widely applied in glycosylation analysis (8–11), are compared (Fig. 1).

For our experiments we prepared a digest of fetal calf serum fetuin using proteinase K, a serine protease from *Tritirachium album*. The enzyme was successfully applied both in solution and after covalent immobilization onto Sepharose beads (see Note 1). The resulting glycopeptides were analyzed by nanoHPLC-ESI-ion trap (IT)-MS/MS. Three different chromatographic methods have been used and the performance of each individual procedure was evaluated. The results of the HILIC separation have recently been reported separately (4). The structures and attachment sites of O- as well as N-glycans have been characterized (Fig. 1), and the glycopeptides identified with the different chromatographic methods are listed in Table 1. The analyses resulted in the following observations and conclusions:

- Only HILIC material allowed the efficient separation of N-glycopeptides from O-glycopeptides.
- In RP-HPLC N- and O-glycopeptides eluted over the same, broad chromatographic range.
- Graphitized carbon HPLC resulted in the very efficient separation of O-glycopeptides, whilst fetuin N-glycopeptides were not detected under the applied conditions. N-glycopeptides

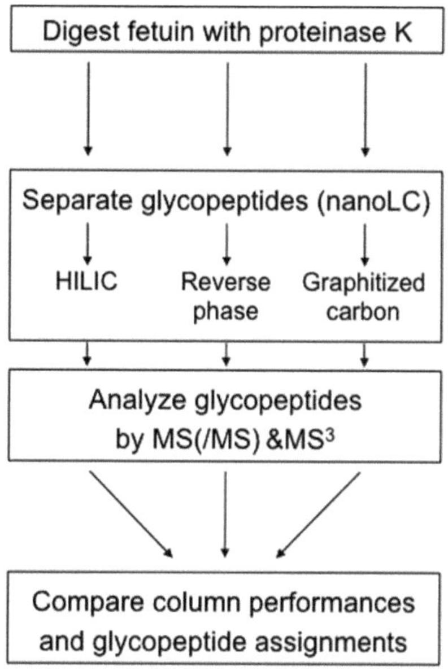

Scheme 1

Fig. 1. Analysis strategy for glycopeptides separation and detection.

were most probably too strongly retained under the applied chromatographic conditions.

- Both HILIC and graphitized carbon nanoLC separated different glycoforms efficiently, i.e., glycopeptides with identical peptide moieties but different glycan moieties. In contrast, reverse phase nanoLC often resulted only in partial separation of glycoforms.

2. Materials

2.1. Preparation of Proteinase K Beads

1. Deionized water (Milli-Q; Millipore) (see Note 2).
2. Coupling buffer: 0.2 M $NaHCO_3$ containing 0.5 M NaCl, pH 8.3.
3. Blocking solution: 0.1 M Tris HCl (pH 8.5).
4. 0.1 M Acetate buffer containing 0.5 M NaCl (pH 4.0).
5. NHS-Activated Sepharose 4 Fast Flow from GE Healthcare Biosciences AB (Uppsala, Sweden).
6. Proteinase K from *T. album* (Sigma-Aldrich, Zwijndrecht, The Netherlands).
7. Fetuin from fetal calf serum (Sigma-Aldrich, Zwijndrecht, The Netherlands).

Table 1
Analysis of proteinase K-generated glycopeptides of fetuin after HILIC, reverse phase and graphitized carbon separation

Glycosylation site	Registered signal	Peptide mass (Da)	Peptide moiety	Glycan moiety	Retention time (min)
(Asialo)fetuin/HILIC					
$S_{296}{}^{A}$	$[1120.5]^{+}$	755.4	291VVGP**S**VV298	$HexNAc_1Hex_1$	1.9
S_{296}	$[1147.6]^{++}$	1345.9	291VVVGP**S**VVAVPLPL304	$HexNAc_1Hex_1NeuAc_2$	3.5
T_{280} or $S_{282}{}^{A}$	$[965.4]^{+}$	600.2	276AAGP**T**P**S**282	$HexNAc_1Hex_1$	3.5
S_{296}	$[1411.6]^{+}$	755.8	291VVVGP**S**VV298	$HexNAc_1Hex_1NeuAc_1$	4.1
$S_{271}{}^{A}$	$[853.4]^{+}$	488.2	271**S**AVPD275	$HexNAc_1Hex_1$	6.1
$S_{271}{}^{A}$	$[1150.4]^{+}$	785.3	268EAP**S**AVPD275	$HexNAc_1Hex_1$	6.9
T_{280} or $S_{282}{}^{A}$	$[823.4]^{+}$	458.2	278GP**T**P**S**282	$HexNAc_1Hex_1$	6.9
T_{280} or S_{282}	$[699.9]^{++}$	742.4	276AAGP**T**PSAA284	$HexNAc_1Hex_1NeuAc_1$	7.5
$S_{271}{}^{A}$	$[1021.5]^{+}$	656.3	269AP**S**AVPD275	$HexNAc_1Hex_1$	7.5
S_{271}	$[721.3]^{++}$	785.4	268EAP**S**AVPD275	$HexNAc_1Hex_1NeuAc_1$	7.8
T_{280} or $S_{282}{}^{A}$	$[701.3]^{++}$	671.4	277AGP**T**P**S**AA284	$HexNAc_2Hex_2$	9.4
T_{280} and S_{282}	$[1282.0]^{++}$	1250.6	276AAGP**T**P**S**AAGPPVAS290	$HexNAc_2Hex_2NeuAc_2$	9.5
T_{280} or $S_{282}{}^{A}$	$[990.0]^{++}$	1250.6	276AAGP**T**P**S**AAGPPVAS290	$HexNAc_2Hex_2$	9.5
S_{271}	$[866.9]^{++}$	785.4	268EAP**S**AVPD275	$HexNAc_1Hex_1NeuAc_2$	9.6
T_{280} or $S_{282}{}^{A}$	$[665.8]^{++}$	600.3	280**T**P**S**AAGP286	$HexNAc_2Hex_2$	10.0
T_{280} or $S_{282}{}^{A}$	$[848.4]^{++}$	600.3	280TPSAAGP286	$HexNAc_3Hex_3$	13.0
$S_{282}{}^{A}$	$[743.8]^{++}$	756.5	282**S**AAGPPVAS290	$HexNAc_2Hex_2$	14.4
$N_{176}{}^{A}$	$[958.3]^{++}$	277.1	175S**N**G177 or 176**N**GS178	$HexNAc_4Hex_5$	17.1

N_{156}^{A}	$[1156.4]^{++}$	689.3	156**N**DSRVV161	$HexNAc_4Hex_5$	19.0
N_{156}^{A}	$[979.4]^{++}$	335.1	156**N**DS158	$HexNAc_4Hex_5$	19.2
N_{176}^{A}	$[1140.9]^{++}$	277.1	175S**N**G177 or 176**N**GS178	$HexNAc_5Hex_6$	19.7
N_{99}^{A}	$[1098.2]^{++}$	204.2	98A**N**99	$HexNAc_5Hex_6$	20.0
N_{156}	$[1184.1]^{+++}$	689.6	156**N**DSRVV161	$HexNAc_5Hex_6NeuAc_3$	22.0
N_{156}^{A}	$[1171.9]^{++}$	335.2	156**N**DS158	$HexNAc_5Hex_6$	23.9
N_{156}	$[1230.2]^{+++}$	826.5	156**N**DSRVVH162	$HexNAc_5Hex_6NeuAc_3$	24.1
T_{280} and S_{282}	$[855.3]^{+++}$	1250.7	276AAGP**T**P**S**AAGPPVAS290	$HexNAc_2Hex_2NeuAc_2$	31.5
Fetuin/reverse phase					
N_{99}	$[942.5]^{4+}$	905.4	98A**N**CSVRQQ105	$HexNAc_5Hex_6NeuAc_3$	3.1
N_{156}	$[965.4]^{+++}$	689.6	156**N**DSRVV161	$HexNAc_4Hex_5NeuAc_2$	9.0
T_{280} or S_{282}	$[957.0]^{++}$	600.4	280**T**P**S**AAGP286	$HexNAc_2Hex_2NeuAc_2$	9.3
N_{176}	$[957.0]^{4+}$	966.4	168LATFAES**N**L176	$HexNAc_5Hex_6NeuAc_3$	9.7
T_{280} or S_{282}	$[1028.0]^{++}$	742.2	276AAGP**T**P**S**AA282	$HexNAc_2Hex_2NeuAc_2$	10.5
N_{156}	$[1184.2]^{+++}$	689.8	156**N**DSRVV161	$HexNAc_5Hex_6NeuAc_3$	11.8
S_{271} or T_{280}	$[819.3]^{++}$	982.6	270P**S**AVPDHAGP**T**281	$HexNAc_1Hex_1NeuAc_1$	11.9
S_{271}	$[721.4]^{++}$	785.6	268EAP**S**AVPD275	$HexNAc_1Hex_1NeuAc_1$	12.4
S_{271}	$[1049.5]^{++}$	785.6	268EAP**S**AVPD275	$HexNAc_2Hex_2NeuAc_2$	14.7
S_{296}	$[1034.5]^{++}$	755.6	291VVVGP**S**VV298	$HexNAc_2Hex_2NeuAc_2$	19.3
S_{296}	$[706.4]^{++}$	755.6	291VVVGP**S**VV298	$HexNAc_1Hex_1NeuAc_1$	23.1
T_{280} or S_{282}	$[899.4]^{+++}$	1092.6	276AAGP**T**P**S**AAGPPV288	$HexNAc_2Hex_2NeuAc_3$	23.0
S_{271}	$[866.9]^{++}$	785.5	268EAP**S**AVPD275	$HexNAc_1Hex_1NeuAc_2$	25.1

(continued)

Table 1
(continued)

Glycosylation site	Registered signal	Peptide mass (Da)	Peptide moiety	Glycan moiety	Retention time (min)
Fetuin/graphitized carbon					
T_{280} or S_{282}	$[628.7]^{++}$	600.2	280**T**P**S**AAGP286	$HexNAc_1Hex_1NeuAc_1$	12.4
S_{271}	$[656.3]^{++}$	656.3	269AP**S**AVPD275	$HexNAc_1Hex_1NeuAc_1$	16.9
S_{271}	$[721.3]^{++}$	785.3	268EAP**S**AVPD275	$HexNAc_1Hex_1NeuAc_1$	19.1
S_{271}	$[732.2]^{++}$	807.2	268EAP**S**AVPD275	$HexNAc_1Hex_1NeuAc_1$	19.5
S_{296}	$[706.3]^{++}$	755.4	291VVGP**S**VV298	$HexNAc_1Hex_1NeuAc_1$	20.7
T_{280} or S_{282}	$[1027.8]^{++}$	742.3	276AAGP**T**P**S**282	$HexNac_2Hex_2NeuAc_2$	20.8
T_{280} or S_{282}	$[866.2]^{++}$	785.3	268EAP**S**AVPD275	$HexNac_1Hex_1NeuAc_2$	22.0
S_{271}	$[821.3]^{++}$	985.3	266EAEAP**S**AVPD275	$HexNAc_1Hex_1NeuAc_1$	23.7
S_{271}	$[956.8]^{++}$	600.2	280**T**P**S**AAGP286	$HexNac_2Hex_2NeuAc_2$	25.0
T_{280} or S_{282}	$[802.3]^{+++}$	1092.4	276AAGP**T**P**S**AAGPPV288	$HexNac_2Hex_2NeuAc_2$	32.5
T_{280} or S_{282}	$[1203.1]^{++}$	1092.3	276AAGP**T**P**S**AAGPPV288	$HexNac_2Hex_2NeuAc_2$	32.5
T_{280} or S_{282}	$[826.0]^{+++}$	1163.7	276AAGP**T**P**S**AAGPPVA289	$HexNac_2Hex_2NeuAc_2$	33.1
S_{296}	$[1034.4]^{++}$	755.5	291VVGP**S**VV298	$HexNac_2Hex_2NeuAc_2$	34.4
T_{280} or S_{282}	$[1031.9]^{++}$	950.3	278GP**T**P**S**AAGPPV288	$HexNac_2Hex_2NeuAc_2$	34.9
T_{280} or S_{282}	$[1010.9]^{++}$	1108.6	274PDAAGP**T**P**S**AAGP286	$HexNac_2Hex_2NeuAc_2$	37.1
T_{280} or S_{282}	$[855.0]^{+++}$	1251.6	280**T**P**S**AAGPPVASVVV293	$HexNac_2Hex_2NeuAc_2$	37.4
T_{280} or S_{282}	$[1282.0]^{++}$	1250.6	276AAGP**T**P**S**AAGPPVAS290	$HexNac_2Hex_2NeuAc_2$	40.8
S_{296}	$[851.2]^{++}$	755.4	291VVGP**S**VV298	$HexNac_1Hex_1NeuAc_2$	42.4

Glycan compositions are given in terms of hexose (Hex), N-acetylhexosamine (HexNAc), and sialic acid (NeuAc). Glycosylation sites are underlined. The results of the HILIC LC-MS experiments are taken from ref. (4)

A, result from an experiment performed on asialofetuin

2.2. Nano HPLC-ESI-MS/MS and Data Analysis

1. NanoHPLC was performed on an Ultimate 3000 system (Dionex/LC Packings, Amsterdam, The Netherlands); in the following, details of the system used by us are given, and possible variations thereof are mentioned.
2. Autosampler with a 20 μL sample loop.
3. Micro-pump system with a switching valve equipped with a guard column (for RP: 5 μm PepMap particles, 300 μm × 5 mm; Dionex; for graphitized carbon separation: Hypercarb-S, 5 μm, 170 μm × 10 mm; packed by Grom Analytik, Rottenburg, Germany). For HILIC nanoHPLC no guard column was used.
4. NanoHPLC (with 1:100 splitter) supplying a flow of approximately 300 nL/min for RP, and 400 nL/min for HILIC and graphitized carbon separation (see Note 3).
5. *NanoHPLC columns*:
   (a) *HILIC*: Nanoscale Amide-80 column (75 μm × 180 mm packed in house; 3 μm, 80 Å, Tosoh Bioscience, Stuttgart, Germany). Mobile phase: solvent A: 80% ACN, 20% 50 mM ammonium formate pH 4.4; solvent B: 50 mM ammonium formate pH 4.4.
   (b) *Reverse phase*: For reverse phase separation a PepMap column (3 μm; 75 μm × 150 mm; Dionex, Amsterdam, The Netherlands) was used. Mobile phase: solvent A: 0.1% formic acid in water; solvent B: 95% ACN.
   (c) *Graphitized carbon*: For separation a Hypersil Hypercarb column (5 μm; 75 μm × 100 mm, Grom) was used. Mobile phase: solvent A: 0.1% formic acid in water; solvent B: 95% ACN.
6. Z-type nano-flow UV-detector (Dionex/LC Packings).
7. Tandem mass spectrometer: Esquire High Capacity Trap (HCTultra) ESI-IT-MS (Bruker Daltonics, Bremen, Germany) equipped with an on-line nano source.
8. Electrospray needles (360 μm OD, 20 μm ID with 10 μm opening; New Objective, Cambridge, MA, USA).
9. Search engine: Obtained peptide masses from the measured MS^2 (tandem mass spectrometry) spectra were used for matching to theoretical masses of peptide moieties generated by random cleavage of fetuin (Uniprot entry P12763) using the FindPept tool (http://www.expasy.org/tools/findpept.html).
10. Database: Swiss-Prot or NCBInr.

3. Methods

Carry out all procedures at room temperature unless otherwise specified.

3.1. Proteolytic Cleavage in Solution

1. For the in solution digests 1.5 mg of fetal calf (asialo)fetuin is dissolved in 300 μL of 50 mM ammonium bicarbonate buffer.
2. 5 mg of proteinase K is dissolved in 1 mL of 20 mM acetic acid and directly added to the (asialo)fetuin solution at an enzyme-substrate ratio of 1:300 or 1:3 (see Note 4).
3. Samples are incubated over night at 37°C.

3.2. Proteolysis with Immobilized Proteinase K

1. Take 100 μL of NHS-activated Sepharose beads and wash twice with ice cold 1 mM HCl solution (10× volume) immediately before use. Before and after each washing step spin the beads down for 2 min at 16.1 rcf—take the supernatant off and discard.
2. Add coupling buffer and wash the beads as above. Repeat until the pH is correct (pH 8.3).
3. Add 0.5 mg of proteinase K dissolved in 500 μL of coupling buffer to the beads (see Note 5).
4. Leave the mixture for 4 h shaking at room temperature.
5. Blocking step: spin the beads down, take the supernatant off and discard. Add 1 mL blocking solution and incubate for 1 h.
6. Wash again with 0.1 M acetate buffer and subsequently with blocking solution (see Note 6). Perform this step three times.
7. 1.5 mg of fetal calf fetuin is dissolved in 300 μL of 50 mM ammonium bicarbonate buffer and directly mix with 10 μL of the proteinase K beads. The digest "suspension" is incubated overnight at 37°C with shaking. The beads are spun down (see above), and an aliquot is directly analyzed by LC-MS. The digest solution may be stored at −20°C (see Note 6).

3.3. Nano-HPLC-ESI-MS/MS

1. Prepare the solvent for the micro-flow pump (0.1% formic acid). Degas with vacuum degasser or by flushing helium through the solution for a minimum of 15 min.
2. Set up the nanoHPLC system with autosampler, micro-pump with switching valve and trap column, HPLC pump with 1:100 splitter,

the nano-column (see Subheading 2.2, item 5), and the Z-type UV-detector (see Note 7; for HILIC separation see Note 8).

3. Switch on the pumps and wash the trap column system and the nano-column.
4. Set up the HPLC method:
   (a) HILIC: trap column: 15 μL/min; nano-column: $t=0$ min, 25% solvent B; $t=30$ min, 40% solvent B; $t=35$ min, 100% solvent B; $t=40$ min, back to 25% solvent B; $t=60$ min, 25% solvent B; flow rate: 400 nL/min.
   (b) RP: trap column: 25 μL/min; nano-column: $t=0$ min, 0% solvent B; $t=15$ min, 25% solvent B; $t=25$ min, 70% solvent B, $t=30$ min, 70% solvent B; flow rate: 300 nL/min.
   (c) Graphitized carbon: trap column: 8 μL/min; nano-column: $t=0$ min, 0% solvent B; $t=15$ min, 25% solvent B; $t=25$ min, 70% solvent B, $t=30$ min, 70% solvent B; flow rate: 400 nL/min.
5. Prepare a dilution (1:10) of the proteinase K digest of fetuin in autosampler vials (stock digest solution is approximately 5 μg/μL) (see Note 9).
6. Run a blank.
7. Put the MS in-line with the nanoESI-source. Choose positive-ion detection mode with the following settings:

   MS scan range: *m/z* 300–1,500 for N- and O-glycans; scan average = 5

   MS/MS scan range: *m/z* 140–2,200; number of precursors = 5; preferred charge state: none; include singly charged ions for O-glycans

   MS^3: number of precursors = 1 (see Note 10).
8. Run your samples using a 1 μL sample injection as default (see Fig. 2 for extracted ion chromatograms (EICs) observed with RP and graphitized carbon nanoHPLC).
9. Analyze the data using processing software. LC-MS/MS data are searched for carbohydrate-derived oxonium fragment ions such as *m/z* 366.14 ($(HexNAc_1Hex_1+H]^+$), 657.23 ($[NeuAc_1HexNAc_1Hex_1+H]^+$), and 292.10 ($[NeuAc_1+H]^+$) (see Note 11).
10. Interpret the obtained MS/MS data to achieve an assignment of the glycan composition and structure, and deduce the mass of the peptide moiety (12–17). Figures 3 and 4 give examples of glycopeptides fragment spectra assignments after RP and graphitized carbon LC-MS/MS analyses. The peptide moiety 291VVVGP<u>**S**</u>VV298 (*m/z* 755.4) is observed with the

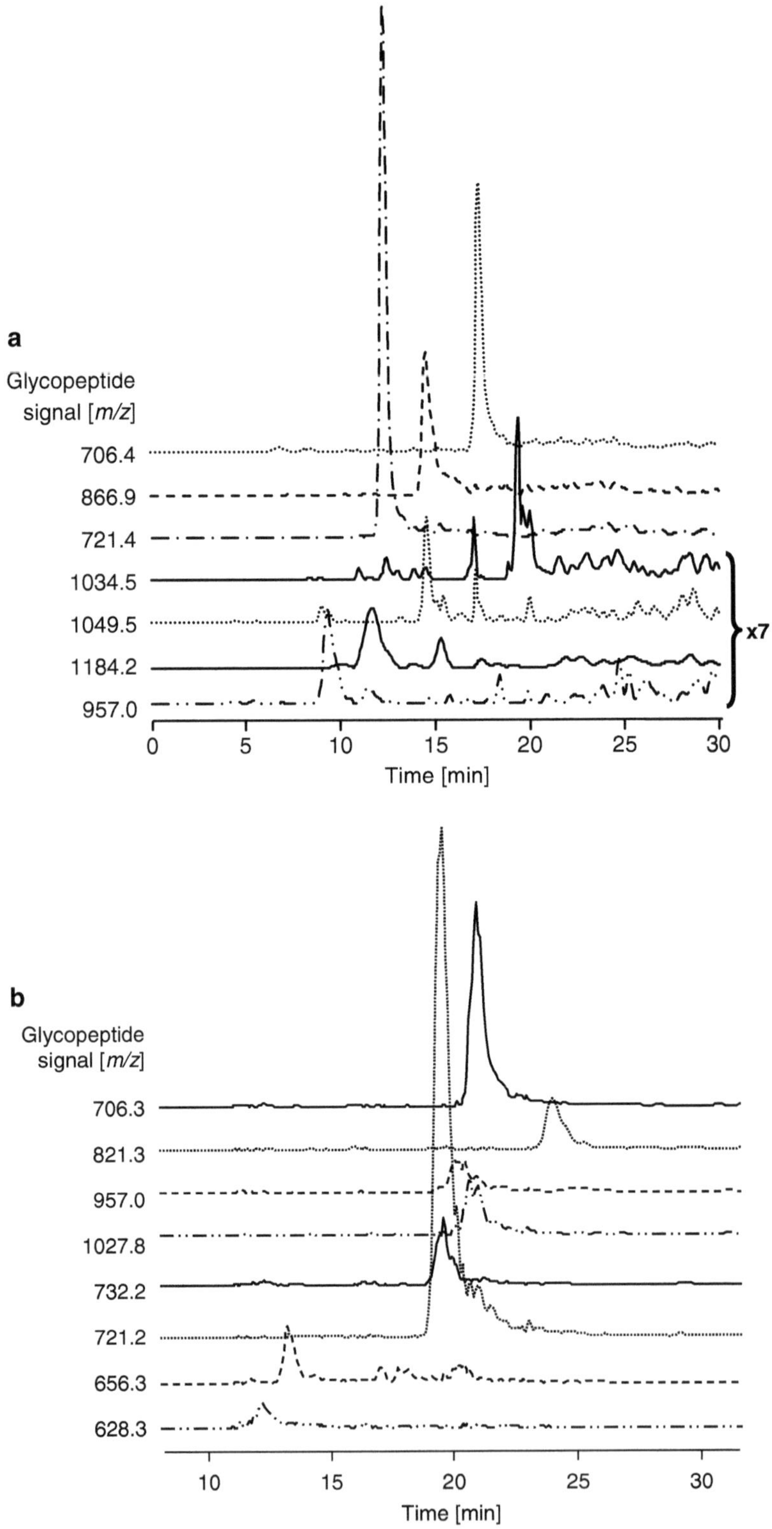

Fig. 2. (**a**) RP extracted ion chromatograms (EICs) for six O-glycosylated peptides and one N-glycosylated peptide from a fetuin proteinase K digest. O-glycopeptides elute at the same time as N-glycopeptides (8–20 min). (**b**) Graphitized carbon EICs for eight O-glycopeptides (11–25 min) from a fetuin proteinase K digest.

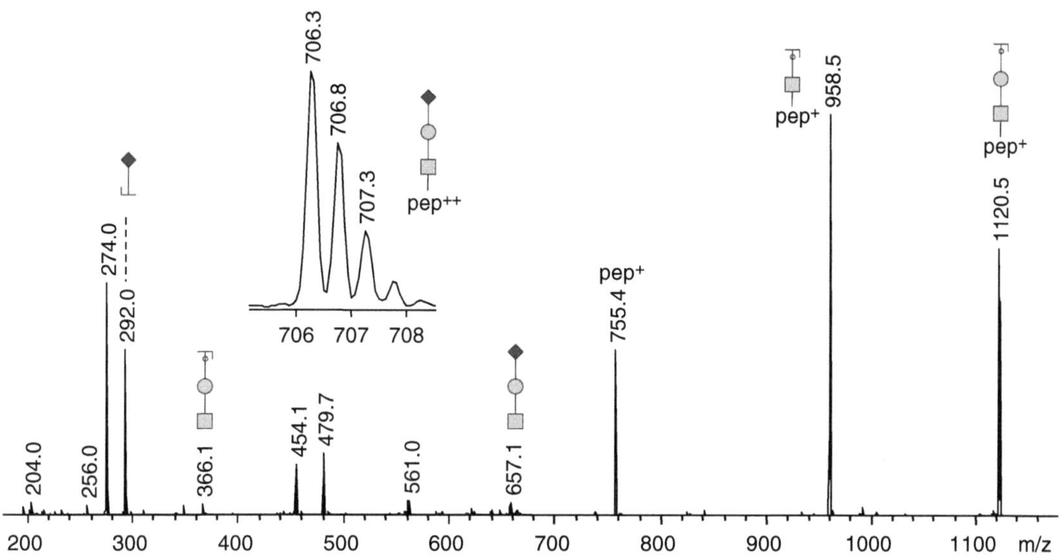

Fig. 3. Proteinase K-generated glycopeptide V291-V298 from fetuin with an O-glycan ($Hex_1HexNAc_1NeuAc_1$) analyzed by graphitized carbon nanoLC-MS/MS. The *inset* shows the mass spectrum of the precursor at *m/z* 706.3 ($[M+2H]^{2+}$). Glycan compositions are given in terms of hexose (Hex), N-acetylhexosamine (HexNAc), and sialic acid (NeuAc). *dark square*, N-acetylglucosamine; *diamond*, N-acetylneuraminic acid; *light circle*, galactose; *dark circle*, mannose.

O-glycan species $HexNAc_1Hex_1NeuAc$ (Fig. 3), and $HexNAc_2Hex_2NeuAc_2$ (Fig. 4b).

3.4. Database Search

1. Peptide masses deduced from the MS/MS data are searched against theoretical masses of randomly cleaved peptide moieties of fetuin (Uniprot entry P12763) using the FindPept tool (http://www.expasy.ch/tools/findpept.html) (see Table 1 and Note 12).
2. The peptide moieties indicated by the Expasy FindPept tool are used for in-silico fragmentation. To this end, the candidate peptide sequences are copied into the program Protein Prospector (http://prospector.ucsf.edu/prospector/mshome.htm) which performs a theoretical fragmentation (see Note 13). The obtained masses can be used for structural assignments of the MS/MS and MS/MS/MS data (see ref. (4) for the latter). As an alternative to Protein Prospector the free on-line tool "Fragment Ion Calculator" (http://db.systemsbiology.net:8080/proteomicsToolkit/FragIonServlet.html) may be used, for example.

4. Notes

1. Immobilization of the enzyme has the following advantages: (1) The enzyme can be easily removed after the reaction via centrifugation of the sample; (2) Immobilization of the enzyme suppresses autoproteolysis; (3) High local concentrations of the proteolytic enzyme may be achieved.

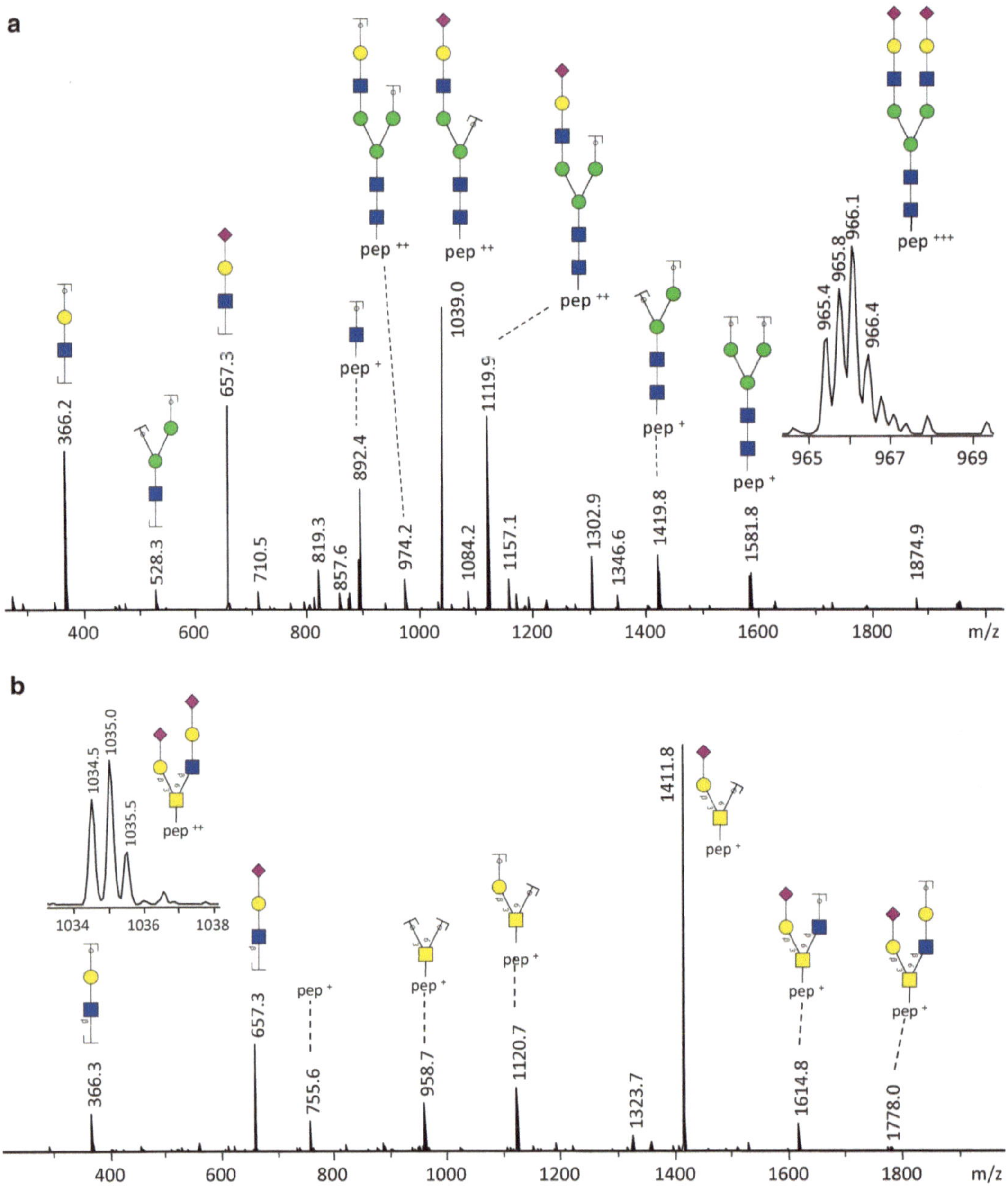

Fig. 4. Fragment ion spectra of proteinase K-generated glycopeptides from fetuin after RP separation. For each identified peptide a fragment ion spectrum (MS^2) of the protonated precursor (*inset*) is shown. (**a**) Glycopeptide N156-V161 with an N-glycan ($HexNAc_4Hex_5NeuAc_2$); precursor at *m/z* 965.4 ($[M+3H]^{3+}$). (**b**) Glycopeptide V291-V298 with an O-glycan ($HexNAc_2Hex_2NeuAc_2$); precursor at *m/z* 1034.5 ($[M+2H]^{2+}$). Glycan compositions are given in terms of hexose (Hex), N-acetylhexosamine (HexNAc), and sialic acid (NeuAc). *dark square*, N-acetylglucosamine; *diamond*, N-acetylneuraminic acid; *light circle*, galactose; *dark circle*, mannose.

2. Use Milli-Q water throughout.
3. Always degas the solvents using helium gas prior to use.
4. By using a broad range of enzyme to substrate ratios glycopeptides with short as well as long peptide moieties were obtained.

This allowed in some cases to find the same glycan structure and glycan attachment site on different peptide moieties making the assignment more reliable. In order to use the most suitable enzyme:substrate ratio for a protein of interest, we suggest to start digesting the protein with various ratios initially trying tenfold incremental steps, i.e., 1:10, 1:100, 1:1,000.

5. Coupling is done very fast at room temperature. It is important to optimize the coupling time to obtain the biological activity of the enzyme. To avoid its autoproteolysis proteinase K was added on ice to the solution before continuation with the following step. Once more to initially 100 μL of bead suspension 0.5 mg of proteinase K dissolved in 500 μL of coupling buffer was added after treating the beads according to Subheading 3.2, steps 1 and 2.
6. At this stage the beads can be stored in 20% ethanol at 4°C. Before (re-)use wash with coupling buffer until a pH of 8.3 is reached. No loss of enzyme activity was noticed after storage of the beads for 1 month.
7. The UV-detector is not essential, but is useful to monitor the performance of the nanoHPLC system.
8. No trap column is used for HILIC separation in order to avoid loss of small O-glycopeptides during trap column wash.
9. Before pipetting the digestion solution into the injection vial it is advisable to centrifuge (16.1 rcf for 1 min) the sample and take the supernatant to avoid getting any kind of particles onto the nanoHPLC system. For RP and graphitized carbon separation 1 μL aqueous sample was injected. To employ the HILIC column 1 μL of sample was brought to 80% ACN prior to use.
10. MS^3 experiments seem particularly helpful if proteinase K is used for the first time to digest a protein of interest in order to identify the peptide moieties obtained after digestion. MS^2 spectra mainly serve for the glycan identification as the oligosaccharide will be fragmented at first (weaker bonds) prior to the peptide portion. Therefore the MS^3 spectra provide mainly information on the peptide nature, and sometimes still on the glycan portion if the oligosaccharide has a large molecular weight (as most N-glycans).
11. MS/MS spectra are screened for glycan marker ions. If spectra contain those, the MS spectrum of the parent ion is checked for the charge state and monoisotopic mass of this particular ion. The MS/MS spectrum is then investigated further for glycan structural assignment and to identify the mass of the peptide moiety.
12. In the "Findpept tool" the UniProtKB ID (P12763 for fetuin), or user-entered sequence are specified as well as the experimentally

identified peptide masses. When specifying the enzyme, proteinase K may NOT be selected, though this option is provided. Instead, "none" should be selected. As mentioned above, under the chosen proteinase K reaction conditions most unmodified stretches of proteins will be cleaved to the level of single amino acids. The glycosylated regions, in contrast, will often be cleaved less efficiently due to steric hindrance, often resulting in peptide tags of 6–8 amino acids. Moreover, optional pretreatment of the protein by reduction/alkylation with acrylamide/iodoacetic acid/iodoacetamide/4-vinyl pyridine has to be specified. In our example the option "nothing" has to be selected. We allowed for a 0.5 Da "Mass tolerance" which is displayed below the selection of pretreatments. Once these details are filled in, one should run the search by clicking on "StartFindPept." The search will provide a list of possible peptide candidates matching the previously entered mass. With the obtained peptides from the list containing a possible glycosylation site (a serine or threonine residue for O-linked oligosaccharides or an asparagine for N-linked glycans) a further search should for some b and y peptide fragments in the MS^2 can be performed (depending on the size of the glycans attached, which will be fragmented first) (see below).

13. In ProteinProspector, the "MS-Product" tool is chosen and the peptide sequence is entered. Of the various options b-ions and y-ions and constant loss sequence of H_2O was ticked. Upon "Induce Fragmentation" a list of possible b- and y-ion masses plus their *m/z* values with additional H_2O loss is provided.

Acknowledgement

Dr. G. Zauner acknowledges financial support by the Netherlands Genomics Initiative (Horizon Breakthrough Project 93518016).

References

1. Hashimoto R, Hirose K, Sato T, Fukushima N, Miura N, Nishimura S (2010) Functional network of glycan-related molecules: glyco-net in glycoconjugate data bank. BMC Syst Biol 4:91
2. Rek A, Krenn E, Kungl AJ (2009) Therapeutically targeting protein-glycan interactions. Br J Pharmacol 157:686–694
3. Varki A, Cummings R, Esko JD, Freeze HH, Hart GW, Marth J (1999) Essentials of glycobiology. Cold Spring Harbour Laboratory Press, Cold Spring Harbor, NY
4. Zauner G, Koeleman CA, Deelder AM, Wuhrer M (2010) Protein glycosylation analysis by HILIC-LC-MS of proteinase K-generated N- and O-glycopeptides. J Sep Sci 33:903–910
5. Dodds ED, Seipert RR, Clowers BH, German JB, Lebrilla CB (2009) Analytical performance of immobilized pronase for glycopeptide footprinting and implications for surpassing reductionist glycoproteomics. J Proteome Res 8:502–512
6. Wuhrer M, Koeleman CA, Hokke CH, Deelder AM (2005) Protein glycosylation analyzed by normal-phase nano-liquid chromatography–mass spectrometry of glycopeptides. Anal Chem 77:886–894

7. Dalpathado DS, Desaire H (2008) Glycopeptide analysis by mass spectrometry. Analyst 133: 731–738
8. Mysling S, Palmisano G, Hojrup P, Thaysen-Andersen M (2010) Utilizing ion-pairing hydrophilic interaction chromatography solid phase extraction for efficient glycopeptide enrichment in glycoproteomics. Anal Chem 82:5598–5609
9. Wuhrer M, Deelder AM, Hokke CH (2005) Protein glycosylation analysis by liquid chromatography-mass spectrometry. J Chromatogr B 825:124–133
10. Ruhaak LR, Deelder AM, Wuhrer M (2009) Oligosaccharide analysis by graphitized carbon liquid chromatography-mass spectrometry. Anal Bioanal Chem 394:163–174
11. Ruhaak LR, Zauner G, Huhn C, Bruggink C, Deelder AM, Wuhrer M (2010) Glycan labeling strategies and their use in identification and quantification. Anal Bioanal Chem 397:3457–3481
12. Mechref Y, Novotny MV (2002) Structural investigations of glycoconjugates at high sensitivity. Chem Rev 102:321–369
13. Zaia J (2004) Mass spectrometry of oligosaccharides. Mass Spectrom Rev 23:161–227
14. Morelle W, Michalski J-C (2005) The mass spectrometric analysis of glycoproteins and their glycan structures. Curr Anal Chem 1:29–57
15. Harvey DJ (1999) Matrix-assisted laser desorption/ionization mass spectrometry of carbohydrates. Mass Spectrom Rev 18:349–450
16. Wuhrer M, de Boer AR, Deelder AM (2009) Structural glycomics using hydrophilic interaction chromatography (HILIC) with mass spectrometry. Mass Spectrom Rev 28: 192–206
17. Viseux N, Costello CE, Domon B (1999) Post-source decay mass spectrometry: optimized calibration procedure and structural characterization of permethylated oligosaccharides. J Mass Spectrom 34:364–376

Part III

Separation Methods

Chapter 10

Glycopeptide Enrichment for MALDI-TOF Mass Spectrometry Analysis by Hydrophilic Interaction Liquid Chromatography Solid Phase Extraction (HILIC SPE)

Pia Hønnerup Jensen, Simon Mysling, Peter Højrup, and Ole Nørregaard Jensen

Abstract

Glycoproteins, and in particular glycopeptides, are highly hydrophilic and are often not retained by reversed phase (RP) chromatography. The separation principle of normal phase (NP) is based on hydrophilic interactions, which in many aspects is complementary to RP separations. Hydrophilic interaction liquid chromatography (HILIC) is a fairly new variation of the NP separations used in the 1970s, the major difference being the use of aqueous solvents. HILIC provides a versatile tool for enrichment of glycopeptides before mass spectrometric (MS) analysis, particularly when used for solid phase extraction (SPE), or in combination with other chromatographic resins or ion-pairing reagents. HILIC SPE can be used for glyco-profiling, i.e., for determining the glycan heterogeneity at one specific glycosylation site, for enrichment of glycopeptides from a complex mixture of peptides, as well as for pre-fractionation of complex samples at the protein or peptide level. In this chapter we present a straightforward HILIC SPE enrichment technique and then combine C18 RP and HILIC enrichment for analysis of glycopeptides. Finally, we demonstrate HILIC enrichment using trifluoroacetic acid as an ion-pairing reagent for the enrichment of glycopeptides prior to mass spectrometry analysis.

Key words: Glycopeptides, Enrichment, HILIC, MALDI-TOF MS, SPE, *N*-glycosylation, Ion-pairing reagent

1. Introduction

Protein glycosylation has many biological functions (1). It is known to be involved in cellular targeting and secretion (2), as well as in regulating enzymatic activity, enhancing stability and solubility of secreted proteins, and it affects the function of proteins in the immune system. Moreover, glycoproteins participate in cell–cell and cell–matrix interactions and mediate complex developmental

Jennifer J. Kohler and Steven M. Patrie (eds.), *Mass Spectrometry of Glycoproteins: Methods and Protocols*,
Methods in Molecular Biology, vol. 951, DOI 10.1007/978-1-62703-146-2_10, © Springer Science+Business Media, LLC 2013

functions (3). Glycosylation is one of the most common types of post-translational modification proteins can undergo. In fact, 13 different monosaccharides and eight amino acids have been reported across species to be involved in glycoprotein linkages (4). The two major types of oligosaccharide attachment to the protein are referred to as N-linked (asparagine) and O-linked glycosylation (serine or threonine). These sites occur in specific domains in proteins. For N-linked glycosylation, there is a glycosylation consensus sequence, (N-X-T/S/(C) (X≠P)); however, this is not the case for O-linked glycans.

Typically glycosylation analysis has to be performed on several levels: intact glycoproteins, glycopeptides, and/or released glycans. Analysis of intact glycoproteins helps visualize the overall glycoforms of a protein population and is often applied to purified proteins with a limited number of different glycans and/or a low number of glycosylation sites. Analysis of the released glycans yields information on the glycan heterogeneity present on the protein and can, together with the known protein sequence, be very helpful in the interpretation of glycopeptide/protein MS data (5). However, neither of these two types of analyses provides information on the actual site of glycosylation. In order to assign glycosylation sites, the protein has to be proteolytically cleaved into peptides. Well-defined peptides with one glycosylation site can be analyzed by MALDI mass spectrometry where the site and attached glycan moieties can often be unambiguously assigned. By use of LC-ESI-MS, peptides with multiple sites can be chromatographically separated and analyzed upon elution. However, peptides with multiple sites remain a difficult challenge. Often the amount of sample available will be the limiting factor on the level of information obtained about the glycosylation of a particular protein.

In general, glycosylation analysis by mass spectrometry is a challenging discipline for a number of reasons: (1) many of the monosaccharides have the same mass and thus cannot be distinguished by soft ionization (MALDI or ESI) mass spectrometry without prior derivatization of the glycan moieties; (2) the stoichiometry of glycopeptides can be much lower than non-glycosylated peptides, due to incomplete site occupancy; (3) heterogeneity due to presence of multiple different glycan forms at a given *N*-glycosylation site in a protein population; (4) the glycopeptide signals are suppressed in the presence of non-glycosylated peptides (6, 7). Enrichment of glycopeptides by hydrophilic interaction liquid chromatography (HILIC) (8, 9), lectins (10, 11), boronic acid (12, 13), graphitized carbon (6, 14), titanium dioxide (15), or combinations hereof (16–18) can be employed to increase the signal intensities of the glycosylated peptides by depleting the far more abundant non-glycosylated peptides.

HILIC is a recent variation of the NP separations of the 1970s, achieving similar retention characteristics while employing

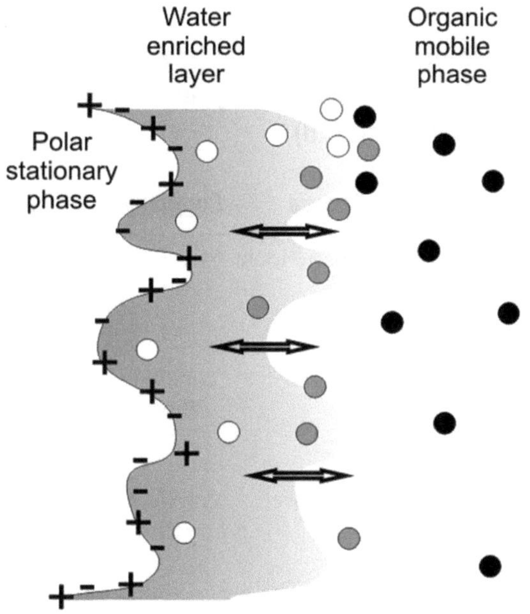

Fig. 1. HILIC principle. HILIC separation is based on passing a mostly organic mobile phase over a polar stationary phase. This forms a water enriched layer next to the stationary phase. The hydrophilic analytes partition into this layer and as the mobile phase becomes more hydrophilic, they are eluted in order of increasing hydrophilicity (*white circle*=hydrophilic analyte; *grey circle*=semi-hydrophilic analyte, *black circle*=hydrophobic analyte).

water-miscible solvents, which is gaining popularity in protein, and particularly glycoprotein, analysis. HILIC is interesting in this context because of its ability to retain and separate hydrophilic compounds, which are not retained or adequately separated by RP chromatography. The principle of HILIC is based on passing a mostly hydrophobic (organic) mobile phase over a hydrophilic stationary phase. Pre-wetting the column forms a water layer around the stationary phase and the separation is based on hydrophilic compounds partitioning into this water layer (Fig. 1). Increasing the water content of the mobile phase will release and elute the compounds in order of increasing hydrophilicity. There are many types of HILIC resins, some with a charged/ionized stationary phase functionality. This adds electrostatic interactions to the phase partitioning separation mechanism (Table 1). This may be advantageous in some applications, but we feel that choosing a neutral HILIC resins provides the most unbiased enrichment of glycopeptides. Obtaining representative relative amounts of different glycoforms in a peptide population is of primary concern if performing glycosylation site profiling for glycoprotein analysis (19).

HILIC is a very versatile separation method and the choice of resin, chemistry/functionality and mobile phase, as well as the

Table 1
Different types of HILIC material. Shown are the most commonly used HILIC resins and some of the applications they have been used for. For a review of these and more materials see ref. (28)

HILIC material	Charge	Company	Application	References
ZIC®-HILIC[a]	Zwitterionic	Merck SeQuant AB	N-glycopeptides	(8, 17–20, 29)
			Xanthines, nucleosides, vitamins	(30)
TSKgel Amide-80[a]	Neutral	TOSOH Bioscience	N-glycopeptides	(19, 29)
			Xanthines, nucleosides, vitamins	(30)
PolySULFOETHYL A™	Negative	PolyLC Inc.	N-glycopeptides	(19)
			Xanthines, nucleosides, vitamins	(30)
PolyHYDROXYETHYL A™	Neutral	PolyLC Inc.	N-glycopeptides	(19, 20)
			O-glycopeptides (mucin-like)	(22)

[a]These materials are not available to buy in bulk. We have obtained material from Merck Sequant AB as a gift and from TOSOH Bioscience by disassembling a guard column

properties of the sample of interest, all have great influence on the result. We here present a glycopeptide enrichment technique using HILIC resin in a solid phase extraction (SPE) micro-column format. SPE facilitates enrichment of smaller amounts of sample very rapidly when compared to HPLC instrumentation. Additionally, SPE allows for easy adjustment of both the type and amount of chromatographic resin used for enrichment as well as various mobile phases to fine-tune the outcome of the enrichment. We find that the speed and adaptability of SPE is advantageous for our glycopeptide enrichment purposes, allowing for both enrichment optimization from novel proteins and rapid enrichment from well-characterized proteins. An overview of glyco-profiling workflows, including the SPE methods mentioned in the chapter, can be found in Fig. 2 and Table 2.

We here illustrate mentioned approaches using HILIC enrichment on a tryptic digest of murine IgG1 (P01868), which contains a single N-glycosylation site. We also show the effects of combining the HILIC enrichment with C18 depletion as well as different ion-pairing reagents in the mobile phase (20). In all cases it needs to be stressed that, while "quick-and-dirty" glycopeptide enrichments using HILIC are often viable, a protein-specific optimization of the protocol is highly recommended when performing accurate glycosylation profiling.

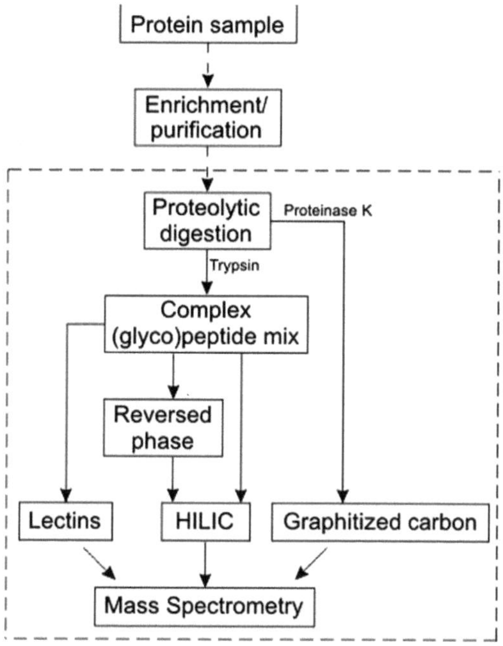

Fig. 2. Flow chart of different glycopeptide enrichment strategies. Shown are some of the different workflows that have been employed to optimize glycopeptide enrichment. Table 2 summarizes strategies, outcomes, and references to this workflow. The *boxed area* indicates the workflows discussed in this chapter. The prior steps are dependent on the sample.

2. Materials

2.1. Reduction, Alkylation, and Proteolytic Digestion

1. Ammonium bicarbonate; 100 mM.
2. Dithiothreitol; 10 mM final concentration in sample.
3. Iodoacetamide; 50 mM final concentration in sample.
4. Trypsin; sequence grade (Promega, Madison/WI, #V5111).

2.2. HILIC Solid Phase Extraction Enrichment

2.2.1. Direct HILIC SPE Glycopeptide Enrichment

1. HILIC resin (see Note 1).
2. Methanol.
3. GELoader® tips (Eppendorf, Hamburg/Germany, #022351656).
4. Optional: Empore™ C18 Disk (3 M, St. Paul/MN, #98060402181).
5. HILIC elution solution (HES): 5.0% (v/v) formic acid.
6. HILIC binding solution 1 (HBS1): 2.0% (v/v) formic acid in 80% (v/v) acetonitrile.

2.2.2. HILIC SPE Glycopeptide Enrichment of C18 Flow-Through Peptides

1. Poros R2 reversed phase resin (C18, 20 μm; Applied Biosystems, Framingham/MA, #1-1128-10).
2. C18 elution solution (CES): 0.1% (v/v) trifluoroacetic acid in 80% (v/v) acetonitrile.
3. C18 binding solution (CBS): 0.1% (v/v) trifluoroacetic acid.

Table 2
Different glycopeptide enrichment strategies

Chromatographic resins	HILIC binding solution	Strategy/result	References
Poros R1/Poros R2/ HILIC	80% MeCN; 0.5% FA	R1 and R2 remove large and hydrophobic peptides and FT was enriched using HILIC	(18)
Poros R2/graphitized carbon/HILIC/TiO_2	80% MeCN; 2% FA	R2 removes hydrophobic peptides and FT was enriched. HILIC and carbon can enrich neutral glycopeptides in a nonbiased manner as long as column capacity is not exceeded	(19)
HILIC	80% MeCN; 2% FA 80% MeCN; 0.1% TFA 80% MeCN; 1% TFA	TFA increased glycopeptide recovery for simple mixtures and decreased non-specific binding for complex mixtures	(20)
Poros R2/graphitized carbon[a]	–	R2 removes hydrophobic peptides and FT was carbon-enriched	(6)
Lectin mix/HILIC	80% MeCN; 0.5% FA	Lectin mix enriches glycoproteins and subsequently lectin mix/HILIC enriches glycopeptides	(16)
Lectin/HILIC	80% MeCN; 0.5% FA	Lectin enriches glycoproteins and subsequently HILIC enriches glycopeptides	(8)

Shown are some of the different combinations of chromatographic resins and ion-pairing reagents that have been employed to optimize glycopeptide enrichment. It is important to stress that the optimal strategy is highly dependent on the sample and has to be optimized for each glycoprotein. The overall workflow of the strategies in the table is illustrated in Fig. 2

MeCN acetonitrile; *FA* formic acid; *TFA* trifluoroacetic acid

[a]This strategy requires proteolytic cleavage with proteinase K or a similar non-specific protease

2.2.3. Ion-Pairing HILIC SPE Glycopeptide Enrichment

1. HILIC binding solution 2 (HBS2): 0.1% (v/v) trifluoroacetic acid in 80% (v/v) acetonitrile.
2. HILIC binding solution 3 (HBS3): 1.0% (v/v) trifluoroacetic acid in 80% (v/v) acetonitrile.

2.3. MALDI-TOF MS

1. MALDI-TOF MS instrumentation.
2. 2,5-Dihydroxybenzoic acid matrix solution: 10 mg/mL in 0.1% (v/v) trifluoroacetic acid and 70% (v/v) acetonitrile.

3. Methods

The method described below requires prior purification of the glycoprotein(s) of interest. This can be achieved by using a variety of techniques, e.g., SDS PAGE or C4/C8 RP-HPLC. For a review on protein purification and fractionation see ref. (21). We illustrate the procedure using a purified murine IgG (P01868).

3.1. Reduction, Alkylation, and Proteolytic Digestion (Estimated Time: 3 h, Plus Overnight Digestion)

1. Reduce and alkylate the glycoprotein using a standard protocol as described in ref. (22).
2. Digest the glycoprotein with trypsin, in a ratio of 1:20 (enzyme:protein; w/w), at 37°C over night (see Note 2).
3. Check that the digestion is complete by ensuring that no signal corresponding to the intact protein mass remains using MALDI-TOF MS (An example of an un-enriched glycoprotein digest can be seen in Fig. 3a).

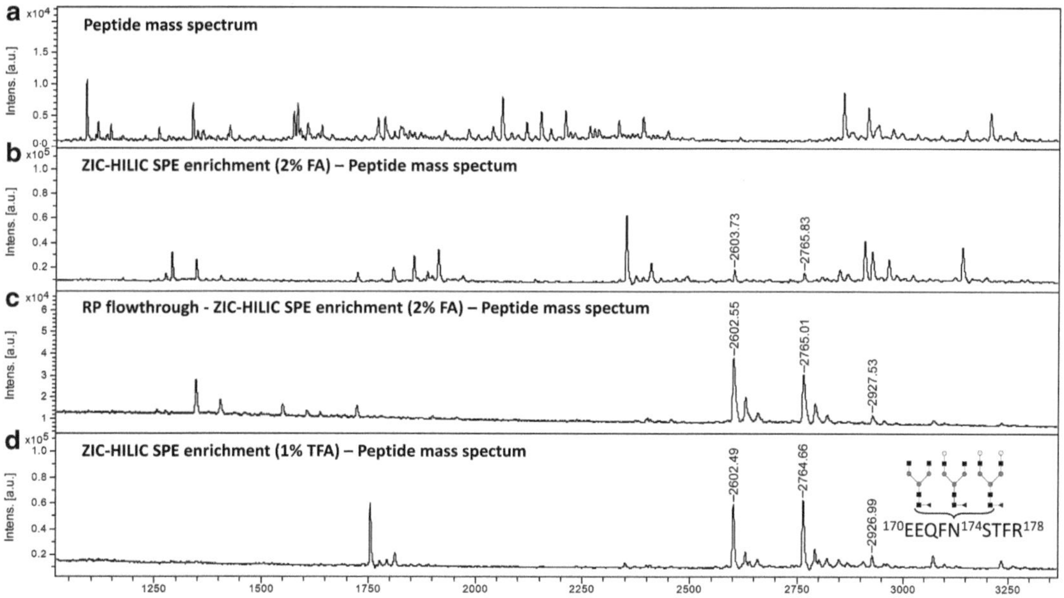

Fig. 3. HILIC enrichment of *N*-glycopeptides from a tryptic digest of murine IgG1. (**a**) MALDI spectrum of the non-enriched tryptic digest, in order to illustrate the effect of the enrichment. (**b**) Standard HILIC enrichment as described in Subheading 3.2.1. The labeled masses correspond to glycopeptides. There is still a significant amount of hydrophilic, non-glycosylated peptides present in the spectrum that suppress the signals from the glycopeptides. (**c**) HILIC enrichment after C18 depletion as described in Subheading 3.2.2. Most of the non-glycosylated peptides were removed by the pre-depletion on the C18 and the glycopeptides in the flow-through were further enriched using HILIC. Without the suppression by non-glycopeptides, the intensity of the glycopeptide signals increases. (**d**) HILIC enrichment using trifluoroacetic acid as an ion-pairing reagent as described in Subheading 3.2.3. Using trifluoroacetic acid as the ion-pairing agent as opposed to formic acid (**b**) yields a more specific enrichment of glycopeptides, although introducing a slight bias in the glyco-profiles when compared to (**b**) and (**c**). Again the removal of non-glycosylated peptides greatly increases the signal intensities for the glycopeptides. The peptide sequence and the different glycans identified on N174 are shown in the inset (*black square* = *N*-acetylglucosamine; *dark grey triangle* = fucose; *grey circle* = mannose; *white circle* = galactose).

3.2. HILIC SPE Enrichment (Estimated Time: 15 min/Sample)

3.2.1. Direct HILIC SPE Glycopeptide Enrichment (Fig. 3b)

1. Make a slurry of your chosen HILIC resin in methanol.
2. Constrict the tip of a GELoader® tip, to act as the base of the micro-column. If available, an Empore™ C18 Disk can also be placed in the tip as an alternative to constriction (see Note 3).
3. Apply the resin slurry to the constricted tip, creating a column with a length appropriate to the desired amount of sample (see Note 4).
4. Wash the column with 20 μL HES.
5. Equilibrate with 20 μL HBS1 (23).
6. Load the sample on the column in 10 μL HBS1 (see Notes 5 and 6).
7. Wash the column with two separate applications of 10 μL HBS1 (see Note 7).
8. Elute glycopeptides with 10 μL HES and (see Note 8).

3.2.2. HILIC SPE Glycopeptide Enrichment of C18 Flow-Through Peptides (Fig. 3c)

1. Make a slurry of Poros R2 reversed phase resin in methanol.
2. Create a Poros R2 SPE micro-column as described in Subheading 3.2.1 steps 2 and 3 (see Note 9).
3. Wash the column with 20 μL CES.
4. Equilibrate with 20 μL CBS.
5. Load the sample on the column in 10 μL CBS and collect the flow-through (see Note 10).
6. Wash the column with 10 μL CBS and pool it with the previously collected flow-through.
7. Lyophilize the flow-through in low-binding eppendorf tubes using a vacuum centrifuge and resuspend it in 10 μL HBS1.
8. Perform a HILIC SPE glycopeptide enrichment as described in Subheading 3.2.1, applying the resuspended flow-through from step 6.

3.2.3. Ion-Pairing HILIC SPE Glycopeptide Enrichment (Fig. 3d)

1. Perform a HILIC SPE glycopeptide enrichment as described in Subheading 3.2.1, replacing the binding solution with HBS2 or HBS3 (see Note 11).

3.3. MALDI TOF MS (Estimated Time: 15 min/Sample)

1. Spot 0.5–1 μL sample (10 pmol) on a MALDI target, add 0.5 μL 2,5-dihydroxybenzoic acid matrix solution, and let it dry (see Note 12).
2. Perform MALDI-TOF MS in positive ionization mode using linear TOF detection. For the authors' notes on relevant analysis parameters, see Note 13. The specific MALDI-TOF parameters employed by the authors are given in Note 14.

3.4. Data Interpretation (Estimated Time: 1 or More h/Sample)

When a MALDI spectrum has been obtained, one looks for characteristic glycan mass differences in the spectrum. These are typically masses corresponding to mono-, di-, or trisaccharide differences between peaks, like 146.06 Da (deoxyhexose), 162.05 Da (hexose; Hex), 203.08 Da (*N*-acetylhexoseamine; HexNAc), 291.10 Da (neuroaminic acid; NeuAc), 365.13 Da (HexHexNAc), or 656.23 Da (HexHexNacNeuAc). These are indicative of the presence of glycopeptides and correspond to different glycoforms of the same glycopeptide (see Note 15). Applying collision-induced dissociation (CID), MS/MS on a glycopeptide will preferentially fragment the glycan, allowing for validation of glycopeptide ions and determination of the glycan composition from the fragment spectrum. Often little or no fragmentation of the peptide backbone occurs, leaving the intact peptide with a single glycan attached at the glycosylation site. In the case of a tryptic digest, it is often possible to identify the peptide from the mass alone. For N-glycosylation, there is often only one possible glycosylation site per peptide, so this can also be assigned (see Note 16).

4. Notes

1. Most resin types can provide unbiased glycopeptide enrichment under standard conditions. ZIC-HILIC (10 μm, 200 Å; Merck SeQuant, UmeÅ/Sweden) was used for the presented trifluoroacetic acid ion-pairing glycopeptide enrichments, but polyHydroxyethyl A (12 μm, 100 Å; PolyLC, Columbia/MD, product number BMHY12) has also been successfully employed. We recommend using HILIC resins with neutral functional groups.
2. The choice of proteolytic agent depends on the protein at hand. Often trypsin will be the first choice as it in many cases produces peptides of an appropriate size for MS analysis (1,000–2,000 Da). In addition, a specific protease like trypsin yields relatively well-defined peptides. Glycopeptides may, however, become too large for optimal analysis when the mass of the glycan is added. And especially O-glycopeptides may have long stretches without lysine or arginine, thus making them unsuitable for cleavage with trypsin. In the case of a known protein, it is advisable to inspect the sequence prior to choosing a proteolytic agent (24). Alternatively, non-specific proteases, such as Proteinase K, can be employed. A 3–5 residue peptide will often remain attached to N-glycans as the steric hindrance it provides will block the protease and prevent digestion in the proximity of the glycan (6).
3. A variety of methods exist for this purpose. The goal is to achieve a constriction that will retain the chromatographic

resin, while allowing solvent to pass through. For resins composed of larger beads (>20 μm), it is often sufficient to constrict by squeezing the tip of the gel-loader tip using the back of a pen. However, when employing 10–20 μm bead resins, it is often necessary to resort to tighter constrictions, such as mashing the tip with the dull side of a scalpel while twisting it 180–360°. It is advised to check that the constriction is not too wide or too narrow before applying resin, in order to avoid wasting resin and buffers. This can be done by generating backpressure inside the tip using a 1 mL plastic syringe, letting air slowly pass through the constriction. In our experience, a constriction which is able to alleviate this backpressure over 5–10 s is able to hold 10 μm bead resins while still allowing solvent through the column.

4. For examples of packed columns, along with column length/resin volume conversion charts and the amount of resin required for unbiased glycopeptide enrichments, consult (19) and associated supplementary figures.
5. It is very important that the final concentration of acetonitrile in the loaded sample is 80% (v/v), otherwise it may not bind to the column. If the sample is dissolved in a volatile buffer, such as ammonium bicarbonate, lyophilizing the sample in low-binding eppendorf tubes (Eppendorf, Hamburg/Germany, #022431081) using a vacuum centrifuge and resuspending it in the appropriate amount of HBS is also a viable option. Collect and analyze the flow-through to check the binding efficiency.
6. During sample loading, and throughout all subsequent washing and elution steps, patience is highly recommended. By only applying very slight amounts of backpressure, allowing the solutions to slowly flow through the column, we observe a marked increase in enrichment efficiency. Working at lower backpressures also aids enrichment by minimizing resin compression. To reduce manual handling, the syringe can be tensioned slightly between two ribs of a rack or similar frame. This also facilitates multiple parallel enrichments.
7. The washing solution may be pooled with the flow-through from the previous step and analyzed together.
8. If an Empore™ C18 Disk was used as an alternative to constriction, also perform a subsequent elution using 5 μL HBS and pool the eluates. The elution with HBS is to ensure none of the glycopeptides eluted from the HILIC with the HES bind to the C18 material of the Empore™ C18 Disk. Performing a second elution step will verify whether or not all glycopeptides have been eluted.

9. The length of the column should be adjusted according to the protein amount and size. Too little capacity and the depletion of hydrophobic peptides will be incomplete. Too great a capacity and depletion of glycans linked to larger peptides may occur. It is recommended to try a few different column lengths for initial experiments to determine the optimal conditions. If no glycopeptides are observed in the final enrichment, the use of C4 or C8 resin can be attempted.
10. The final composition of the loaded sample should be aqueous with no organic content. If the sample is dissolved in an organic buffer, lyophilizing the sample in low-binding eppendorf tubes using a vacuum centrifuge and resuspending it in the appropriate amount of CBS is recommended.
11. Using higher concentrations of trifluoroacetic acid, especially 1% trifluoroacetic acid, as an ion-pairing agent during HILIC SPE appears to result in effects similar to those observed when the capacity of the HILIC resin is reduced, increasing competition between bound species and potentially inducing an enrichment bias toward glycopeptides containing larger glycans. It is important to note that resin amounts which can provide sufficient capacity for unbiased glycopeptide enrichment under standard HILIC conditions may induce a bias when 1% trifluoroacetic acid, or even 0.1% trifluoroacetic acid, is present in the HBS. As a rule of thumb, try to increase column size when using a trifluoroacetic acid ion-pairing HBS and verify that the enrichment was unbiased, if performing glycopeptide enrichment for the purpose of glyco-profiling. For a more detailed discussion of the subject, please consult (20).
12. The 2,5-dihydroxybenzoic acid matrix and positive ionization mode is used for the analysis of glycopeptide populations containing the neutral high-mannose and complex-type structures without terminal sialic acid. Under these conditions, the labile terminal sialic acids fragment, making them difficult to observe. For MALDI-TOF analysis of sialic acid containing glycopeptides, it is recommended to use a 2,4,6-trihydroxyacetophenone matrix (1 mg/mL in 50% (v/v) acetonitrile; 50% (v/v) 20 mM aqueous ammonium citrate) and performing the analysis in negative ionization mode (25).
13. Linear detection is used, as post-source fragmentation of the glycan moiety is very pronounced during MALDI-TOF MS, hence reflector detection will skew the relative glycan ratios towards the less glycosylated fragmented forms, introducing a bias in the glyco-profile. The use of "warmer" matrices, such as alpha-cyano-4-hydroxycinnamic acid, will generally result in greater degrees of glycan fragmentation. When acquiring spectra, low laser powers and short delay times (50–100) ns were

used to minimize in-source fragmentation (26). Continuous extraction (no delay time) has been attempted by the authors, but the resulting loss in resolution was too great for it to be a viable option. For our analysis, 500–1,000 laser shots, spread over the entire spot, are generally summed. This minimizes the effect of inhomogenic sample distribution within the matrix spots, which can cause a bias in the resulting glyco-profile. Moreover, summing more shots also yields better overall ion statistics to facilitate a more accurate quantitation. For a more thorough evaluation of important parameters when performing unbiased glyco-profiling using MALDI-TOF of glycopeptides, please consult (19). If performing analysis of sialylated glycopeptides, be aware that sialic acids have been reported to be lost if the pH of the solution becomes too low, or the temperature of the solution becomes too high (27).

14. For our analyses, we used different instruments and parameters for non-sialylated and sialylated glycopeptides. Non-sialylated glycopeptides: Bruker Ultraflex (Bruker Daltronics) in positive ionization mode, linear detection, Pulsed Ion Extraction at 100 ns, 25 kV source 1 voltage, 23.5 kV source 2 voltage, 6 kV lens voltage. Sialylated glycopeptides: Voyager-DE STR (Applied Biosystems) in negative ionization mode, linear detection, Extraction delay time at 50 ns, 20 kV extraction voltage. Both instruments were equipped with a pulsed nitrogen laser (337 nm) and ionization performed at the lowest power able to yield a discernable spectrum to minimize in-source fragmentation. Choice of instruments for positive and negative ionization was a personal judgment of spectrum quality.
15. The different glycoforms can also arise from post-source fragmentation of the glycans, but as long as linear ion mode detection is applied, this should not be reflected in the spectrum.
16. The glycosylation consensus for N-glycosylation is N-X-T/S/(C) (X≠P; N-X-C is rarely observed). There is no such consensus for *O*-glycosylation, which can essentially occur at any serine or threonine in a peptide, making assignment of these even more challenging.

Acknowledgments

PHJ was supported by a postdoctoral fellowship from "Annie og Otto Johs. Detlefs' Almennyttige Fond."

References

1. Varki A (1993) Biological roles of oligosaccharides: all of the theories are correct. Glycobiology 3:97–130
2. Gagneux P, Varki A (1999) Evolutionary considerations in relating oligosaccharide diversity to biological function. Glycobiology 9:747–755
3. Varki A, Lowe JB (1999) Biological roles of glycans. In: Varki A, Cummings R, Esko J, Freeze H, Hart G, Marth J (eds) Essentials of glycobiology, 2nd edn. Cold Spring Harbor Laboratory Press, Woodbury, NY, pp 57–68
4. Spiro RG (2002) Protein glycosylation: nature, distribution, enzymatic formation, and disease implications of glycopeptide bonds. Glycobiology 12:43R–56R
5. Deshpande N, Jensen PH, Packer NH, Kolarich D (2009) GlycoSpectrumScan: fishing glycopeptides from MS spectra of protease digests of human colostrum sIgA. J Proteome Res 9:1063–1075
6. Larsen MR, Højrup P, Roepstorff P (2005) Characterization of gel-separated glycoproteins using two-step proteolytic digestion combined with sequential microcolumns and mass spectrometry. Mol Cell Proteomics 4:107–119
7. Mirgorodskaya E, Krogh TN, Roepstorff P (2000) Characterization of protein glycosylation by MALDI-TOFMS. In: Chapman J (ed) Methods in molecular biology: mass spectrometry of proteins and peptides. Humana Press, Totawa, NJ, pp 273–292
8. Hägglund P, Bunkenborg J, Elortza F, Jensen ON, Roepstorff P (2004) A new strategy for identification of N-glycosylated proteins and unambiguous assignment of their glycosylation sites using HILIC enrichment and partial deglycosylation. J Proteome Res 3:556–566
9. Kieliszewski MJ, O'Neill M, Leykam J, Orlando R (1995) Tandem mass spectrometry and structural elucidation of glycopeptides from a hydroxyproline-rich plant cell wall glycoprotein indicate that contiguous hydroxyproline residues are the major sites of hydroxyproline O-arabinosylation. J Biol Chem 270:2541–2549
10. Bunkenborg J, Pilch BJ, Podtelejnikov AV, Wi niewski JR (2004) Screening for N-glycosylated proteins by liquid chromatography mass spectrometry. Proteomics 4:454–465
11. Hirabayashi J (2004) Lectin-based structural glycomics: glycoproteomics and glycan profiling. Glycoconj J 21:35–40
12. Rawn JD, Lienhard GE (1974) Binding of boronic acids to chymotrypsin. Biochemistry 13:3124–3130
13. Sparbier K, Koch S, Kessler I, Wenzel T, Kostrzewa M (2005) Selective isolation of glycoproteins and glycopeptides for MALDI-TOF MS detection supported by magnetic particles. J Biomol Tech 16:407–413
14. Alley WR, Mechref Y, Novotny MV (2009) Use of activated graphitized carbon chips for liquid chromatography/mass spectrometric and tandem mass spectrometric analysis of tryptic glycopeptides. Rapid Commun Mass Spectrom 23:495–505
15. Larsen MR, Jensen SS, Jakobsen LA, Heegaard NHH (2007) Exploring the sialiome using titanium dioxide chromatography and mass spectrometry. Mol Cell Proteomics 6:1778–1787
16. Calvano CD, Zambonin CG, Jensen ON (2008) Assessment of lectin and HILIC based enrichment protocols for characterization of serum glycoproteins by mass spectrometry. J Proteomics 71:304–317
17. Kondo A, Miyamoto T, Yonekawa O, Giessing AM, Østerlund EC, Jensen ON (2009) Glycopeptide profiling of beta-2-glycoprotein I by mass spectrometry reveals attenuated sialylation in patients with antiphospholipid syndrome. J Proteomics 73:123–133
18. Kondo A, Thaysen-Andersen M, Hjernø K, Jensen ON (2010) Characterization of sialylated and fucosylated glycopeptides of β2-glycoprotein I by a combination of HILIC LC and MALDI MS/MS. J Sep Sci 33:891–902
19. Thaysen-Andersen M, Mysling S, Højrup P (2009) Site-specific glycoprofiling of N-linked glycopeptides using MALDI-TOF MS: strong correlation between signal strength and glycoform quantities. Anal Chem 81:3933–3943
20. Mysling S, Palmisano G, Højrup P, Thaysen-Andersen M (2010) Utilizing ion-pairing hydrophilic interaction chromatography solid phase extraction for efficient glycopeptide enrichment in glycoproteomics. Anal Chem 82:5598–5609
21. Stasyk T, Huber LA (2004) Zooming in: fractionation strategies in proteomics. Proteomics 4:3704–3716
22. Christiansen MN, Kolarich D, Nevalainen H, Packer NH, Jensen PH (2010) Challenges of determining O-glycopeptide heterogeneity: a fungal glucanase model system. Anal Chem 82:3500–3509

23. Hägglund P, Matthiesen R, Elortza F, Højrup P, Roepstorff P, Jensen ON, Bunkenborg J (2007) An enzymatic deglycosylation scheme enabling identification of core fucosylated N-glycans and O-glycosylation site mapping of human plasma proteins. J Proteome Res 6:3021–3031
24. Højrup P (2009) Peptide mapping for protein characterization. In: Walker JM (ed) The protein protocols handbook, 3rd edn. Humana Press, Totawa, NJ, pp 965–983
25. Papac DI, Wong A, Jones AJS (1996) Analysis of acidic oligosaccharides and glycopeptides by matrix-assisted laser desorption/ionization time-of-flight mass spectrometry. Anal Chem 68:3215–3223
26. Tsarbopoulos A, Bahr U, Pramanik BN, Karas M (1997) Glycoprotein analysis by delayed extraction and post-source decay MALDI-TOF-MS. Int J Mass Spectrom Ion Process 169–170:251–261
27. Harvey DJ (1999) Matrix-assisted laser desorption/ionization mass spectrometry of carbohydrates. Mass Spectrom Rev 18:349–450
28. Yoshida T (2004) Peptide separation by hydrophilic-interaction chromatography: a review. J Biochem Biophys Methods 60:265–280
29. Wohlgemuth J, Karas M, Eichhorn T, Hendriks R, Andrecht S (2009) Quantitative site-specific analysis of protein glycosylation by LC-MS using different glycopeptide-enrichment strategies. Anal Biochem 395:178–188
30. Lämmerhofer M, Richter M, Wu J, Nogueira R, Bicker W, Lindner W (2008) Mixed-mode ion-exchangers and their comparative chromatographic characterization in reversed-phase and hydrophilic interaction chromatography elution modes. J Sep Sci 31:2572–2588

Chapter 11

Separation and Identification of Glycoforms by Capillary Electrophoresis with Electrospray Ionization Mass Spectrometric Detection

Alina D. Zamfir, Corina Flangea, Alina Serb, Ana-Maria Zagrean, Andreas M. Rizzi, and Eugen Sisu

Abstract

Capillary electrophoresis (CE) is a resourceful and versatile separation method for the analysis of complex carbohydrate mixtures. In combination with electrospray ionization (ESI) mass spectrometry (MS), CE enables fast, sensitive, and efficient separations for the accurate identification of a large variety of glycoform mixture types. In this chapter several reliable off- and on-line CE-based methods for the analysis of glycoforms with ESI MS/MS are presented. The first part of this chapter is dedicated to the application of off-line CE/ESI MS to complex mixtures of O-glycopeptides and mixtures of proteoglycan-derived O-glycans, i.e., glycosaminoglycans such as depolymerized hybrid chains of chondroitin sulfate (CS) and dermatan sulfate (DS). Procedures for off-line fractionation of these heterogeneous mixtures followed by ESI MS screening and sequencing of single glycoforms by collision-induced dissociation (CID) at low energies are also described. Ample sections are further devoted to on-line CE/ESI MS technique and its application to separation and identification of O-glycopeptides and CS/DS oligosaccharides. The concept and construction principles of two different sheathless CE/ESI MS interfaces together with the protocols to be applied for successful on-line analysis of O-glycopeptides and CS/DS oligosaccharides are presented in details in the last part of the chapter.

Key words: Glycopeptides, Glycosaminoglycans, Capillary electrophoresis, Electrospray ionization, Tandem mass spectrometry

1. Introduction

Capillary electrophoresis (CE) is an instrumental evolution of traditional slab gel electrophoretic techniques and is based on differences in solute velocity in an electric field. In CE the electromigration of analytes takes place in narrow-bore capillaries. The application

Jennifer J. Kohler and Steven M. Patrie (eds.), *Mass Spectrometry of Glycoproteins: Methods and Protocols*, Methods in Molecular Biology, vol. 951, DOI 10.1007/978-1-62703-146-2_11,

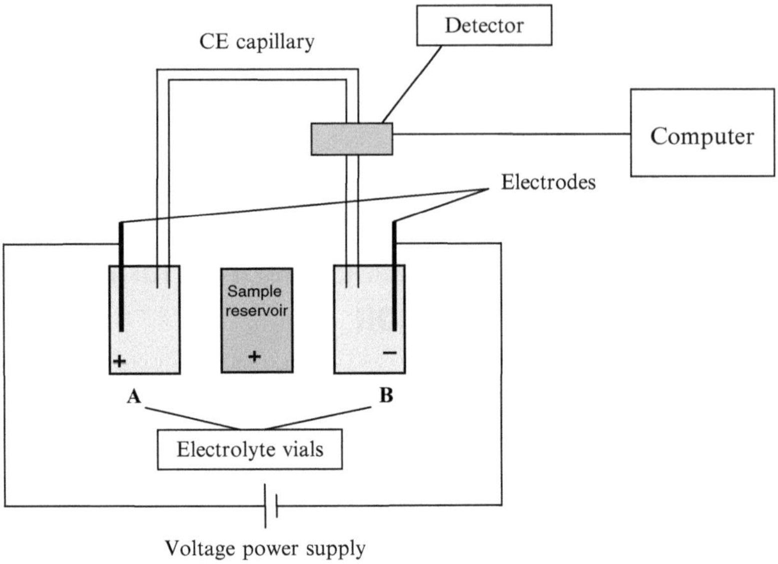

Fig. 1. Basic CE setup.

of a high electric field, up to 0.6 kV/cm, enhances the efficiency of the separation over traditional high performance liquid chromatography (HPLC) methods predominately due to reduced mass transfer and laminar flow characteristics. In principle, a CE setup (Fig. 1) consists of a fused silica separation capillary, two buffer vials A and B and a reservoir containing the solution of analyte, a high-voltage power supply delivering up to 30–40 kV, and a detector which can be of various types: UV detector, electrochemical detector, laser-induced fluorescence (LIF), or mass spectrometry (MS).

The sample, usually dissolved in buffer/electrolyte, is injected into the capillary by either: (1) hydrodynamic injection using pressure or vacuum application while the injection end of the capillary is inserted in the vial containing the analyte solution or (2) electrokinetic injection induced by voltage application.

In CE the separation is usually carried out at constant potential in direct or normal polarity (injection at anode and detection at cathode) or reverse polarity (injection at cathode and detection at anode). Although not commonly used, gradients or steps in the voltage may be useful in simultaneous analysis of compounds having very different electrophoretic mobility.

The variation in solute velocity is given by the different electrophoretic mobility of the analytes, which depends on the charge of the analyte, pH of the solution, viscosity, temperature inside the capillary, *m/z* ratio, the applied electric field, and dimensions of the capillary. Therefore, besides instrumental parameters, optimization of electrolyte pH, ionic strength, chemical composition, and concentration are key to performing an efficient CE separation (1, 2).

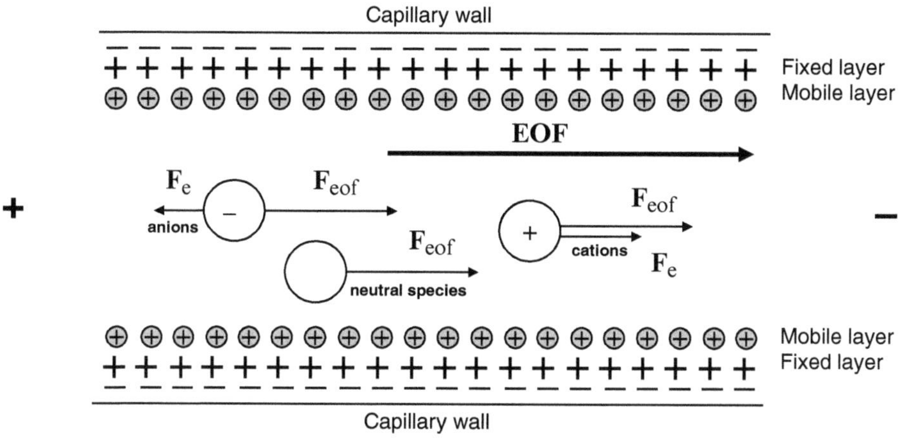

Fig. 2. Schematic of the CE process in direct polarity.

Under certain solution conditions (pH > 3.0), fused silica surface possesses an excess of negative charges due to the ionization of silanol (Si-OH) to negatively charged silanoate ($Si\text{-}O^-$). Attracted to the negatively charged silanoate groups, the cations of the buffer solution, which balance the surface charge, will form two inner layers called the diffuse or electrical double layer (Fig. 2) and create a potential difference. The first layer is fixed to the silanoate groups, while the outer layer is mobile. When a voltage is applied across the capillary, the mobile layer of the electrical double layer is attracted toward the cathode. Due to solvation, the ionic movement transports the bulk flow solution creating an electroosmotic flow (EOF) under the electric field. $\mathbf{F}_{eof}$, the force of the EOF, is one order of magnitude higher than $\mathbf{F}_e$ the electrophoretic force; therefore, EOF causes the movement of all species regardless the charge in the same direction (Fig. 2). The separation occurs under the action of different $\mathbf{F}_e$ so that if the mixture contains positive and negative ions as well as neutral species, in direct polarity the first will elute the positive ions, followed by the molecules which did not undergo ionization. The negative ions, drifted by $\mathbf{F}_{eof}$, will also migrate toward the cathode but will elute much later. In reverse polarity, the EOF is oriented against the desired direction of ion motion. It may be suppressed by careful reconsideration of solution parameters (pH < 3.0). In this case, the $\mathbf{F}_e$ becomes concomitantly the drift and separation force giving rise to an elevated separation efficiency and resolution.

In comparison to CE, HPLC gained more in popularity primarily due to the development of routine operational procedures. Additionally, in CE, only about 1% of the capillary can be filled with sample, which corresponds to injected volumes of tens of nL. Although stacking or on-line preconcentration techniques (3) are capable to increase this amount by an order of magnitude, this is still far from the μL-range injection volumes possible in HPLC.

Nevertheless, compared to chromatographic methods, which separate components of a mixture based on their polarity or size, by employing high electric fields to separate charged and neutral species, CE has the ability to achieve even one million theoretical plates. As a consequence, CE exhibits a number of advantages (1–5) including high analysis speed, reproducibility of experiments, ease of automation, miniaturization, and most importantly, the unsurpassed separation efficiency and resolution.

Among all biopolymers found in nature, carbohydrates form a special category, which exhibits the highest level of structural diversity. A reliable analysis of glycoform mixtures, particularly those of biological origin, calls for the development and optimization of efficient separation protocols (5–7). Being characterized also by high versatility, CE has emerged as one of the most resourceful methods in glycomics. Lately, CE performance in this field was considerably augmented by its combination with electrospray ionization (ESI) mass spectrometry (MS) for detection and identification of separated glycoforms in complex mixtures (8–15). By CE/ESI MS technique, molecular masses of the glycans separated by CE can be directly measured with high accuracy; additionally, in one and the same experiment, specific fragment ions of single components may be generated by tandem MS (MS/MS) or even multistage MS (MS^n) to deduce in detail the molecular structures and glycosylation site(s).

When dealing with simple glycoform mixtures, an accessible method is the off-line CE/ESI MS in which CE separation accompanied by collection of fractions and their off-line ESI MS analysis can be accomplished (15). Even for complex mixtures such an approach is recommended for a partial separation, since it may eliminate the effects of the ion suppression in ESI MS, with beneficial consequences upon the detection of single species in a multicomponent samples. Unlike on-line coupling, off-line method provides higher flexibility toward system optimization since the CE instrument and the mass spectrometer can be optimized separately. Furthermore, post-separation treatments of the fractionated samples prior to MS analysis, like concentration by solvent evaporation, modification of buffer composition, dialysis, centrifugation, etc., are possible. However, since fractions of a few nanoliters are collected into tens of microliter electrolyte volumes, lack of sensitivity is a particular downside of the method.

Due to its sensitivity, on-line CE/ESI MS gained in popularity as a more convenient approach for the analysis of complex carbohydrate mixtures (8–15). On-line CE/ESI MS coupling requires an interface able to provide an efficient transfer of the analyte from the CE capillary into MS without affecting the separation efficiency. Besides the interface design and the performance of the formed CE/MS electric circuit, other numerous parameters influence the results of CE/ESI MS analysis and are to be taken into account for optimization (16). These are: (1) the choice of an electrolyte compatible with both the ionic species formation/separation by CE and electrospray process; (2) range and polarity of the applied CE and ESI potentials; (3) fine positioning of the sprayer with respect to the MS sampling orifice, and (4) general solution and instrumental parameters such as buffer and sample concentration, pH, injection time and pressure, and capillary temperature. From the separation point of view, a fundamental and general concern is that the best suited CE electrolytes such as borate- or phosphate-based buffers are usually inappropriate for the electrospray process being non-volatile and causing unstable behavior with the ESI source (12, 15, 16). Therefore, the compatibility of the CE electrolyte is one of the major challenges in interfacing CE to ESI MS. From this perspective, carbohydrates as either oligosaccharides or glycoconjugates provide even more limited number of options because of the restrictive conditions for ion formation, separation, and detection (15).

Nowadays in proteomics and glycomics, the most widely used on-line CE/ESI MS setup is the sheath liquid interface based on a make-up liquid that ensures the electrical contact between the CE capillary and ESI sprayer (17–19). The sheath or make-up liquid is usually a mixture of acetic acid or ammonium acetate with an organic solvent such as methanol. In principle, the make-up liquid picks up the analyte eluting from the CE capillary in a solvent appropriate for ESI and the whole resulting mixture is then sprayed into the mass spectrometer. Consequently, in all sheath flow-based configurations, the make-up liquid is mixed with the CE buffer and the sample; therefore, the major drawback of this setup is the low sensitivity as a consequence of the inherent analyte dilution.

A system eliminating the addition of sheath liquid employs a conductive sprayer for electrical contact (15, 16, 20). In such a sheathless design, the sample is sprayed directly into the mass spectrometer from either the separation capillary etched as a sprayer (Figs. 3 and 4) or from a tapered micro- or nano-emitter connected via joints (Fig. 5) to the CE column (16, 21–25). Employment of these types of emitters results in superior sensitivity, lower spraying potential, closer positioning of the sprayer to the orifice of the mass spectrometer, and considerable improvement of ion transfer into MS. Due to all these advantages, immediately after its introduction, sheathless setups became very popular in proteomics and more recently have become intensively used also in glycomics and glycoproteomics as well (21–24, 26).

In this chapter efficient procedures for the analysis of glycoforms by CE in off- and on-line combination with ESI MS and tandem MS are described. The first sections are devoted to the applicability of off-line CE/ESI MS and tandem MS by collision-induced dissociation (CID) at low energy for screening of

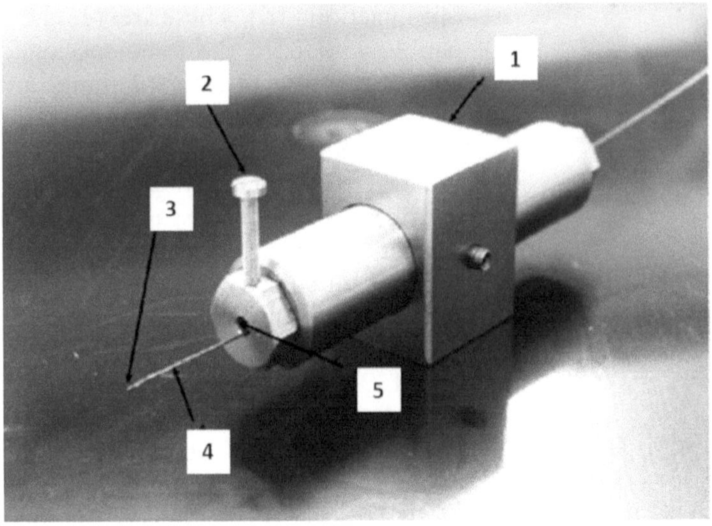

Fig. 3. Sheathless on-line CE/ESI MS interface with one-piece CE column having the terminus etched as a microsprayer and copper-coated. *1* Stainless steel clenching device; *2* capillary fixing screw; *3* microsprayer tip; *4* copper coating; *5* capillary insertion orifice (reproduced from ref. (16) with permission from Elsevier).

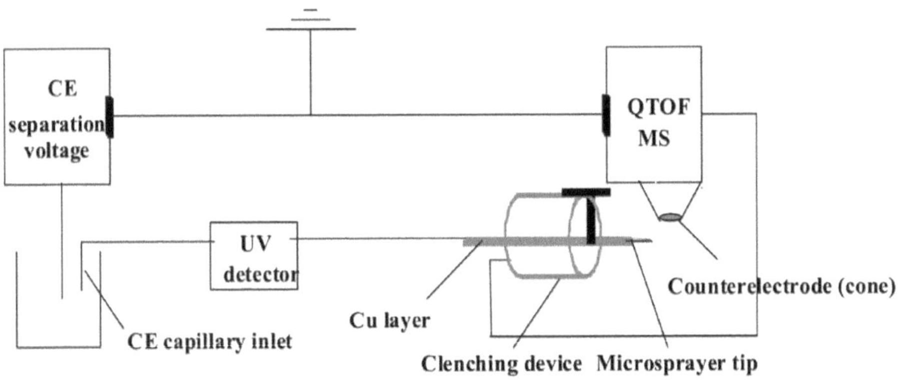

Fig. 4. Schematic of the on-line CE/ESI MS coupling via sheathless interface with one-piece CE column (reproduced from ref. (21) with permission from Wiley).

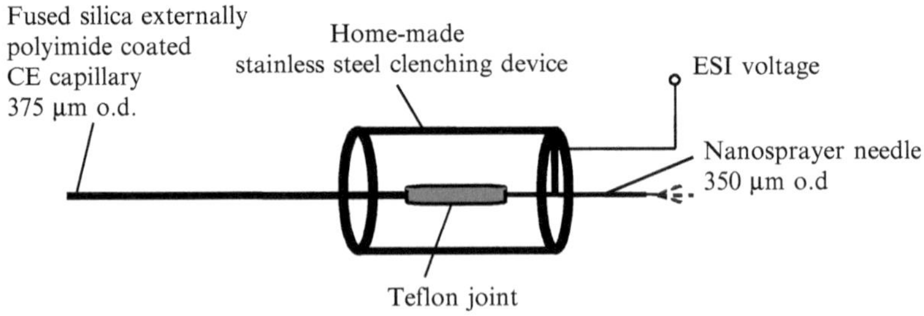

Fig. 5. Sheathless CE/ESI MS interface with electrospray emitter attached to the outlet of the CE column via a Teflon tubing joint (adapted from ref. (23) with permission from Wiley).

O-glycosylated sialylated peptides and chondroitin sulfate/dermatan sulfate (CS/DS) glycosaminoglycans (GAGs). The second part of the chapter describes the techniques for manufacturing of two different sheathless interfaces and shows their implementation in glycan analysis with on-line CE/ESI MS. The last part is dedicated to a comprehensive description of the protocols for application of on-line CE/ESI MS for the separation and identification of glycoforms in the mixtures of O-glycopeptides and CS/DS oligosaccharides.

2. Materials

2.1. Capillary Electrophoresis Instrument

Beckman Coulter P/ACE™ 5000 series instrument with hydrodynamic injection, UV detector, deuterium lamp, 2 nm wavelength accuracy, 190–380 nm wavelength range with filter selection controlled by the System Gold dedicated software package or Beckman Coulter P/ACE™ MDQ system with hydrodynamic injection, selectable-wavelength UV/Vis (200, 214, 254, and 280 nm filters included) detector and 32 Karat™ Software configured on personal computer (see Note 1).

2.2. Mass Spectrometry

Orthogonal hybrid quadrupole time-of-flight (QTOF) mass spectrometer (Micromass, Manchester, UK or Waters, Milford, MA, USA) with nanoelectrospray ion source in Z-spray geometry and personal computer running MassLynx software system under Windows to control the QTOF MS instrument, acquisition, and processing of the MS data. Instrument performances: (1) resolution >5,000 (full width at half maximum, FWHM); (2) mass accuracy >10 ppm for *m/z* (150–900); (3) scan rate from 2 to 10 scan/s; (4) mass range: *m/z* 7,000–20,000 for MS, *m/z* 3,000–4,000 for MS/MS (see Note 2).

2.3. Off-Line CE/ESI MS

2.3.1. Analysis of O-Glycopeptides

1. Analytical grade methanol (MeOH), ammonium acetate, ammonium hydroxide, 32% ammonia, acetonitrile from Merck (Darmstadt, Germany) to be used without further purification.
2. pH-meter, model 766, Calimatic (Knick, Germany).
3. Disposable filter units with 0.2 μm membrane from Schleicher & Schuell (Dassel, Germany).
4. Digital SpeedVac system SPD111V (Thermo Electron, Asheville, NC, USA) coupled to a vacuum pump PC 2002 Vario with CVC 2000 Controller from Vaccubrand GmbH, (Wertheim, Germany).
5. Deionized water from Milli-Q water system (Millipore, Bedford, MA, USA).

6. Laboratory desktop Eppendorf 5804R centrifuge (Eppendorf AG, Hamburg, Germany).
7. Weighing digital balance, Sartorius, ED 153 (Sartorius AG Göttingen, Germany).
8. Fused-silica tubing CE capillaries: 50 μm ID × 375 μm OD, overall length 57 cm, externally coated with polyimide (BGB Analytik Vertrieb, Schloßböckelheim, Germany).
9. Omega glass capillaries (Hilgenberg, Germany).
10. Vertical pipette puller, model 720 (David Kopf Instruments, Tujunga, CA, USA).
11. Calibration standard G2421A electrospray "tuning mix" (Agilent Technologies, Santa Rosa, CA, USA).
12. Nitrogen for CE and QTOF mass spectrometer, purity ≥99.999% vol.
13. Argon for tandem MS by CID, purity = 99.9995% vol.
14. Sample: mixture of sialylated *O*-glycopeptides.

2.3.2. Analysis of Chondroitin and Dermatan Sulfate (CS/DS) Oligosaccharides

1. Analytical grade MeOH, ammonium acetate, ammonium hydroxide, 32% ammonia from Merck (Darmstadt, Germany) to be used without further purification.
2. pH-meter, filters, Digital SpeedVac, weighing digital balance, fused-silica tubing CE capillaries, deionized water, centrifuge, omega glass capillaries, vertical pipette puller, calibration standard, nitrogen and argon as described at Subheading 2.3.1.
3. CS/DS oligosaccharides released, partially depolymerized, fractionated by size exclusion chromatography (SEC) and purified for CE/ESI MS as described in refs. (27, 28).

2.4. On-Line CE/ESI MS

2.4.1. Sheathless Interface with One-Piece CE Column

1. Fused-silica tubing CE capillary 50 μm ID × 375 μm OD, minimum length 130–150 cm, externally coated with polyimide (BGB Analytik Vertrieb, Schloßböckelheim, Germany).
2. 40% Hydrofluoric acid (HF) from Sigma-Aldrich.
3. Copper suspension or spray in dimethylether (Gutter Supply, Lake Bluff, IL, USA or Rust-Oleum, Vernon Hills, IL, USA).
4. Meiji EMT3-P—×20/×40 dual-power stereo microscope (Meiji Techno, Santa Clara, CA, USA).

2.4.2. Sheathless Interface with Butted Electrospray Emitter

1. Fused-silica tubing CE capillary 75 μm ID × 375 μm OD, minimum length 130–150 cm, externally coated with polyimide (BGB Analytik Vertrieb, Schloßböckelheim, Germany).
2. Externally distal coated Pico Tip™ nanospray needles 75 μm ID × 350 μm OD (15 ± 1.5 μm ID of the tip) from New Objective (Cambridge, MA, USA).

3. Teflon® PTFE Tubing, 350 μm ID (Professional Plastics, Inc., Fullerton, CA, USA).
4. Meiji EMT3-P—×20/×40 dual-power stereo microscope (Meiji Techno, Santa Clara, CA, USA).

2.4.3. On-Line CE/ESI MS of O-Glycopeptides in Normal Polarity

1. Analytical grade MeOH, ammonium acetate, ammonium hydroxide, 32% ammonia from Merck (Darmstadt, Germany).
2. pH-meter, filters, Digital SpeedVac, weighing digital balance, deionized water, centrifuge, calibration standard, nitrogen as described at Subheading 2.3.1.
3. Sample: mixture of sialylated O-glycopeptides (see Note 3).

2.4.4. On-Line CE/ESI MS of O-Glycopeptides in Reverse Polarity

1. Analytical grade MeOH, 98% formic acid, and 32% ammonia from Merck (Darmstadt, Germany).
2. pH-meter, filters, Digital SpeedVac, weighing digital balance, deionized water, centrifuge, calibration standard, nitrogen as described at Subheading 2.3.1.
3. Sample: mixture of sialylated O-glycopeptides (see Note 3).

2.4.5. On-Line CE/ESI MS of CS/DS Oligosaccharides

1. Analytical grade MeOH, ammonium acetate, and 32% ammonia from Merck (Darmstadt, Germany).
2. pH-meter, filters, Digital SpeedVac, weighing digital balance, deionized water, centrifuge, calibration standard, nitrogen, and argon as described at Subheading 2.3.1.
3. CS/DS oligosaccharides released, partially depolymerized and purified for CE/ESI MS as described in ref. (23).

3. Methods

3.1. Off-Line CE/ESI MS

Off-line CE/ESI MS assumes collection in different vials of fractions separated by CE, their transfer into the glass nanoESI capillary, and subsequent analysis by ESI MS and tandem MS (27–29). Fraction collection might be carried out by either calculating the time window when a compound has migrated to the end of the capillary or using a prerun to obtain a CE/UV or laser-induced fluorescence (LIF) profile and estimate the migration time of different carbohydrate compounds. Collected fractions are to be screened by nanoESI MS to determine the molecular mass and identify the glycoforms in the mixtures. Some species of interest may be further subjected to tandem MS by CID to characterize in detail the glycan and peptide moieties and, under certain sequencing conditions, the glycosylation site(s).

3.1.1. O-Glycopeptides: CE Separation and Fraction Collection

1. Prepare 50 mL CE electrolyte (buffer): 50 mM aq. ammonium acetate buffered to pH 12.0 with 32% ammonia solution.
2. Prepare 50 μL sample/buffer stock solution of 1 μg/μL.
3. Centrifuge the sample/buffer solutions for 1 h and collect the supernatant.
4. Prepare 50 mL rinsing solution: 19 M aq. ammonium hydroxide.
5. Filter all solutions through a 0.2 μm disposable filter unit and store them at 4°C.
6. Cut by the aid of the ceramic cutter from Beckman instrument supplies, a fused-silica capillary piece of 57 cm.
7. At 7 cm from one end (outlet), create a transparent window by removing the external polyimide capillary coat on a segment of approximately 0.5 cm length (see Note 4).
8. Prepare the CE capillary cartridge according to the description provided by Beckman supplier and place the transparent window in front of the UV detector casement (see Note 5).
9. Rinse the CE capillary for 10 min with MeOH and dry it by flushing it for 10 min with air.
10. Condition the capillary by 10–15 min rinsing with the CE buffer, 50 mM ammonium acetate buffered to pH 12.0 with 32% ammonia solution.
11. Load 10 μL sample/buffer solution in a clean CE vial A and place it in the autosampler.
12. Prepare several buffer vials (B, C, D, E, etc.) containing each 10 μL CE buffer solution and place them in the autosampler.
13. Set the detection at 214 nm and the separation time at 30 min.
14. Set the temperature of the capillary cartridge at 20°C.
15. Inject the sample from vial A in the CE capillary inlet by applying a constant nitrogen pressure of 0.5 psi for 4 s, which will result in approximately 13 nL injected sample volume.
16. Separate the injected mixture with inlet vial B, outlet vial C, for 30 min in normal polarity under 20 kV CE voltage, while recording the CE/UV profile at 214 nm.
17. After the separation is completed, rinse the CE capillary for 20 min with 19 M aq. ammonium hydroxide.
18. Condition again the capillary by rinsing for 5 min with the CE buffer.
19. Repeat the steps 11–13 to check the reproducibility of migration times (t_1, t_2, t_3, etc.) of the components (UV absorption peaks) recorded on the electropherogram.

20. Rinse again the CE capillary for 20 min with 19 M aq. ammonium hydroxide.
21. Condition again the capillary by 5 min rinsing with the CE buffer.
22. Proceed with the steps 11 and 12.
23. Separate the injected mixture in normal polarity under 20 kV CE voltage with inlet vial B, outlet vial D.
24. Stop the separation between time t_1 and time t_2 and change through the software the autosampler position for: inlet vial B and outlet vial C.
25. Continue the separation under 20 kV.
26. Stop the separation between time t_2 and time t_3 and change for: inlet vial B and outlet vial D.
27. Proceed the same for all succeeding fractions.
28. At the end of the separation, collect the fractions from vials C, D, E, etc. and transfer them in different Eppendorf tubes. Store them at 4°C.
29. Prepare the CE capillary for the next runs by rinsing it for 30 min with 19 M aq. ammonium hydroxide.
30. Prepare the CE capillary for the next day by rinsing it for 20 min with MeOH and dry it by flushing it for 10 min with air (see Note 6).

3.1.2. Off-Line NanoESI MS and Tandem MS Analysis of the Separated O-Glycopeptide Fractions

1. Prepare the nanoESI capillaries with tapered tip by pulling the omega glass tubes in the vertical pipette puller.
2. Transfer 5–10 μL of the CE fraction solution into the capillary by the aid of a micropipette loader.
3. Set up the loaded capillary on the nanoESI source of the QTOF mass spectrometer by inserting into the source stainless steel wire.
4. Tune the QTOF mass spectrometer for operating in negative ion mode ESI with a capillary voltage of 800–1,000 V, sampling cone potential (cone voltage) of 30–50 V, source block temperature 80–100°C, desolvation gas pressure at a nominal flow rate around 50 L/h (see Note 7).
5. Optimize these nanoESI parameters and adjust the capillary tip position with respect to the counterelectrode (cone) to obtain a stable spray and avoid the *in-source* fragmentation.
6. Acquire the signal at a scan speed of 2.1 scans/s until a fair signal-to-noise ratio is obtained and record the total ion chromatogram (TIC). Recommended acquisition time is above 5 min (see Note 8).

7. Import the TIC and generate the spectrum by combining it in progress across all TIC scans.
8. Identify in the MS run the ion, which according to the calculated mass corresponds to a structure of interest for further detailed investigation.
9. Leave the MS settings unaltered, choose MS/MS option in the tune page of MassLynx, set HM and LM at values of 10.0 and 10.0, respectively.
10. Acquire the CID MS/MS signal for at least 5–10 min at a scan speed of 2.1 scans/s and argon pressure of 12 psi while varying gradually the collision energy from 20 to 50 eV (see Notes 8 and 9). Stop the acquisition, import the TIC, and generate the spectrum by combining it in progress across over all TIC scans (see Note 10).
11. Apply the steps 1–10 to all collected fractions using a new nanoESI capillary for each fraction.
12. Load in a new nanoESI capillary a tune mix solution (1:100) diluted in acetonitrile. Generate the calibration spectrum in the negative ion mode by repeating the steps 3–7. Apply the calibration file to the sample spectra acquired in MS and MS/MS modes. On a Waters QTOF mass spectrometer, by proper calibration an average mass accuracy of at least 10 ppm can be achieved.

3.1.3. Chondroitin and Dermatan Sulfate Oligosaccharides, CE Separation, and Fraction Collection

Most of CS/DS GAG species exhibit a regular sulfation pattern, i.e., one sulfate group per disaccharide repeat, typically situated at GalNAc; however, oversulfated and undersulfated CS/DS glycoforms were also reported (23, 24, 26–28, 30, 31). By CE fractionation, CS/DS species with high molar sulfate content can be separated from the non-sulfated ones existent in the mixture and characterized by ESI MS in terms of epimerization, sulfate distribution along the chain, and sulfation site(s) within the monomer ring (30). A CE partial separation in fractions according to sulfation degree is helpful in particular for delineating the real under- and non-sulfated species from the possible artifacts induced by the *in-source* decay of the sulfate groups in the ESI MS mode.

1. Prepare 50 mL CE electrolyte (buffer): 50 mM aqueous/MeOH (40:60) ammonium acetate buffered to pH 12.0 with 32% ammonia solution.
2. Prepare 50 μL sample/buffer stock solution of 0.5 μg/μL.
3. Proceed with the steps 3–9 described at the Subheading 3.1.1.
4. Condition the capillary by 10–15 min rinsing with the CE buffer, 50 mM aqueous/MeOH ammonium acetate buffered to pH 12.0 with 32% ammonia solution.

5. Proceed with the steps 11–15 described at the Subheading 3.1.1.
6. Separate the injected mixture with inlet vial B, outlet vial C, for 30 min in normal polarity under 25 kV CE voltage, while recording the CE/UV profile at 214 nm.
7. Proceed with the steps 17–22 described at the Subheading 3.1.1.
8. Separate the injected mixture in normal polarity under 25 kV CE voltage with inlet vial B, outlet vial D.
9. Proceed with the step 24 described at the Subheading 3.1.1.
10. Continue the separation under 25 kV.
11. Proceed with the steps 26–30 described at the Subheading 3.1.1.

3.1.4. Off-Line NanoESI MS and Tandem MS Analysis of the Separated CS/DS Fractions

1. Evaporate to complete desiccation the CE fractions in SpeedVac and dissolve each dried sample in 10–15 μL analytical grade methanol.
2. Proceed with the steps 1–3 described at the Subheading 3.1.2.
3. Tune the QTOF mass spectrometer for operating in negative ion mode ESI with a capillary voltage of 800 V, sampling cone potential (cone voltage) of 18–20 V, source block temperature 80–100°C, desolvation gas pressure at a nominal flow rate around 50 L/h (see Note 7).
4. Proceed with the steps 5–9 described at the Subheading 3.1.2.
5. Acquire the CID MS/MS signal for at least 5–10 min at an argon pressure of 8–10 psi while varying gradually the collision energy from 10 to 40 eV (see Notes 8 and 9). Stop the acquisition, import the TIC, and generate the spectrum by combining it in progress across over all TIC scans (see Note 10).
6. Proceed with the step 11 described at the Subheading 3.1.2.
7. Apply the calibration file to the sample spectra acquired in MS and MS/MS modes.

3.2. On-Line CE/ESI MS

In sheathless on-line CE/ESI MS, the glycoforms separated by CE are sprayed directly into MS from the tip of either the CE separation capillary (Figs. 3 and 4) or a butted needle (Fig. 5), at flow rates in the range of a few microliters per minute to a few nanoliters per minute. As compared to the off-line method, the on-line CE/ESI MS exhibits superior separation efficiency, sensitivity, and ionic transfer into MS. Although a significant number of configurations were produced in research laboratories, currently, there is still a lack of commercial sheathless interfaces or completely standardized procedures for smooth and fast in-house interface production. Therefore, in the next part of this chapter the protocols for construction and application to glycan analysis of two effective sheathless CE/ESI MS interfaces are presented.

3.2.1. Construction of the Sheathless Interface with One-Piece CE Column

This interface is based on coating the CE capillary terminus tapered as a sprayer tip with a copper layer to provide the electrical contact needed for both CE and ESI (21). Copper deposition at the CE column terminus is a simple, optimal, and feasible solution for handling the problem of aggressive highly alkaline media, which are required for CE separation of carbohydrates in normal (direct) polarity.

1. Cut by the aid of the ceramic cutter from Beckman instrument supplies, a fused-silica capillary piece of 130–150 cm.
2. Heat up locally the capillary, 4–5 cm from one end, in a flame of corresponding melting temperature, while gently pulling manually the pieces apart.
3. Under visual inspection with the microscope, remove the tiny long wire resulted by pulling under flame, using the ceramic cutter.
4. To further reduce the outer diameter of the tapered tip and to smooth the edge, immerse the tip in 40% HF for 15–30 min (see Notes 11 and 12).
5. After etching, rinse the external capillary surface for at least 2 min with distilled water and purge it with nitrogen to clean up the surface from any trace of HF.
6. On a length of 5–6 cm from the capillary tapered end, smear the surface of the tip with a liquid layer of copper suspension in dimethylether, by the aid of a thin paint brush (see Notes 13 and 14).
7. Leave the tip to dry in air at room temperature for 3–5 min (see Note 15).
8. Prepare with this column a CE capillary cartridge with a capillary inlet of regular length and the rest of the column (100–130 cm) with tapered and coated end as the outlet. Set up the cartridge in the CE instrument with the capillary inlet in the dedicated slot and the long outlet column kept outside.
9. Introduce the CE column outlet with the tapered and coated emitter in a stainless steel clenching device shaped as shown in Fig. 3 (see Note 16).
10. Position the CE instrument as close as possible to the QTOF MS. Mount the interface onto the ESI high-voltage plate of the QTOF MS source and position the CE emitter in the vicinity of the entrance hole of the cone, at a distance less than 5 mm (Fig. 4, see Note 17).

3.2.2. Analysis of O-Glycopeptides in Normal Polarity

1. Prepare 50 mL CE electrolyte (buffer): 50 mM ammonium acetate (in 60% water and 40% methanol) buffered to pH 12.0 with 32% ammonia solution.
2. Prepare 50 μL sample/buffer stock solution of 0.75 μg/μL.

3. Centrifuge the sample/buffer solutions for 1 h and collect the supernatant.
4. Prepare 50 mL rinsing solution: 19 M aq. ammonium hydroxide.
5. Filter each solution through a 0.2 μm disposable filter unit and store them at 4°C.
6. Rinse the CE capillary 10 min with MeOH and dry it by flushing it 10 min with air (see Note 18).
7. Condition the capillary by 10–15 min rinsing with the CE buffer, 50 mM ammonium acetate buffered to pH 12.0 with 32% ammonia solution (see Note 18).
8. Load 10 μL sample/buffer solution in a clean CE vial A and place it in the autosampler.
9. Load 10 μL buffer solution in a clean CE vial B and place it in the autosampler.
10. Set the temperature of the capillary cartridge to 20°C.
11. Set the CE separation time at 30 min. If in-line UV monitoring is also chosen, set the wavelength at 214 nm.
12. Tune the QTOF mass spectrometer for operating in negative ion mode ESI with a capillary voltage of 1.2 kV, cone voltage 40 V, source block temperature 70–80°C, desolvation gas pressure at a nominal flow rate around 100 L/h.
13. Inject the sample from vial A in the CE capillary inlet by applying a constant nitrogen pressure of 0.5 psi for 8 s, which will result in 25 nL injected volume.
14. Start the option "separate" on the CE instrument in normal polarity under 25 kV CE voltage, applied on the inlet electrode, with vial B at the inlet and no vial at the outlet.
15. Start the acquisition on QTOF tune page and record the TIC at a scan speed of 2.1 scans/s (example in Fig. 6, see Note 19).
16. Acquire the signal for 30 min. Stop the acquisition, import the TIC, and generate the spectra of the separated components by combining in progress across the scans corresponding to each peak (examples in Figs. 7 and 8).
17. Apply the calibration file to each spectrum.
18. Prepare the CE capillary for the next runs by rinsing it for 30 min with 19 M aq. ammonium hydroxide.
19. Prepare the CE capillary for the next day by rinsing it for 20 min with MeOH and dry it by flushing it for 10 min with air (see Note 6).

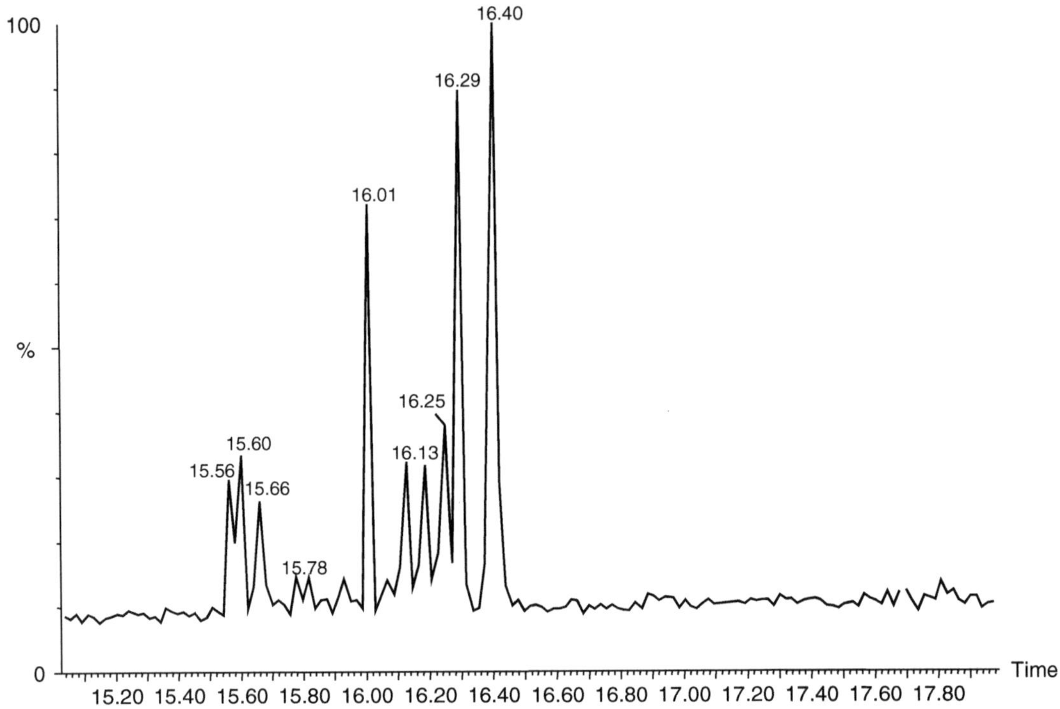

Fig. 6. On-line CE/ESI-MS TIC in normal polarity and negative ion mode of a mixture of O-glycosylated aminoacids and peptides from urine of a patient suffering from Schindler's disease. Fused silica CE column: 125 cm length; CE potential 25 kV; CE buffer: 50 mM aqueous +40% MeOH ammonium acetate with 32% ammonia (pH 12); sample concentration: 0.75 μg/μL buffer (in 5 μL); 8 s injection by pressure; 25 nL injected; ESI potential −1.2 kV; ESI cone potential 40 V (reproduced from ref. (21) with permission from Wiley).

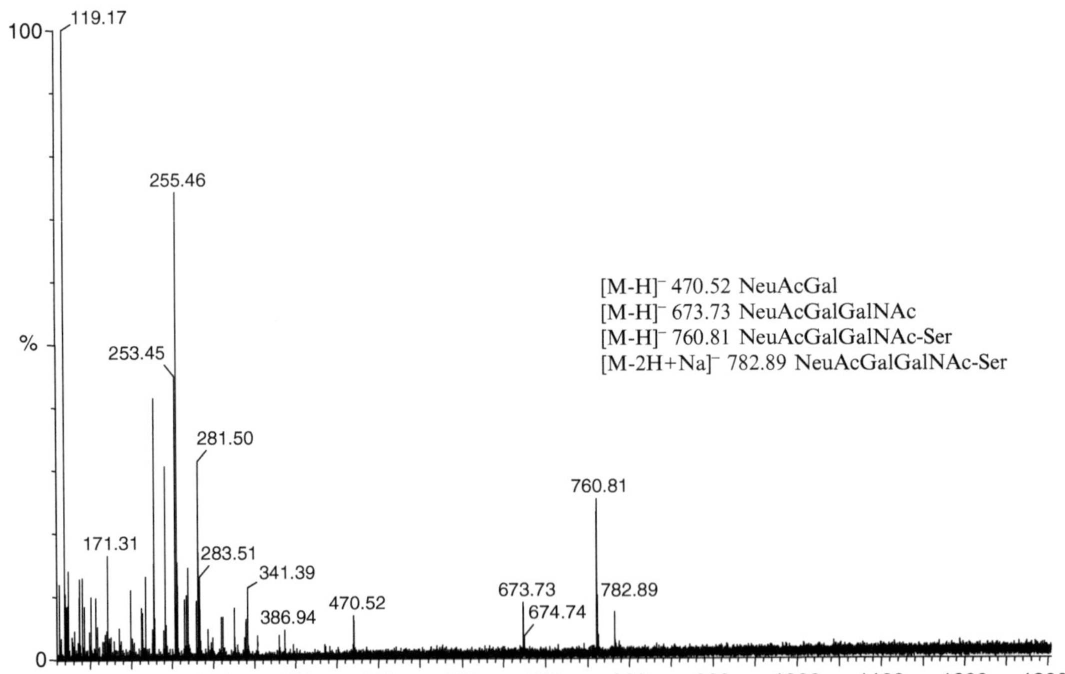

Fig. 7. Spectrum derived by combining in progress across the TIC peak at 16.01 min depicted in Fig. 6 (reproduced from ref. (21) with permission from Wiley).

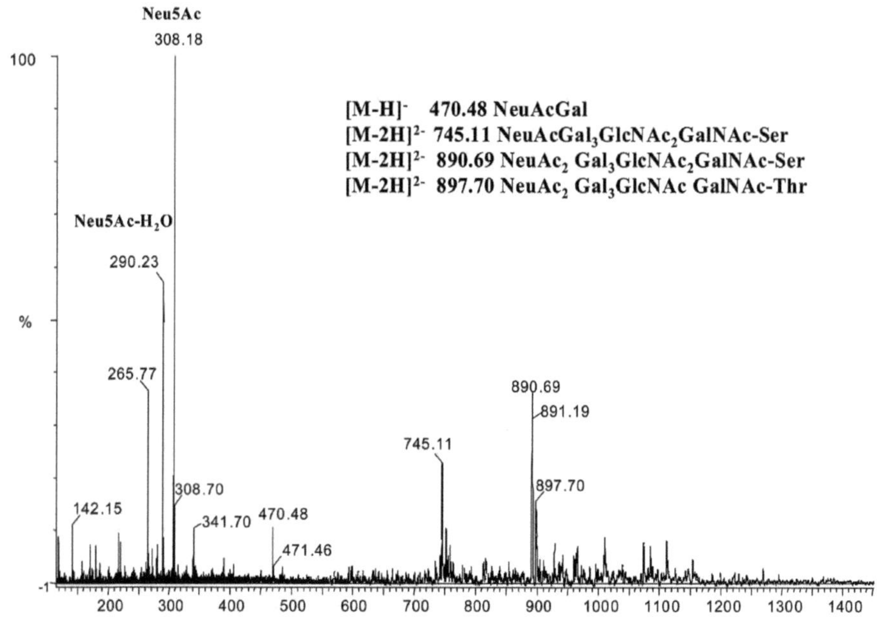

Fig. 8. Spectrum derived by combining in progress across the TIC peak at 16.40 min depicted in Fig. 6 (reproduced from ref. (21) with permission from Wiley).

3.2.3. Construction of the Sheathless Interface with Butted Needle

Different techniques for obtaining a long-lasting metal deposition on the CE capillary were reported (16); however, in all these studies difficulties in obtaining a durable deposition at the end of a one-piece CE column were encountered. In the case of copper-coated sprayer, the major disadvantage is that, although several refurbishments of the coating are possible, after many measurements the electrical contact is irreversibly degraded. To produce a new interface, the relatively laborious procedure of capillary tapering and coating must be again carried out. A solution to this problem is the development of two-piece-CE columns (22, 23), with commercially available, disposable emitters as described below.

1. Cut by the aid of the ceramic cutter of Beckman instrument supplies, a fused-silica capillary piece of 130–150 cm.
2. Prepare the CE capillary cartridge with a capillary inlet of regular length and the rest of the column (100–130 cm) as the outlet. Set up the cartridge in the CE instrument with the capillary inlet in the dedicated slot and the long outlet kept outside.
3. Cut by the aid of the ceramic cutter a Teflon tubing piece of 1 cm (see Note 20).
4. Insert the CE capillary outlet and the externally distal coated nanospray needle into the Teflon joint as shown in Fig. 5. Adjust the Teflon ends by filling and inspect at the microscope the connection and alignment of the capillary and needle in the joint (see Note 21).

5. Introduce the CE column outlet with butted emitter in the stainless steel clenching device as shown in Fig. 5 (see Note 16).
6. Position the CE instrument as close as possible to the QTOF MS. Mount the interface onto the ESI high-voltage plate of the QTOF MS source, assemble the system as depicted in Fig. 4 and position the nanosprayer in the vicinity of the entrance hole of the cone (see Note 17).

3.2.4. Analysis of O-Glycopeptides in Reverse Polarity

1. Prepare 50 mL CE electrolyte (buffer): 0.1 M formic acid (in 60% water and 40% methanol) buffered to pH 2.8 with 32% ammonia solution (scc Notc 22).
2. Prepare 50 μL sample/buffer stock solution of 0.75 μg/μL.
3. Centrifuge the sample/buffer solutions for 1 h and collect the supernatant.
4. Filter each solution through a 0.2 μm disposable filter unit and store them at 4°C.
5. Rinse the CE capillary for 30 min with MeOH and dry it by flushing it 15 min with air (see Note 18).
6. Condition the capillary by 15–20 min rinsing with the CE buffer, 0.1 M formic acid (in 60% water and 40% methanol) buffered to pH 2.8 with 32% ammonia solution (see Note 18).
7. Load 10 μL sample/buffer solution in a clean CE vial A and place it in the autosampler.
8. Load 10 μL buffer solution in a clean CE vial B and place it in the autosampler.
9. Set the temperature of the capillary cartridge to 20°C.
10. Set the CE separation time at 150 min (see Note 23). If in-line UV monitoring is also chosen, set the wavelength at 214 nm.
11. Tune the QTOF mass spectrometer for operating in negative ion mode ESI with a capillary voltage of 800–900 V, cone voltage 40 V, source block temperature 40–60°C, desolvation gas pressure at a nominal flow rate around 50 L/h, scan speed 2.1 scans/s.
12. Inject the sample from vial A in the CE capillary inlet by applying a constant nitrogen pressure of 0.5 psi for 5 s, which will result in approximately 15 nL injected volume.
13. Start the option "separate" on the CE instrument in reverse polarity under −25 kV CE voltage applied on the inlet electrode, with vial B at the inlet and no vial at the outlet.
14. Start the acquisition on QTOF tune page and record the TIC (example in Fig. 9).
15. Start acquiring the ESI/MS signal approximately 30–40 min after sample injection (see Note 23).

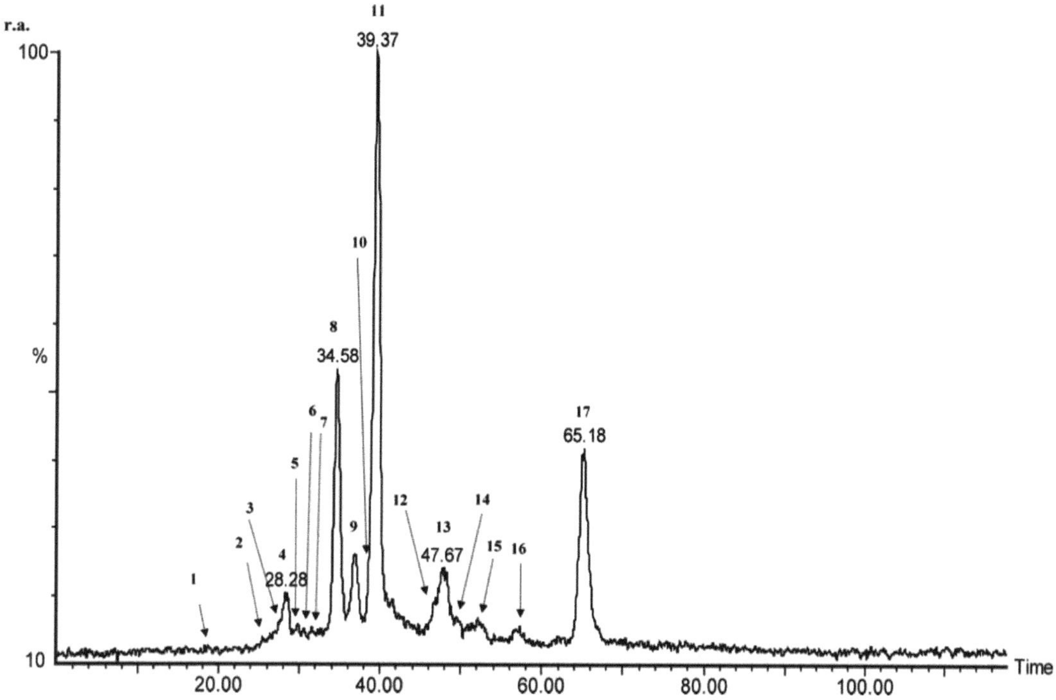

Fig. 9. On-line CE/ESI-MS TIC in reverse polarity and negative ion mode of a mixture of O-glycosylated aminoacids and peptides from urine of patient suffering from Schindler's disease. Fused silica CE column: 130 cm length; CE potential −25 kV; CE buffer: 0.1 M methanol/water (6:4% v/v) formic acid (pH 2.8); sample concentration; 0.75 µg/µL (5 pmol injected); 5 s injection by pressure; ESI potential −0.8 kV; ESI cone potential 40 V (reproduced from ref. (22) with permission from Wiley).

16. Acquire the signal for 80–100 min at a scan speed of 2.1 scans/s (see Note 23). Stop the acquisition, import the TIC, and generate the spectra of the separated components by combining in progress across the scans corresponding to each peak. In the case of highly complex mixtures, process from TIC the extracted ion chromatogram (XIC) of the ion of interest and its derived mass spectrum (example in Fig. 10a, b).
17. Apply the calibration file to each spectrum.
18. Prepare the CE capillary for the runs and the next day by rinsing it for 30 min with MeOH and dry it by flushing it for 15 min with air (see Note 6).

3.2.5. Analysis of Chondroitin and Dermatan Sulfate Oligosaccharides

1. Prepare 50 mL CE electrolyte (buffer): 50 mM aqueous/MeOH (40:60) ammonium acetate buffered to pH 12.0 with 32% ammonia solution.
2. Prepare 50 µL sample/buffer stock solution of 0.1 µg/µL.
3. Centrifuge the sample/buffer solutions for 1 h and collect the supernatant.
4. Filter all sample and buffer solutions through a 0.2 µm disposable filter unit and store them at 4°C.

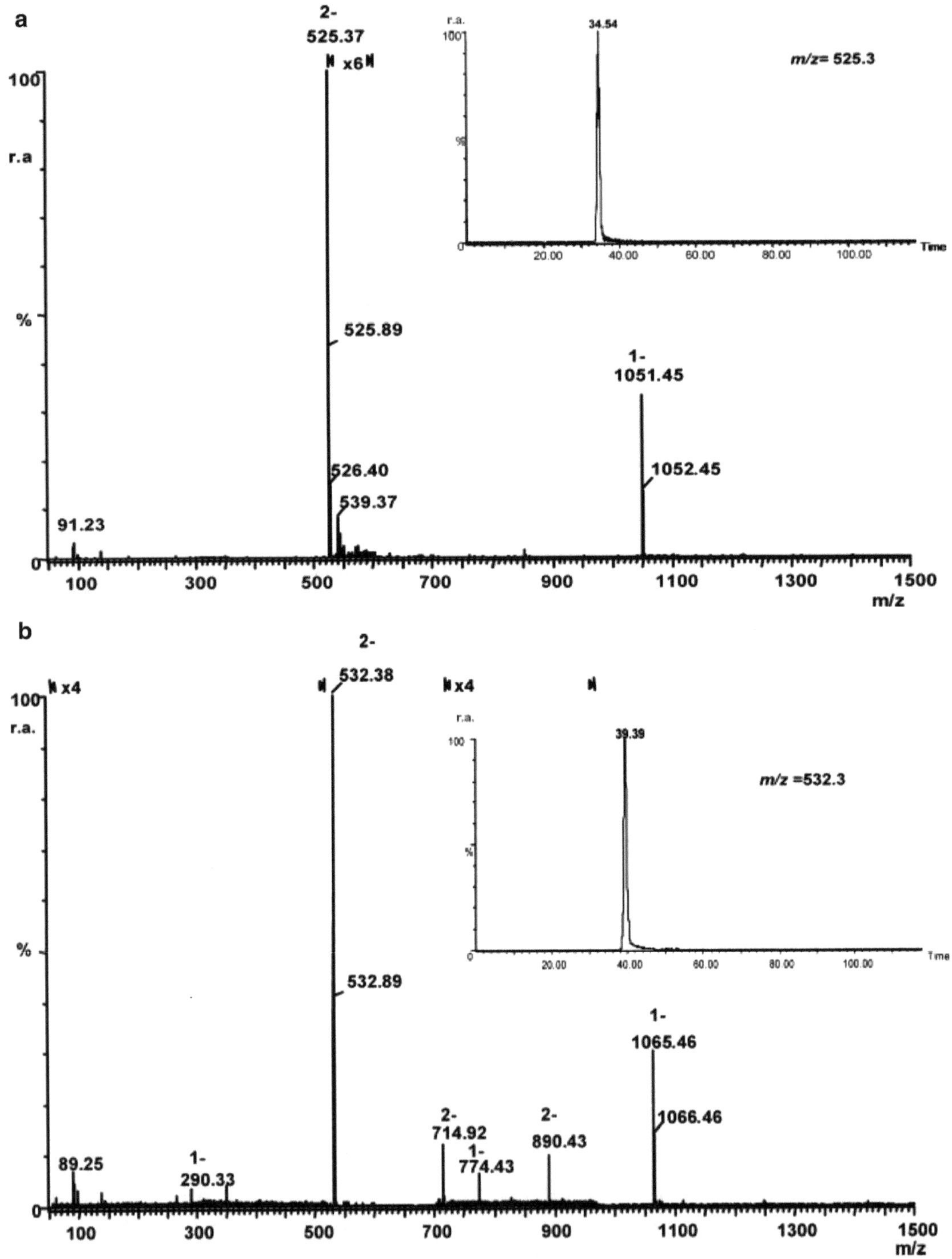

Fig. 10. Reverse polarity CE/(−)nanoESI-QTOF mass spectra obtained by combining across the extracted ion cromatogram (XIC) of the doubly deprotonated ions at (**a**) *m/z* 525.3 corresponding to $Neu5Ac_2HexHexNAc$-Ser and (**b**) *m/z* 532.3 corresponding to $Neu5Ac_2HexHexNAc$-Thr. Insets: XIC of the ions at *m/z* 525.3 and *m/z* 532.3 processed from the TIC MS in Fig. 9 (reproduced from ref. (22) with permission from Wiley).

5. Rinse the CE capillary for 30 min with MeOH and dry it by flushing it 15 min with air (see Note 18).
6. Condition the capillary by rinsing it for 15–20 min with the CE buffer, 50 mM aqueous/MeOH (40:60) ammonium acetate buffered to pH 12.0 with 32% ammonia solution (see Note 18).
7. Load 10 μL sample/buffer solution in a clean CE vial A and place it in the autosampler.
8. Load 10 μL buffer solution in a clean CE vial B and place it in the autosampler.
9. Set the temperature of the capillary cartridge to 15–20°C.
10. Set the CE separation time at 30 min. If in-line UV monitoring is also chosen, set the wavelength at 214 nm.
11. Tune the QTOF mass spectrometer for operating in negative ion mode ESI with a capillary voltage of 700 V, cone voltage 15 V, source block temperature 40–60°C, desolvation gas pressure at a nominal flow rate around 50 L/h, scan speed 2.1 scans/s.
12. Inject the sample from vial A in the CE capillary inlet by applying a constant nitrogen pressure of 0.5 psi for 6 s, which will result in approximately 20 nL injected volume.
13. Start the option "separate" on the CE instrument in normal polarity under 30 kV CE voltage applied on the inlet electrode, with vial B at the inlet and no vial at the outlet.
14. Start the acquisition on QTOF tune page and record the TIC.
15. Acquire the signal for 30 min at a scan speed of 2.1 scans/s. Stop the acquisition, import the TIC, and generate the spectra of the separated components by combining in progress across the scans corresponding to each peak.
16. Apply the calibration file to each spectrum.
17. Prepare the CE capillary for the runs and the next day by rinsing it for 30 min with MeOH and dry it by flushing it for 15 min with air (see Note 6).

4. Notes

1. Agilent 7100, Agilent G1600 3D capillary electrophoresis systems from Agilent Technologies (worldwide sales and support) or PrinCE systems (400–700 series) from Prince Technologies (Emmen, The Netherlands) are also suitable.
2. High resolution instruments with CID capabilities such as Orbitrap or Fourier transform ion cyclotron (FTICR) MS may be also used.

3. Sialylated O-glycopeptides obtained after digestion of fetuin from fetal calf serum (Sigma Aldrich, St. Louis, MO, USA) are recommended as control.
4. The external polyimide capillary coat can be removed properly by local burning in a flame. After coat removal, the transparent area should be several times rinsed with distilled water to eliminate impurities and purged with nitrogen for drying.
5. Manipulate carefully the capillary after the removal of the 0.5 cm polyimide sector. Bare fused silica is particularly fragile; therefore, it may easily break in the region of the uncovered segment.
6. Condition daily the capillary and change it for a new one after 20–30 runs. It is also strongly recommended to change the capillary from sample to sample.
7. In the case of sialylated, fucosylated, or sulfated glycans, since sialic acid, fucose, and sulfate group(s) are very labile, the cone voltage should not exceed the indicated maximal values. Higher cone voltages induce *in-source* fragmentations with particular loss of these moieties.
8. The low *signal-to-noise* ratio, generated by the reduced concentration of analytes in the CE-collected fraction, may be compensated by longer signal accumulation; acquisition of even 500–800 scans or more combined in a single mass spectrum is occasionally needed.
9. In the case of sialylated, fucosylated, or sulfated glycans, because of sialic acid, fucose, and sulfate group(s) lability under CID conditions, collision energy and gas pressure should not exceed the indicated maximal values. Higher values result in molecule exhaustive desialylation, desulfation, or fucose detachment, which impedes the determination of sialylation, sulfation, or fucosylation pattern and sites.
10. The following computer software, databases, and data analysis tools to assist the interpretation of the MS and MS/MS of glycopeptides and/or GAGs are recommended:
    SimGlycan®
    (http://www.premierbiosoft.com/glycan/index.html)
    GlycoWorkbench
    (http://www.eurocarbdb.org/applications/ms-tools)
    GlycoMod Tool
    (http://www.expasy.org/tools/glycomod/)
    GlycoFragment
    (www.dkfz.de/spec/projekte/fragments/)
    GlycoSearchMS
    (www.dkfz.de/spec/glycosciences.de/sweetdb/ms/)

11. Hydrofluoric acid is highly toxic and especially hazardous to handle. It should be handled with extreme care, exceeding that accorded to other chemicals. Carry out this procedure only under ventilation (properly functioning chemical fume hood). Use chemical splash spectacles together with a face shield. It is obligatory to wear: (1) laboratory coat with a chemical splash apron made out of natural rubber, neoprene, or viton; (2) medium or heavy-weight viton, nitrile, or natural rubber gloves; (3) a second pair of nitrile gloves under the gloves for protection against leaks. Do not wear shorts or open-toed shoes. Prior to this procedure read carefully the safety instructions and consult the literature (32–34).
12. During etching with hydrofluoric acid, pump water or air through the capillary (at a flow rate of 2–3 μL/min) using a regular laboratory syringe pump. This rinsing procedure is compulsory in order to prevent hydrofluoric acid from entering the capillary where it would chemically attack the interior of the surface and widen the tip orifice.
13. For higher precision in surface smearing, perform this operation under visual inspection with the microscope.
14. To avoid clogging of the tip by fine copper particles, during deposition rinse continuously the capillary with distilled water (flow rate 2 μL/min) using the syringe pump.
15. The copper-coated microsprayer tip is not everlasting; the electrical contact can be maintained around 30 h of functioning under ESI voltage, without the need of supplementary copper deposition. Once the electrical contact degraded, do not change the column. The copper coat of the same tip may be several times refurbished by fresh depositions.
16. The sprayer should extend 1–2 cm over the holder.
17. If simultaneous in-line UV detection is required, a transparent window is to be produced as described at Subheading 3.1.1. The window must be placed in the front of UV detector casement as illustrated in Fig. 4.
18. CE emitter tip unit may be left connected to the mass spectrometer during the rinse of the capillary.
19. In on-line CE/ESI MS method, QTOF MS instrument will record the TIC as a set of migrating peaks (CE profile) vs. time/scan axis; each peak corresponds to the component or components which have eluted at the respective moment. By combining in progress across each peak, the spectrum of the component(s) eluted at that time will be generated.
20. Prior to repeated couplings and between runs, purge the Teflon joint by distilled water to remove possible impurities and purge it with air or nitrogen to dry it and hinder any air bubble formation.

21. An inappropriate connection or a misalignment of the capillary and needle in the joint results in dead-volume formation, which severely deteriorates the separation efficiency.
22. At lower concentrations of ammonium formate, Corona discharge at the ESI tip and air bubbles occur. They are attributable to the redox reaction at the cathode. 0.1 M methanol/ water (6:4% v/v) formic acid at pH 2.8 is an optimal value for suppression of the EOF and generation of a stable spray.
23. For the non-pressurized reverse polarity CE/ESI MS, the use of a buffer at a pH 2.8 in combination with internally uncoated capillary walls results in almost total suppression of EOF. Under these conditions, the migration of the negatively charged analytes toward the anode is driven only by the electrical force. The separation efficiency is certainly much higher than in the case of pressure-assisted reverse polarity CE/ESI MS; however, the migration time is much longer.

Acknowledgments

The authors are grateful to the Romanian National Authority for Scientific Research, CNCS-UEFISCDI (Grants No. PN-II-ID-PCE-2011-3-0047, PN-II-RU-TE-2011-2-0008 and POSDRU/89/1.5/S/60746) for financial support.

References

1. Lazar IM, Lazar AC, Cortes DF, Kabulski JL (2011) Recent advances in the MS analysis of glycoproteins: theoretical considerations. Electrophoresis 32:3–13
2. Suntornsuk L (2010) Recent advances of capillary electrophoresis in pharmaceutical analysis. Anal Bioanal Chem 398:29–52
3. Breadmore MC, Dawod M, Quirino JP (2011) Recent advances in enhancing the sensitivity of electrophoresis and electrochromatography in capillaries and microchips (2008–2010). Electrophoresis 32:127–148
4. Haselberg R, de Jong GJ, Somsen GW (2011) Capillary electrophoresis-mass spectrometry for the analysis of intact proteins 2007–2010. Electrophoresis 32:66–82
5. Honda S (2003) Capillary electrophoresis of monosaccharides. Methods Mol Biol 213:81–92
6. Tryfona T, Stephens E (2010) Analysis of carbohydrates on proteins by offline normal-phase liquid chromatography MALDI-TOF/TOF-MS /MS. Methods Mol Biol 658:137–151
7. Smejkal P, Szekrényes A, Ryvolová M, Foret F, Guttman A, Bek F, Macka M (2010) Chip-based CE for rapid separation of 8-aminopyrene-1,3,6-trisulfonic acid (APTS) derivatized glycans. Electrophoresis 31:3783–3786
8. Neusüss C, Pelzing M (2009) Capillary zone electrophoresis-mass spectrometry for the characterization of isoforms of intact glycoproteins. Methods Mol Biol 492:201–213
9. Sanz-Nebot V, Balaguer E, Benavente F, Neusüss C, Barbosa J (2007) Characterization of transferrin glycoforms in human serum by CE-UV and CE-ESI-MS. Electrophoresis 28:1949–1957
10. Schweda EK, Richards JC (2010) Profiling LPS glycoforms of non-typeable Haemophilus influenzae by multiple-stage tandem mass spectrometry. Methods Mol Biol 600:79–92
11. Sisu E, Flangea C, Serb A, Zamfir AD (2011) Modern developments in mass spectrometry of chondroitin and dermatan sulfate glycosaminoglycans. Amino Acids 41: 235–256

12. Amon S, Zamfir AD, Rizzi A (2008) Glycosylation analysis of glycoproteins and proteoglycans using capillary electrophoresis-mass spectrometry strategies. Electrophoresis 29:2485–2507
13. Zaia J (2009) On-line separations combined with MS for analysis of glycosaminoglycans. Mass Spectrom Rev 28:254–272
14. Shin I, Zamfir AD, Ye B (2008) Protein carbohydrate analysis: gel-based staining, liquid chromatography, mass spectrometry, and microarray screening. Methods Mol Biol 441:19–39
15. Zamfir AD, Peter-Katalinić J (2004) Capillary electrophoresis-mass spectrometry for glycoscreening in biomedical research. Electrophoresis 25:1949–1963
16. Zamfir AD (2007) Recent advances in sheathless interfacing of capillary electrophoresis and electrospray ionization mass spectrometry. J Chromatogr A 1159:2–13
17. Martin GB, Mansion F, Servais AC, Debrus B, Rozet E, Hubert P, Crommen J, Fillet M (2009) CE-MS method development for peptides analysis, especially hepcidin, an iron metabolism marker. Electrophoresis 30:2624–2631
18. Maxwell EJ, Zhong X, Zhang H, van Zeijl N, Chen DD (2010) Decoupling CE and ESI for a more robust interface with MS. Electrophoresis 31:1130–1137
19. Lau SS, Stainforth NM, Watson DG, Skellern GG, Wren SA, Tettey JN (2008) CE hydrogen deuterium exchange-MS in peptide analysis. Electrophoresis 29:393–400
20. Haselberg R, Ratnayake CK, de Jong GJ, Somsen GW (2010) Performance of a sheathless porous tip sprayer for capillary electrophoresis-electrospray ionization-mass spectrometry of intact proteins. J Chromatogr A 1217:7605–7611
21. Zamfir AD, Dinca N, Sisu E, Peter-Katalinić J (2006) Copper-coated microsprayer interface for on-line sheathless capillary electrophoresis electrospray mass spectrometry of carbohydrates. J Sep Sci 29:414–422
22. Bindila L, Peter-Katalinić J, Zamfir AD (2005) Sheathless reverse-polarity capillary electrophoresis-electrospray mass spectrometry for analysis of underivatized glycoconjugates. Electrophoresis 26:1488–1499
23. Zamfir AD, Seidler DG, Schönherr E, Kresse H, Peter-Katalinić J (2004) On-line sheathless capillary electrophoresis/nanoelectrospray ionization-tandem mass spectrometry for the analysis of glycosaminoglycan oligosaccharides. Electrophoresis 25:2010–2016
24. Zamfir AD, Peter-Katalinić J (2001) Glycoscreening by on-line sheathless capillary electrophoresis/electrospray ionization-quadrupole time of flight-tandem mass spectrometry. Electrophoresis 22:2448–2457
25. Shi LH, Jin YX, Moon DC, Kim SK, Park SR (2009) A sheathless CE/ESI-MS interface with an ionophore membrane-packed electroconduction channel. Electrophoresis 30:1661–1669
26. Seidler DG, Peter-Katalinić J, Zamfir AD (2007) Galactosaminoglycan function and oligosaccharide structure determination. Scientific World Journal 19:233–241
27. Zamfir AD, Seidler DG, Kresse H, Peter-Katalinić J (2002) Structural characterization of chondroitin/dermatan sulfate oligosaccharides from bovine aorta by capillary electrophoresis and electrospray ionization quadrupole time-of-flight tandem mass spectrometry. Rapid Commun Mass Spectrom 16:2015–2024
28. Zamfir AD, Seidler DG, Kresse H, Peter-Katalinić J (2003) Structural investigation of chondroitin/dermatan sulfate oligosaccharides from human skin fibroblast decorin. Glycobiology 13:733–742
29. Zamfir AD, König S, Althoff J, Peter-Katalinić J (2000) Capillary electrophoresis and off-line capillary electrophoresis-electrospray ionization quadrupole time-of-flight tandem mass spectrometry of carbohydrates. J Chromatogr A 895:291–299
30. Zamfir AD, Flangea C, Sisu E, Serb AF, Dinca N, Bruckner P, Seidler DG (2009) Analysis of novel over- and under-sulfated glycosaminoglycan sequences by enzyme cleavage and multiple stage MS. Proteomics 9:3435–3444
31. Flangea C, Schiopu C, Sisu E, Serb A, Przybylski M, Seidler DG, Zamfir AD (2009) Determination of sulfation pattern in brain glycosaminoglycans by chip-based electrospray ionization ion trap mass spectrometry. Anal Bioanal Chem 395:2489–2498
32. Hoffman RS, Nelson LS, Howland MA, Lewin NA, Flomenbaum NE, Goldfrank LR (2007) Goldfrank's manual of toxicologic emergencies. McGraw-Hill, New York, p 1333
33. Roblin I, Urban M, Flicoteau D, Martin C, Pradeau D (2006) Topical treatment of experimental hydrofluoric acid skin burns by 2.5% calcium gluconate. J Burn Care Res 27:889–894
34. Matsuno K (1996) The treatment of hydrofluoric acid burns. Occup Med 46:313–317

Chapter 12

Structural Separations by Ion Mobility-MS for Glycomics and Glycoproteomics

Larissa S. Fenn and John A. McLean

Abstract

This chapter describes the utility of ion mobility-mass spectrometry (IM-MS) for the detection and characterization of glycoproteins and associated glycoconjugates. IM-MS provides separations in two dimensions; one on the basis of molecular surface area or structure, and the other on molecular mass which creates the ability to differentiate biomolecular classes and isobaric species. When applied to the characterization of glycoproteins, IM-MS separates peptides from the associated glycans in the same digest without purification, and can also be used to separate different isomeric glycans which is a significant challenge in current glycomic studies. The chapter details the methodologies to use IM-MS for the study of glycans and glycoproteins for an audience ranging from new and potential practitioners to those already utilizing the technique.

Key words: Ion mobility, Ion mobility-mass spectrometry, IM-MS, Structural separations, MALDI, IM-MS/MS, Glycomics, Glycoproteomics

1. Introduction

Glycomics has progressed into a critical area of study due to the implications of carbohydrate participation in many biological functions, and variations in glycosylation being associated with many disease states (1–3). Protein glycosylation is one of the more intricate forms of post-translational modification (PTM) and is estimated to be present on over 50% of eukaryotic proteins (4). Glycoproteins have vital functions inside various organisms, and their associated glycans assist in the structure, function, and stability of proteins. Glycoproteins are involved in many important biological functions (e.g., embryonic development and the recognition of hormones, toxins, and other signals on the cell surface) and

Jennifer J. Kohler and Steven M. Patrie (eds.), *Mass Spectrometry of Glycoproteins: Methods and Protocols*, Methods in Molecular Biology, vol. 951, DOI 10.1007/978-1-62703-146-2_12, © Springer Science+Business Media, LLC 2013

processes (e.g., coordination of immune function, cell division, and protein regulation and interactions) (5). With all of these important tasks of glycosylation, detrimental effects may occur from variation or defects in glycosylation patterns. Several disease states such as Alzheimer's disease, HIV, cancer, and diabetes have characteristic defects in glycosylation patterns or unique glycoproteins associated with the disease (3). Further information about the functions of glycans and glycoconjugates can be explored in several excellent texts (6, 7) along with other chapters in this book. Overall, the function of carbohydrates are derived from their composition and structure necessitating rapid and efficient structural determination from complex mixtures, including glycoconjugates such as glycoproteins and glycolipids.

Collectively these challenges motivate the development of higher-throughput, more accurate, and minimal sample manipulation strategies for carbohydrate structure elucidation. Recently, 2D ion mobility-mass spectrometry (IM-MS) has been applied to the field of biological analysis (8–10). Ion mobility (IM) separates ions based on their apparent surface area or ion-neutral collision cross section (11). When merged with MS, IM can separate gas-phase ions in one dimension based on their structure, and a second dimension related to their mass to charge (*m/z*). The advantages provided by IM-MS would likely be of great utility in the field of glycoproteomics.

This chapter focuses on using IM-MS technologies for the study of carbohydrates and glycoproteins in the pursuit of combining omics (e.g., simultaneous glycomics, proteomics, lipidomics, etc.). Identification and conformational characterization of glycoproteins is pursued through studies of carbohydrate standards and separation of glycoprotein digests provided by the structural dimension of IM-MS. In this introductory chapter, IM-MS structural characterization is summarized along with the theoretical background and instrumentation. The following sections describe an overview of IM-MS instrumentation, the theory of IM separations, different types of IM separations, and data interpretation in IM-MS conformation space. Previous studies of carbohydrates and glycoproteins using IM-MS, and methods for the characterization of glycoproteins using IM-MS are also discussed.

1.1. Ion Mobility Applications to the Life Sciences

Ion mobility (IM), which has existed for over a century, is a well-developed separation technique that has been used extensively in the rapid detection of drugs and warfare agents due to its ease of use, low cost, speed, and sensitivity (12, 13). The coupling of IM to MS was first performed in the early 1960s (14, 15), but the utility of IM-MS for biomolecular separations was not fully realized until combined with soft ionization techniques, such as electrospray ionization (ESI) and matrix-assisted laser desorption/ionization (MALDI) (16, 17) which were not developed until the late

1980s. The first pioneering studies, which used IM-MS to determine peptide and protein structures, were performed in the late 1990s (18–20). Following these early studies, research over the past decade has extended IM-MS techniques to the study of complex biological samples, such as whole cell lysates (21), plasma (22–25), homogenized tissue (21, 26, 27), non-covalent complexes (28–30), or directly from thin tissue sections (31, 32). However, IM-MS was essentially limited to a few laboratories where custom instrumentation was constructed. The recent introduction of commercially available IM-MS instruments, in several forms, has further stimulated the use of IM-MS for life sciences research. The following sections describe an overview of IM-MS separations (Subheading 1.2), IM-MS instrumentation (Subheading 1.3), the theory of IM separations (Subheading 1.4), and data interpretation in 2D IM-MS conformation space (Subheading 1.4). Materials and methods for characterizing carbohydrates and glycoproteins using IM-MS are then detailed (Subheadings 2 and 3).

1.2. Overview of Ion Mobility Separations

Most IM-MS instruments have the same general layout. They are similar to mass spectrometers with the IM region inserted between the source and mass analyzer; hence, IM is a post-ionization separation technique (Fig. 1a). From this general layout, instruments can vary due to the type of IM used, the choice of mass analyzer or ionization source (i.e., ESI, nESI, MALDI), the insertion of a quadrupole for mass selection before IM or MS analysis, etc. There are two main methods for differentiating ions using ion mobility, either separating the ions using space or time. The main focus of this chapter is on time-dispersion, but we also highlight several carbohydrate studies utilizing space-dispersion.

The types of IM that use a time-dispersion of the ions are drift tube and traveling wave ion mobility spectrometers (DTIM and TWIM, respectively). DTIM and TWIM separate ions on the basis of molecular surface area due to interactions with a neutral buffer gas present in the IM drift cell. These interactions are not like high energy ion-neutral gas-phase collisions used in collision-induced dissociation (CID) but are low energy gas-phase elastic collisions akin to the collisions of billiard balls. Ions are injected into the IM drift tube and migrate under the influence of a weak electrostatic field gradient (Fig. 1b) where they interact with the neutral drift gas. This field is electrostatic for drift tube and electrodynamic for traveling wave separations, respectively. Smaller ions have a higher mobility than larger ions which result in shorter drift times vs. longer drift times. While the ions traverse the drift cell, their migration is impeded by collisions with the neutral drift gas, typically helium or nitrogen, to a degree that is proportional to apparent surface area or collision cross section. The actual experimental parameter obtained from IM separations is the ion arrival time distribution (t_{ATD}), or the time between ion injection and ion detection.

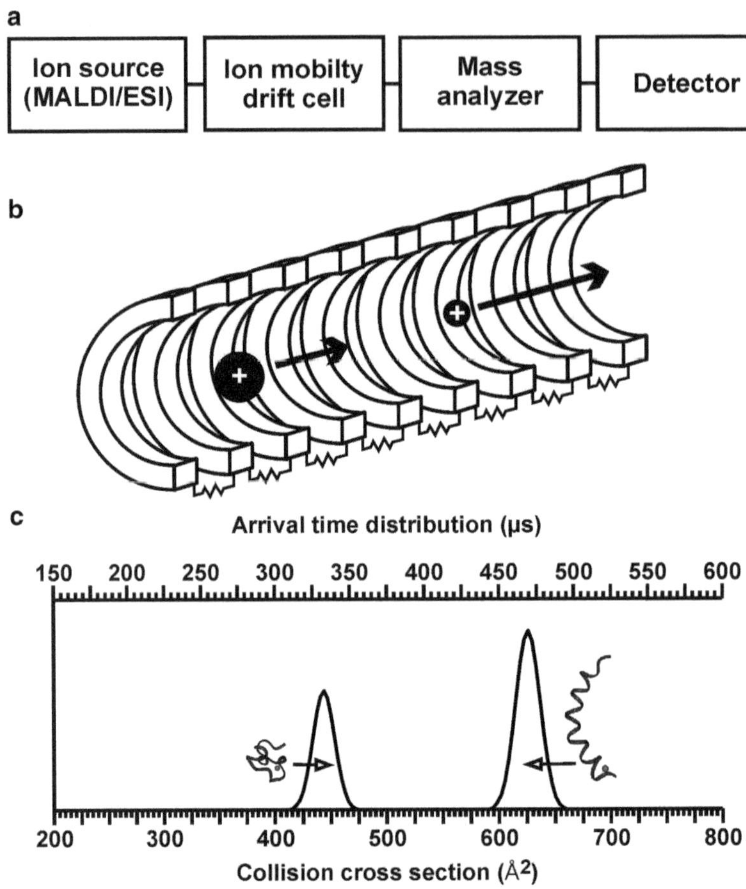

Fig. 1. (**a**) A block diagram of the primary components of biological IM-MS instrumentation. (**b**) A conceptual depiction of an IM drift cell. A stack of ring electrodes are connected via resistors in series to form a voltage divider, which is typically designed to generate a relatively uniform electrostatic field along the axis of ion propagation. Ions of larger apparent surface area experience more collisions with the neutral drift gas and therefore elute slower than ions of smaller apparent surface area. (**c**) A hypothetical IM separation for peptide ions exhibiting two distinct structural subpopulations corresponding to globular (*left*) and to helical (*right*) conformations. The arrival time distribution data (*top axis*), or what is measured, can be transformed to a collision cross-section profile (*bottom axis*) via (4) and described in Subheadings 1.4 and 3.1. Adapted with kind permission from Springer Science+Business Media: (43), p. 906, Fig. 1.

It can be converted to collision cross section or apparent surface area as illustrated in Fig. 1c for DTIM. The difference between the separations for DTIM and TWIM is attributed to their instrumental design which is examined in the next sections.

1.3. IM-MS Instrumentation

There are two main methods of separating ions with IM that have been mentioned, through space or time dispersion. The most common techniques used for separation through space are differential mobility and field asymmetric ion mobility spectrometry (FAIMS) whereas the most common methods for separating ions through time-dispersion is with drift tube or traveling wave IM (DTIM or TWIM, respectively). The methods presented here for the characterization of glycoproteins concentrate on the use of TWIM and DTIM.

1.3.1. Drift Tube Ion Mobility

The first IM instruments utilized a drift tube (33). DTIM-MS has the basic design described previously and depicted in Fig. 1b in that it has a series of stacked ring electrodes that create an electrostatic field to create a forward force that is impeded by collisions with a buffer gas. IM resolution typically ranges from 30 to 50 ($r = t/\Delta t$ at FWHM), whereas longer, cryogenically cooled, or higher pressure drift tubes have been reported with resolutions exceeding 100 (34–36). The drift time can be understood based on the kinetic theory of gases and used to calculate the ions absolute collision cross section without the need for standards (37–40). The calculation of collision cross sections is detailed in Subheadings 1.4 and 3.1. For a derivation of ion-neutral collision cross section theory, the reader is directed to several excellent texts and reviews (11, 41, 42).

1.3.2. Traveling Wave Ion Mobility

TWIM is a newly developed technique in comparison to DTIM. The recent commercial availability of TWIM instrumentation (Waters, Corp.) has made IM-MS accessible to the glycobiology community, not just those labs capable of building the instrumentation. Similar to drift tube instruments, TWIM separates ions by time dispersion through collisions with a background buffer gas, but in contrast, it uses electrodynamic fields rather than electrostatic fields (43, 44). This is accomplished by transmitting voltage pulses sequentially across a stack of ring electrodes (similar to Fig. 1b), which creates the traveling wave (45). Conceptually, TWIM separations are performed based on the susceptibility of different ions to the influence of the specific wave characteristics and have been described as the ability of ions to "surf" on waves (44). Since traveling wave separations utilize dynamic electric fields, presently TWIM measurements can only provide estimated collision cross sections based on internal standards from DTIM absolute collision cross sections (46, 47).

The first commercial platform (Synapt HDMS, now referred to as G1) comprises an interchangeable ESI and MALDI source, a mass resolving quadrupole, a trapping region for injecting pulses of ions into the TWIM, the TWIM drift cell, an ion transfer region, and an orthogonal TOFMS ($r = m/\Delta m$ at FWHM of >17,500). Adjustable wave parameters include: traveling wave pulse height, wave velocity, and ramping either of these variables. CID can be performed in the regions before and after the TWIM drift cell (see Note 1). Generally resolution in the TWIM of the G1 is <15, but this is sufficient for the separation of many molecular classes of interest. For example, TWIM has been used to separate biomolecular signals from complex samples (48) and to study the structure of peptides following CID in the trapping region (49). Recently, the Synapt G2 HDMS was released which has improved TWIM resolution (≥40) and improved mass resolution (≥50,000). It also can be easily interfaced with many different ionization sources and combined with other separation techniques prior to ionization (high-performance liquid chromatography (HPLC), ultra-performance liquid chromatography (UPLC), etc.)

1.4. Ion Mobility Theory: Converting Drift Time to Collision Cross Section

This section details the methodology currently used to determine ion-neutral collision cross sections from data acquired with uniform electrostatic field DTIM. For estimating collision cross sections from TWIM data, see procedures described elsewhere (46, 47). Directions for implementing these measurements and equations experimentally are detailed in Subheading 2.

In order to calculate the collision cross section of an ion, the ion has to traverse the drift cell under the influence of a weak electrostatic field (*E*, ca. 20–30 V/cm/Torr) which provides "low-field" conditions (i.e., constant IM proportionality constant, *K*). The separation of the ions is measured as ion drift velocity which is determined by:

$$v_{\mathrm{d}} = KE \tag{1}$$

In order to standardize the value of *K* for comparison across different instruments, the pressure (*p*, Torr) of the neutral drift gas and the temperature (*T*, Kelvin) of separation must be considered which results in a standard or reduced mobility (K_0):

$$K_0 = K \frac{p}{760} \frac{273}{T} \tag{2}$$

This reduced mobility value is normalized to standard temperature and pressure (i.e., 0°C and 760 Torr) and can be related to the ion-neutral collision cross section through the kinetic theory of gases:

$$K_0 = \frac{(18\pi)^{1/2}}{16} \frac{ze}{(k_{\mathrm{B}}T)^{1/2}} \left[\frac{1}{m_i} + \frac{1}{m_n} \right]^{1/2} \frac{760}{p} \frac{T}{273} \frac{1}{N_0} \frac{1}{\Omega} \tag{3}$$

where *ze* is the charge of the ion, m_i and m_n are the mass of the ion and neutral, respectively, k_{B} is Boltzmann's constant, N_0 is the number density of the drift gas at STP ($2.69 \times 10^{19}/\mathrm{cm}^3$), and Ω the ion-neutral collision cross section. This assumes that the collisions are completely elastic. For an IM drift cell of fixed length (*L*), the drift time (t_{d}) of the packet across the cell can be used to solve for the ion-neutral collision cross section by substituting for K_0 in (3) and rearranging:

$$\Omega = \frac{(18\pi)^{1/2}}{16} \frac{ze}{(k_B T)^{1/2}} \left[\frac{1}{m_i} + \frac{1}{m_n} \right]^{1/2} \frac{t_{\mathrm{d}} E}{L} \frac{760}{p} \frac{T}{273} \frac{1}{N_0} \tag{4}$$

which is the form used to solve for collision cross sections from IM data (see Subheading 3.1). Further, since the collision cross sections follow the hard sphere model, molecular dynamics simulations can be performed to interpret structures consistent with the empirical data (50–53).

1.5. Data Interpretation for Glycomics in Conformation Space

Typical data for an IM-MS experiment is acquired in three dimensions; *m/z*, IM arrival time distribution, and relative abundance of the signal. However, to analyze the data and centroid the peaks, the data is presented in a 2D plot with intensity being represented by false coloring. This conversion from 3D figure to 2D plot is presented in Fig. 2a, b for the separation of glycans and peptides from a glycoprotein digest. We refer to the 2D IM-MS plot as

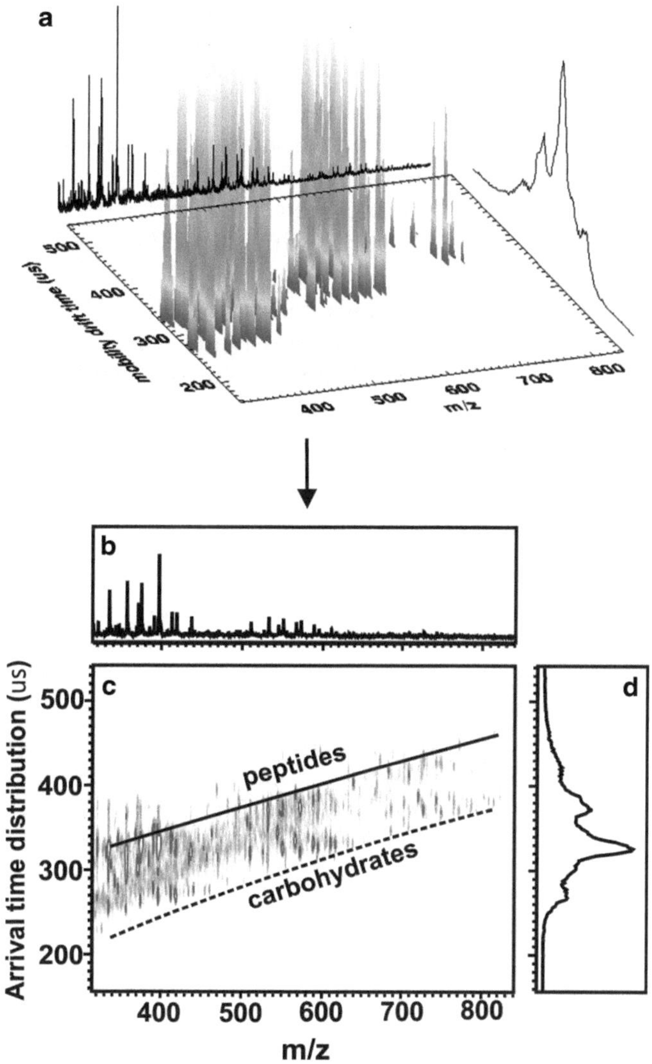

Fig. 2. (**a**) A 3D representation of IM-MS data obtained for human glycoprotein (HGP) digested with Pronase. (**b**) A 2D IM-MS conformation space plot for the analysis of the HGP digest. This data illustrates the variation of gas-phase packing efficiencies for different types of biomolecules. Even though the glycans may still have amino acids attached, a clear differentiation between the peptides and glycans can be noted. (**c**) An integrated mass spectrum over all arrival time distributions. (**d**) The integrated arrival time distribution over the full mass range which would be obtained if a detector was placed after the IM drift cell.

conformation space because it represents biomolecular structure, or conformation, as a function of *m/z* (see Note 2). An integrated mass spectrum over all arrival time distributions is shown in Fig. 2c, which is what would be observed in the absence of IM. An integrated IM arrival time distribution is illustrated by the curve of Fig. 2d which would be obtained by placing the detector directly after the IM drift cell. By plotting the data in 2D conformation space two distinct correlations are observed, one for peptides and one for carbohydrates, respectively. Note that either extracted mass spectra or arrival time distributions can be derived from conformation space data.

One of the main challenges in glycomics is the high probability of carbohydrates and glycans with different structures having the same mass, therefore being isobaric. When using MS alone, these isobaric molecules cannot be differentiated by the intact mass. However, with the addition of the structural separations of IM, some isobaric carbohydrates can be differentiated. This has been demonstrated with DTIM (54) and TWIM (55) and is shown in Fig. 3. In this figure, three pairs of isobaric structural and positional isomers were separated using the additional dimension of IM with MS.

In parallel with the separation of different carbohydrates, the structural separations provided by IM can also be used to differentiate isobaric species belonging to different biomolecular classes (Fig. 4). Although biomolecules are generally composed of a limited combination of elements (e.g., C, O, H, N, S, and P), different biomolecular classes preferentially adopt structures at a given *m/z* correspondent to the prevailing intermolecular folding forces for that class. A representative plot delineating regions of conformation space for which different biomolecular classes (e.g., nucleotides, carbohydrates, peptides, lipids) are predicted to occur is presented in Fig. 4a. These separations are a result of the different gas-phase packing efficiencies of the different classes (nucleotides > carbohydrates > peptides > lipids) (56). This plot is reinforced through the calculation of collision cross sections for standards of each biomolecular class (Fig. 4b). The separation of different biomolecular classes can be utilized in glycoproteomics through the ability to identify peptides and glycans present in a complex sample simultaneously which is further discussed in Subheading 2.

1.6. IM-MS for the Characterization of Glycans and Glycoproteins

Although the use of MS to characterize glycoproteins has been performed extensively for many years, the use of IM-MS for the characterization of carbohydrates and glycoproteins has only recently become increasingly prominent. This again is attributed to the limitation of IM-MS instrumentation to those labs which could construct it. Most early IM-MS studies concentrated on peptides and proteins. However there were a few experiments evaluating

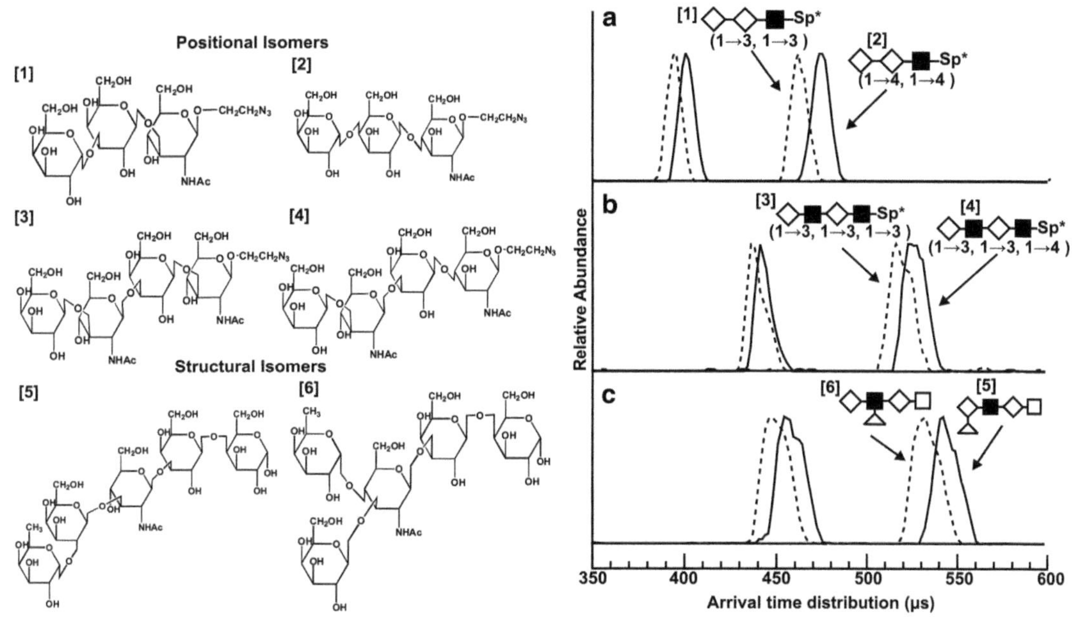

Fig. 3. Structures of the isobaric sets of positional and structural isomers (*left*) and the associated drift time profiles (*right*). (*Left*) Note the difference in structures between glycans 1 and 2 are two 1 ⟶ 3 glycosidic linkages being replaced with two 1 ⟶ 4 linkages. Glycans 3 and 4 have one linkage variation, and glycans 5 and 6 (LNFP1 and LNFP2) vary in the location of fucose from galactose to *N*-acetylglucosamine. (*Right*) Drift time profiles at an electrostatic field strength of 20.6 V/cm/Torr within the ion mobility drift cell. Structures of the oligosaccharides are replaced with shape representations. Drift times are related to the ion structure in that larger, more elongated ions experience more collisions with the neutral buffer gas present in the drift cell causing a longer drift time than more compact structures. (**a**) In the comparison between glycans 1 (*dotted line*) and 2 (*solid line*), the 1 ⟶ 3 linkages of glycan 1 cause it to have a shorter drift time which indicates a more compact structure that glycan 2, which is more elongated. (**b**) Glycans 3 (*dotted line*) and 4 (*solid line*) have differing drift times due to the change in one glycosidic linkage. The 1 ⟶ 3 linkages allow glycan 3 to adopt a more compact conformation when compared to its positional isomer, which has one 1 ⟶ 4 linkage. (**c**) Drift time profiles for glycans 5 (*solid line*) and 6 (*dotted line*) are compared. LNFP2 has a shorter drift time than LNFP1 at both voltages. This is attributed to increased branching in LNFP2 that allows the glycan to adopt a more compact structure. Individual monosaccharide representations are as follows: *open diamond*—galactose; *filled square*—*N*-acetylglucosamine; *open triangle*—fucose; with linkage information for the positional isomers provided in the *parenthesis* below each representation (54). Reproduced by permission of the PCCP Owner Societies.

the use of IM-MS for glycomic studies. The first carbohydrate analyses conducted in the late 1990s aimed at examining short linear polysaccharides and cyclodextrins using DTIM-MS and comparing their collision cross sections to those obtained from molecular dynamic simulations (57). These studies also investigated at the interaction of carbohydrates with Na^+ along with the resulting effect of metal coordination on the overall carbohydrate structure. Additional studies examined ways to enhance the ionization and improve sensitivity for oligosaccharides in ESI-IM-MS instruments by utilizing an ion trap interface and different injection energies (58, 59). These were followed by studies to examine variations in conformation of hexose complexes with zinc ligands (60) through collision cross section determinations and theoretical computational interpretation.

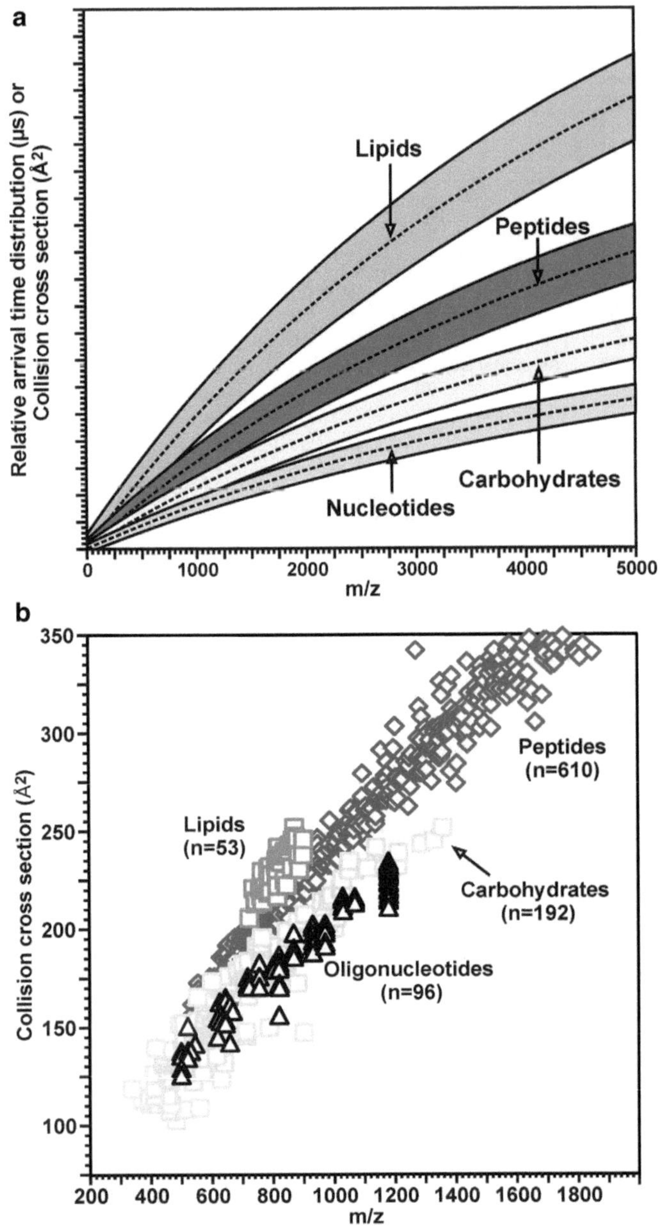

Fig. 4. (**a**) A hypothetical plot illustrating the differences in IM-MS conformation space for different molecular classes based on different gas-phase packing efficiencies. (**b**) A plot of collision cross section as a function of *m/z* for different biologically relevant molecular classes, including: oligonucleotides ($n=96$), carbohydrates ($n=192$), peptides ($n=610$), and lipids (53). All species correspond to singly-charged ions generated by using matrix-assisted laser desorption/ionization (MALDI), where error $\pm 1\sigma$ is generally within the data point. Values for peptides species are from ref. (40). (**a**) Adapted with kind permission from Springer Science + Business Media: (43), p. 906, Fig. 2(a). (**b**) Adapted with kind permission from Springer Science + Business Media: (56), p. 235, Fig. 1(a).

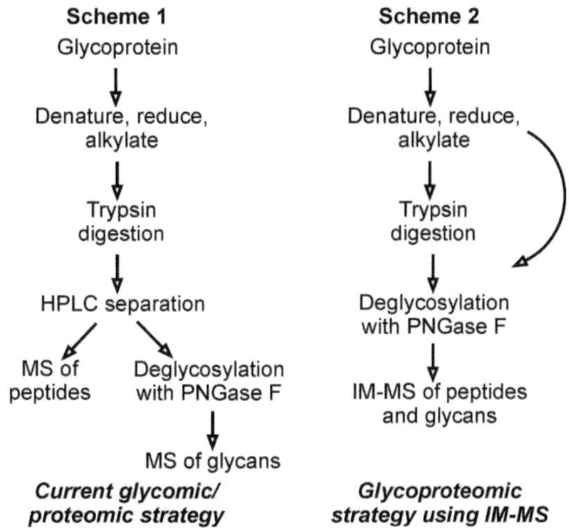

Fig. 5. Schemes 1 and 2 present a comparison of current glycomic and proteomic protocols vs. the glycoproteomic strategy using IM-MS (88). Reproduced by permission of The Royal Society of Chemistry (RSC).

More recent studies have centered on structurally differentiating and determining stereochemical information about monomeric or small di- and trisaccharide structures using DTIM (54, 55, 61–64), TWIM (55, 65, 66), and FAIMS (67). N-linked and O-linked glycans removed from glycoproteins have also been characterized by IM-MS from purified samples (68, 69) after separation and extensive purification from serum (22) or urine (48). In addition, sulfated glycans were resolved by IM-MS through the differentiation of isomers (70), interpretation of collision cross sections with molecular modeling (71), and interactions with defensin-inspired peptides (72). N-glycan structure and glycosylation sites for IgG have been determined using IM-MS/MS (68), and intact glycosylated IgG antibodies have also been analyzed using IM-MS to differentiate IgG1 and two different isoforms of IgG2 (73). Most of these studies removed the glycans from the glycoprotein and purified prior to the carbohydrates prior to analysis which is similar to other contemporary MS methodologies for the characterization of glycoproteins which analyze the glycans, peptides, or glycopeptides separately. In this work, we focus on the rapid characterization of biomolecules in complex samples without time-consuming purification steps before analysis.

1.6.1. Simultaneous Glycoproteomics Using IM-MS

The use of IM combined with MS allows for the simultaneous detection of different biomolecular classes (i.e., lipids, peptides, carbohydrates, and oligonucleotides) with little or no purification needed (Fig. 5). In these methods, we concentrate on the characterization of carbohydrate standards using DTIM-MS for the determination of the collision cross sectional area and region of 2D IM-MS space occupied by carbohydrates when compared to other

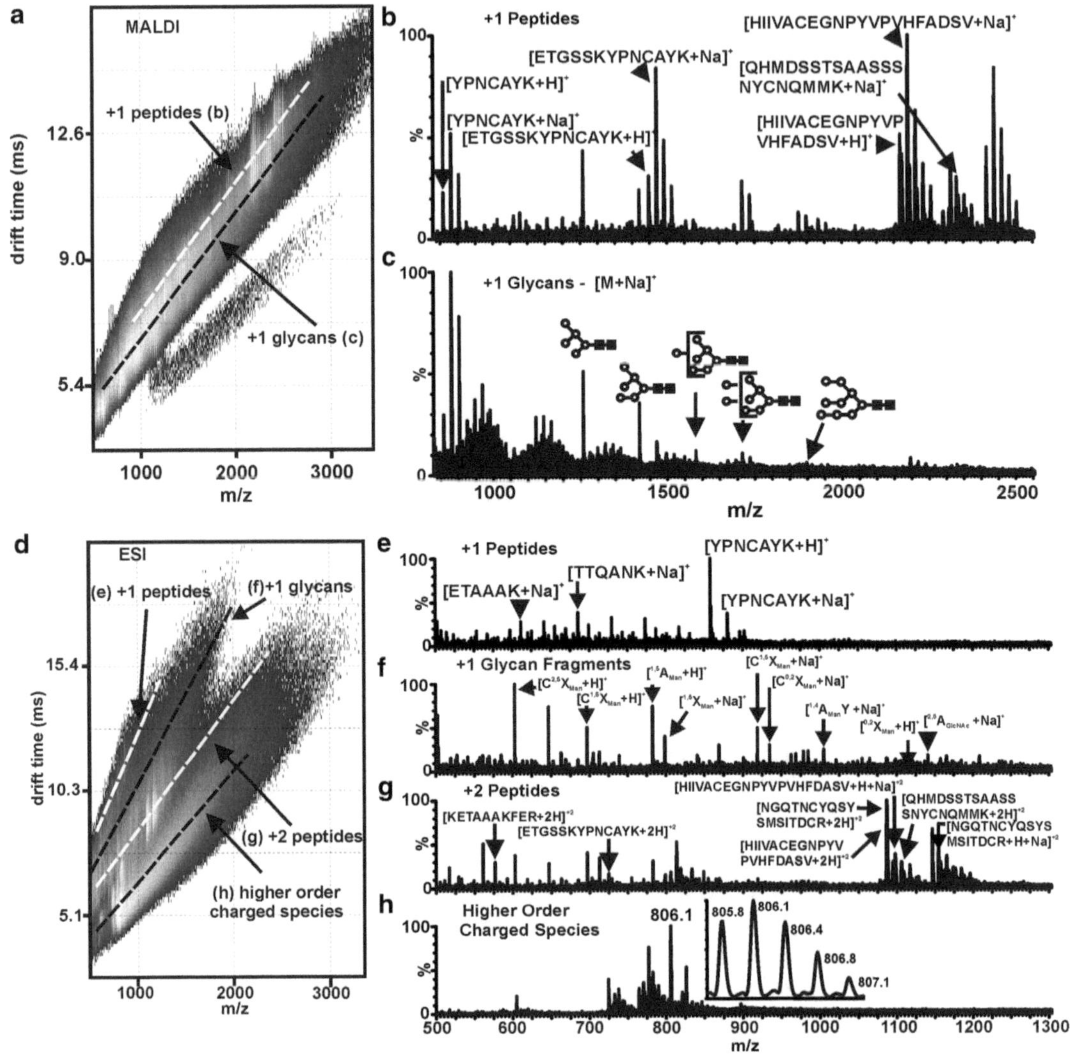

Fig. 6 MALDI- and ESI-IM-MS plot and extracted mass spectra from RNAse B digested and deglycosylated with trypsin and PNGase F, respectively. (**a**) A 2D MALDI-IM-MS plot of conformation space. Structural separations are observed for peptides (labeled (b)) and glycans (labeled (c)). Since MALDI is used, all identified peaks correspond to singly-charged species as sodium-coordinated glycans and protonated peptides. (**b**) An extracted mass spectrum corresponding to peptides (along *white dashed-line* in (**a**)). (**c**) An extracted mass spectrum corresponding to glycans (along *black dashed-line* in (**a**)). *Open circles* and *filled boxes* correspond to mannose and *N*-acetylglucosamine, respectively. Unidentified peaks seen at lower masses are due to in-source fragmentation of the glycans present. (**d**) A 2D ESI-IM-MS plot of conformation space. Structural separations are observed for singly-charged peptides (labeled (e)), singly-charged glycans (labeled (f)), doubly-charged peptides (labeled (g)), and higher order charged species (labeled (h)). (**e**) An extracted mass spectrum corresponding to singly-charged peptides (along *top white dashed-line* in (**d**)). (**f**) An extracted mass spectrum corresponding to singly-charged glycans with identification of fragments by Domon and Costello nomenclature (along *top black dashed-line* in (**d**)). (**g**) An extracted mass spectrum corresponding to doubly-charged peptides (along *bottom white dashed-line* in (**d**)). (**h**) An extracted mass spectrum corresponding to higher order charged species (along *bottom black dashed-line* in (**d**)). The *inset* illustrates the isotopic pattern for a triply-charged analyte (88). Reproduced by permission of The Royal Society of Chemistry (RSC).

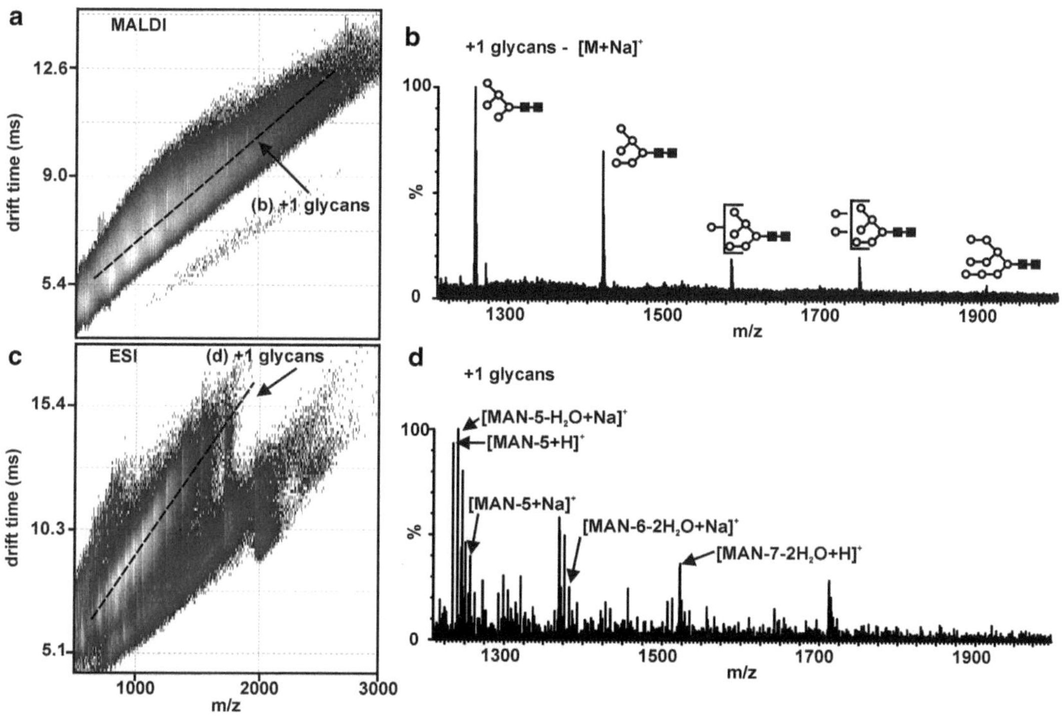

Fig. 7. Plots and extracted mass spectra from intact RNAse B that has been deglycosylated with PNGase F and analyzed using MALDI-IM-MS (**a**, **b**) and ESI-IM-MS (**c**, **d**). Note that the protein was not proteolytically digested and remained intact (M_r = 13,700 Da). (**a**) A 2D MALDI-IM-MS plot of conformation space. Structural separations are observed for singly-charged glycans (labeled (b)) which are then compared to those identified in Fig. 6. (**b**) An extracted mass spectrum corresponding to singly-charged glycans (along *dashed-line* in (**a**)). (**c**) A 2D ESI-IM-MS plot of conformation space. Structural separations are noted for singly-charged glycans (labeled (d)) which are then compared to those identified in Fig. 6. (**d**) An extracted mass spectrum corresponding to singly-charged glycans (along *dashed-line* in (**c**)) (88). Reproduced by permission of The Royal Society of Chemistry (RSC).

biomolecules such as lipids, peptides, and oligonucleotides. We then describe the simultaneous separation and characterization of peptides and glycans from glycoprotein digests (Figs. 6 and 7) and glycans from lipids in a human milk sample without the need for extensive purification (Fig. 8).

In Fig. 6, the simultaneous characterization of glycans and peptides in a digest of RNAse B using ESI and MALDI is presented. The protein is first digested with trypsin and subsequently with PNGase F to produce the peptides and glycans. In Fig. 7, the confidence in the identifications can be increased through performing only the deglycosylation of RNAse B to analyze glycans only in the mass range of interest. In both of these figures, the peptides and glycans are ionized using both ESI and MALDI. There are several differences between ESI and MALDI that would guide an investigator to select one source vs. the other. Firstly, ESI generally produces ions of multiple charge states (e.g., $[M+nH]^{n+}$)

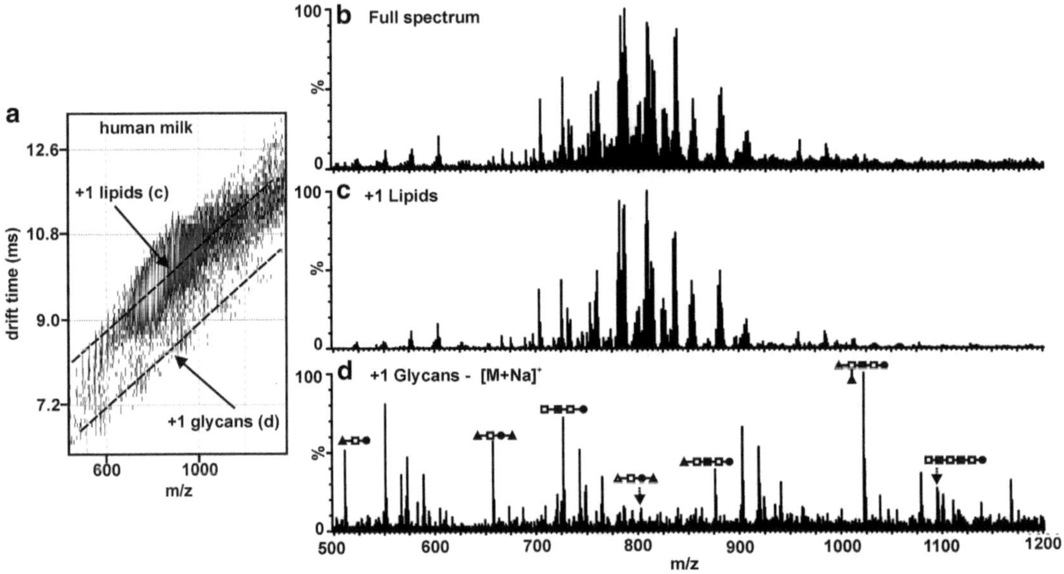

Fig. 8. MALDI-IM-MS plot and extracted mass spectra from human milk with no prior purification. (**a**) A 2D IM-MS plot of conformation space. Structural separations are observed for lipids (labeled (c)) and glycans (labeled (d)). Since MALDI is used, all identified peaks correspond to singly-charged species. (**b**) An integrated mass spectrum for all of conformation space. This is what would be seen if using MS alone to characterize the human milk sample. (**c**) An extracted mass spectrum corresponding to lipids (along *top dashed-line* in (**a**)). (**d**) An extracted mass spectrum corresponding to glycans (along *bottom dashed-line* in (**a**)). Carbohydrate structure representations are as follows: *filled circle*—glucose; *open triangle*—sialic acid; *filled square*—*N*-acetylglucosamine; *open square*—galactose; *filled triangle*—fucose. Adapted from (89).

while MALDI generally produces singly-charged species (e.g., $[M+H]^+$). For fragmentation-based MS/MS studies, multiply-charged species are advantageous; however, by partitioning the signal into multiply-charged channels can result in complicated spectra. Secondly, MALDI is more generally tolerant of salts than is ESI. Thirdly, MALDI is inherently an off-line ionization source, while ESI is more easily coupled with additional separation steps such as liquid chromatography. A more detailed description of the advantages and limitations of MALDI and ESI for glycoproteomics is presented in ref. (88).

1.6.2. IM-MS/MS Measurements for Confident Glycan Identification

Most current MS methodologies for carbohydrate characterization require the use of CID or tandem MS (MS/MS or MS^n) to confirm the carbohydrate sequence and branching patterns (74–77). Similar to traditional MS/MS, structural information can be obtained through the use of IM-MS/MS. The collision cell for fragmentation can be inserted before, or after, the IM drift cell determined by the desired information. An advantage to fragmenting after IM separation is that all fragment ions will have the same drift time as the parent. Commercial IM-MS instruments currently available have the ability to isolate a certain mass through the use of a resolving quadrupole, to fragment ions before or after the IM separation region, and to fragment ions up to MS^5 (45). These abilities are very useful when characterizing complex samples like those encountered in glycomics research. In addition to parent ion selection by mass, the parent ion can be selected by mobility or structure using time dispersion in the drift cell (8).

2. Materials

2.1. Collision Cross Section Measurements for MALDI-IM-MS

1. *Sample for which collision cross section is desired.* This can be a pure compound or within a mixture but must be abundant enough to obtain sufficient signal for five measurements at different voltage settings in the IM dimension. For carbohydrates in Figs. 3 and 4, standards were used at a concentration of 1 mg/mL in DDI water prepared for MALDI analysis. Standards can be purchased from several companies including those that specialize in carbohydrates such as Dextra (Reading, UK) and V-Labs, Inc. (Covington, LA). Many of the carbohydrates used for the presented collision cross section database were obtained from the Consortium for Functional Glycomics.
2. *Drift tube IM standards/calibrants.* Mass standards are species, usually peptides and proteins that span the mass range of interest. For DTIM structural standards, C_{60} and C_{70} (fullerenes) are typically used due to their existence in one structural form. Additionally, fullerenes are sometimes used as mass standards since they are structurally separated from biomolecules in 2D IM-MS space and provide numerous gas-phase reaction products resulting in peaks spanning a large mass range in increments of 24 Da. These standards can also be used to evaluate DTIM resolution and instrument performance.

   Generally, the standard peptide bradykinin (RPPGFSPFR) is used to validate gas pressure in DTIM by comparison of the collision cross section measurement with the accepted value of 242 ± 2 Å^2 (20). Bradykinin can be mixed with matrix of choice or a standard solution of 1 mg/mL in H_2O can be combined 1:1 v/v with 20 mg/mL α-cyano-4-hydroxycinnamic acid in 50% methanol.
3. *Traveling wave IM standards/calibrants.* As discussed in Subheading 1.3, TWIM provides relative collision cross sections therefore requiring internal standards with corresponding DTIM obtained absolute collision cross section values. Published absolute collision cross sections can be obtained from several databases, including: (a) peptide collision cross sections determined by ESI (78, 79), (b) intact protein collision cross sections determined by ESI (80), (c) peptide collision cross sections determined by MALDI (40), and (d) biologically

relevant carbohydrate, lipid, and oligonucleotide collision cross sections determined by MALDI (56). For these comparative measurements, it is necessary to have standards in the same biomolecular class as the samples being measured (81).

2.2. Simultaneous Glycomics and Proteomics Using IM-MS

1. *Purified glycoprotein containing N-linked glycans or glycan-containing sample of interest* (i.e., human milk).
2. *Drift tube IM standards/calibrants. See* Subheading 2.1 for details.
3. *Peptide-N4-(acetyl-b-glucosaminyl)-asparagine amidase F* (PNGase F), usually from *Chryseobacterium meningosepticum*, for the removal of N-linked glycans can be obtained through Prozyme Glyko, Calbiochem, or other vendors.

3. Methods

3.1. Performing Collision Cross Section Measurements Using MALDI-DTIM-MS

1. In order to take measurements, the samples should be prepared for MALDI analysis. In these experiments, the 1 mg/mL carbohydrate standards were combined 1:1 by volume (200:1 molar ratio) with saturated 2,5-dihydroxybenzoic acid (DHB) in 50% ethanol:DDI water.
2. Following insertion of the sample target into the instrument, mass and IM standard/calibrants are measured. In particular, in using MALDI-IM-MS methods, the laser pulse serves as the start signal (t_0) for measuring the IM arrival time distribution (t_{ATD}). These time distinctions are necessary for the calculations in step 4.
3. Following structural separation in the IM drift cell filled with an inert gas (usually 1–10 Torr of He or N_2; see Note 3) ions are directed through a skimming and differential pumping region where the pressure is reduced from 1 to 10 Torr to ~10^{-8} Torr for mass analysis in the orthogonal TOFMS. The stop time for t_{ATD} corresponds to the ion injection time for the TOFMS measurement.
4. To perform the collision cross section calculations as described in Subheading 1.3 (e.g., (4)), the arrival time distribution must be corrected for time spent in regions outside of the drift cell (i.e., time spent traversing from the MALDI plate into the drift cell, in skimming and differential pumping regions, and ion optic regions prior to insertion into the TOFMS). This will result in the drift time (t_d) of the ions within the IM drift cell used in the calculation of collision cross section:

$$t_d = t_{ATD} - t_{dtc} \quad (5)$$

To determine the value of t_{dtc}, IM separations are performed by varying the voltage across the drift cell while maintaining all other experimental parameters constant. The t_{ATD} measured at each drift voltage and are then plotted vs. the inverse of drift voltage ($1/V$). Provided the range of voltages used maintains ion separations under low field conditions, this plot will result in a linear correlation. If nonlinearity is observed, a calculation of the low field limit should be performed, because curvature in this plot indicates that mobility is not constant over the voltage range used. A linear regression of this data results in a y-intercept corresponding to t_{dtc} (see Note 4). Preferably at least five voltages should be used to define this line although for high precision measurements as many voltages as is practical should be used.

5. After the t_d has been determined, it can now be used to calculate the collision cross section, Ω, of the ion of interest through Equation (4) (see Notes 5 and 6) (11).
6. After calculation of the collision cross section, the value can be further related to the structure using molecular dynamic simulations. Detailed information about these computational methods can be found in other resources (50–53).
7. For calculating relative collision cross sections using TWIM-MS, the two main procedures used can be found in the literature (46, 47).

3.2. Performing Simultaneous Glycoproteomics Using IM-MS (see Note 7)

1. The N-linked glycan-containing glycoprotein is prepared by making a 1 mg/mL solution in DDI water or 50 mM ammonium acetate at pH ~7.5 (see Note 8). An aliquot (~1 nmol) is pipetted into a microcentrifuge tube.
2. Thermal denaturation is performed by heating the sample at 90°C for 15 min (see Note 9) (82). To quench denaturation, the sample is placed in a –20°C freezer for ~15 min.
3. To reduce disulfide bonds in the protein, dithiothreitol (DTT) is added to make the final concentration 5 mM and reacts at 60°C for 30 min (83) (see Note 10).
4. To alkylate free cysteines, iodoacetamide is added to a final concentration of 50 mM and reacts for 1 h at 37°C in the dark (84) (see Note 10).
5. For proteolytic digestion of the glycoprotein, trypsin is added to the sample (approximately 20:1 weight of substrate per weight of trypsin) and allowed to digest at 37°C for 24 h. This step is omitted for control digests where only glycans are in the mass range of interest. The sample is placed in a –20°C freezer for at least 15 min to stop the enzyme activity (can be stored overnight).

6. For removal of the N-linked glycans of interest, 1 μU (microunit) of PNGase F is added to the sample and incubated at 37°C for at least 12 h (see Note 11). The sample is then placed in a –20°C freezer until analysis.
7. To prepare for MALDI analysis, the glycoprotein digests are combined 1:1 by volume (200:1 M ratio) with saturated DHB in 50% ethanol:DDI water and then spotted on a MALDI plate.
8. In preparation for ESI analysis, the glycoprotein digest was dissolved in 50:50 (v/v) water:methanol to a final concentration of 25 mM (see Note 12).
9. The samples are then analyzed using a Synapt HDMS G1 or G2 (Waters Corp.) equipped with TWIM drift cell and operated with MassLynx software. For both ionization sources, the ion guide T-wave is operated at 300 m/s and linearly ramped in amplitude from 5 to 20 V over each experiment. The transfer guide T-wave is operated at 248 m/s and with a constant 3 V amplitude. Ion injection voltages in the Trap and Transfer were set at 6 and 4 V, respectively.

   For MALDI and ESI, it is advantageous to optimize all source settings for the sample of interest paying particular attention to the optimization of glycan signal. Cone voltage in ESI along with laser energy in MALDI should be tuned to suppress carbohydrate in-source fragmentation.
10. For data analysis, MassLynx software is also used along with Driftscope for the visualization of 2D data. In Driftscope, the regions associated with different biomolecular classes, in particular carbohydrates and peptides, can be selected and extracted in order to identify the peaks associated with each class (see Figs. 6 and 7). In order to increase confidence in identifications, IM-MS/MS can be used (see Subheading 1.6.2).

3.3. Performing Simultaneous Glycolipidomics Using IM-MS

1. Dilute sample with DDI water. For the human milk example presented, the optimal dilution was 1:10 milk to DDI water by volume.
2. Mix diluted sample with matrix as described in Subheading 3.2, step 6. The type of matrix, concentration, or matrix to analyte ratio can be varied to optimize for carbohydrate signal in the complex mixture if desired.
3. Analyze using MALDI-IM-MS for best results using same settings as above (optimized for glycan signal). An example of the plot obtained is presented in Fig. 8.

4. Notes

1. The Synapt G1 HDMS has activation/dissociation regions to perform up to MS^5, but usually MS^3 is the practical maximum.
2. IM-MS 2D data is presented in one of the two ways with either *m/z* on the abscissa and arrival time distribution on the ordinate axes (in which all the data in this work is presented) or the reverse.
3. Our DTIM drift cells use He due to its low polarizability and low mass relative to other inert gases. However, most TWIM drift cells utilize N_2, and other drift gases or drift gas additives can be used to promote interactions between the ion and drift gas.
4. The plot of arrival time distribution vs. $1/V$ has a *y*-intercept that correlates to t_{dtc} or drift time correction because it represents the limit of $t_d \rightarrow 0$ at infinite drift cell voltage.
5. When high accuracy collision cross section measurements are needed, the drift time correction should be evaluated for each species. This is due to correcting for the additional ion-neutral collisions in the differential pumping regions at the exit and/or entrance of the IM drift cell.
6. When calculating collision cross section, much care should be taken in the dimensionality of the units used. This is due to the equation being derived from classical electrodynamics, and the units for *E* should be expressed in cgs Gaussian units, i.e., statvolts per centimeter, where 1 statvolt equals 299.79 V. Note that statvolts per centimeter is equivalent to statcoulombs per square centimeter and that elementary charge, *e*, is 4.80×10^{-10} statcoulombs.
7. Simultaneous glycoproteomics methodologies are focused on the characterization of N-linked glycans. However, these experiments can be adapted for the characterization of O-linked glycans and glycoproteins treated with other enzymes such as Pronase.
8. There has been some debate against using ammonium-containing buffers (85). However, for these studies, ammonium acetate was used and acceptable glycan signal was obtained.
9. Other methods of protein denaturation can be used (i.e., chemical (86, 87)) but have not been evaluated here.
10. Reduction and alkylation are performed in typical proteomic digestions and directions are given here if the procedure is desired. However, for the presented examples (Figs. 6 and 7), reduction and alkylation were not performed.

11. One unit is defined as the amount of enzyme that will completely catalyze the release of N-linked oligosaccharides from 1 μmol denatured ribonuclease in 1 min at 37°C, pH 7.5.
12. ESI analysis is performed based on high-throughput methodology (minimum time per sample) and does not utilize derivatization strategies known to enhance ionization and decrease fragmentation of glycans. These can be done if time allows.

Acknowledgements

Financial support for this work was provided by the National Institutes of Health (1R01GM092218-01 and RC2DA028981), the US Defense Threat Reduction Agency (HDTRA-09-1-0013), Vanderbilt University College of Arts and Sciences, Vanderbilt Institute of Chemical Biology, and Vanderbilt Institute for Integrative Biosystems Research and Engineering. We thank Nichole M. Lareau for critical comments and assistance with this manuscript. Richard M. Caprioli (Vanderbilt University, Department of Biochemistry) and the Vanderbilt University Mass Spectrometry Research Center for use of the Synapt HDMS. The carbohydrate compounds, LacNAc, Le^{c}, Lac, B tetra type 1, 2′F-B type 2, LNT, Le^{A}Le^{x}, Di-Le^{c}, Di-Le^{A}, LNnT, Galα3-type1, B2-tri, Pk, P1 tri, Tri-LacNAc, H-type2-LN-LN, P1 penta, P1 antigen, GNLNLN, and 3′SLN-Lec, were provided by the Carbohydrate Synthesis/Protein Expression Core of The Consortium for Functional Glycomics funded by the National Institute of General Medical Sciences grant GM62116.

References

1. Bahl OP (1992) Glycoconjugates: composition, structure, and function. Marcel Dekker, New York, NY
2. Gorelik E, Galili U, Raz A (2001) On the role of cell surface carbohydrates and their binding proteins (lectins) in tumor metastasis. Cancer Metastasis Rev 20:245–277
3. Montreuil J, Vliegenthart JFG, Schachter H (1995) Glycoproteins I. Elsevier Science, New York, NY
4. Apweiler R, Hermjakob H, Sharon N (1999) On the frequency of protein glycosylation, as deduced from analysis of the SWISS-PROT database. Biochim Biophys Acta 1473:4–8
5. Rudd PM, Elliott T, Cresswell P, Wilson IA, Dwek RA (2001) Glycosylation and the immune system. Science 291:2370–2376
6. Taylor ME, Drickamer K (2006) Introduction to glycobiology. Oxford University Press, Oxford, UK
7. Varki A, Cummings RD, Esko JD, Freeze HH, Stanley P, Bertozzi CR, Hart GW, Etzler ME (eds) (2009) Essentials of glycobiology, 2nd edn. Cold Spring Harbor Laboratory Press, Woodbury, NY
8. McLean JA, Ruotolo BT, Gillig KJ, Russell DH (2005) Ion mobility-mass spectrometry: a new paradigm for proteomics. Int J Mass Spectrom 240:301–315

9. Clemmer DE, Jarrold MF (1997) Ion mobility measurements and their applications to clusters and biomolecules. J Mass Spectrom 32:577–592
10. Kanu AB, Dwivedi P, Tam M, Matz L, Hill HH Jr (2008) Ion mobility-mass spectrometry. J Mass Spectrom 43:1–22
11. Mason EA, McDaniel EW (1988) Transport properties of ions in gases. Wiley, Indianapolis, IN
12. Baumbach J (2006) Process analysis using ion mobility spectrometry. Anal Bioanal Chem 384:1059–1070
13. Eiceman GA, Karpas Z (2005) Ion mobility spectrometry, 2nd edn. CRC Press, Boca Raton, FL
14. McAfee KB Jr, Edelson D (1963) Identification and mobility of ions in a Townsend discharge by time-resolved mass spectrometry. Proc Phys Soc Lond 81:382–384
15. Barnes WS, Martin DW, McDaniel EW (1961) Mass spectrographic identification of the ion observed in hydrogen mobility experiments. Phys Rev Lett 6:110–111
16. Gieniec J, Mack LL, Nakamae K, Gupta C, Kumar V, Malcolm D (1984) Electrospray mass spectroscopy of macromolecules: application of an ion-drift spectrometer. Biomed Mass Spectrom 11:259–268
17. von Helden G, Wyttenbach T, Bowers MT (1995) Inclusion of a MALDI ion source in the ion chromatography technique: conformational information on polymer and biomolecular ions. Int J Mass Spectrom Ion Process 146–147:349–364
18. Shelimov KB, Clemmer DE, Hudgins RR, Jarrold MF (1997) Protein structure in vacuo: gas-phase confirmations of BPTI and cytochrome c. J Am Chem Soc 119:2240–2248
19. von Helden G, Wyttenbach T, Bowers MT (1995) Conformation of macromolecules in the gas-phase—use of matrix-assisted laser-desorption methods in ion chromatography. Science 267:1483–1485
20. Wyttenbach T, vonHelden G, Bowers MT (1996) Gas-phase conformation of biological molecules: bradykinin. J Am Chem Soc 118:8355–8364
21. Myung S, Lee YJ, Moon MH, Taraszka J, Sowell R, Koeniger S, Hilderbrand AE, Valentine SJ, Cherbas L, Cherbas P, Kaufmann TC, Miller DF, Mechref Y, Novotny MV, Ewing MA, Sporleder CR, Clemmer DE (2003) Development of high-sensitivity ion trap ion mobility spectrometry time-of-flight techniques: a high-throughput nano-LC-IMS-TOF separation of peptides arising from a Drosophila protein extract. Anal Chem 75:5137–5145
22. Isailovic D, Kurulugama RT, Plasencia MD, Stokes ST, Kyselova Z, Goldman R, Mechref Y, Novotny MV, Clemmer DE (2008) Profiling of human serum glycans associated with liver cancer and cirrhosis by IMS-MS. J Proteome Res 7:1109–1117
23. Liu X, Plasencia M, Ragg S, Valentine SJ, Clemmer DE (2004) Development of high throughput dispersive LC-ion mobility-TOFMS techniques for analysing the human plasma proteome. Brief Funct Genomic Proteomic 3:177–186
24. Liu XY, Valentine SJ, Plasencia MD, Trimpin S, Naylor S, Clemmer DE (2007) Mapping the human plasma proteome by SCX-LC-IMS-MS. J Am Soc Mass Spectrom 18:1249–1264
25. Valentine SJ, Plasencia MD, Liu XY, Krishnan M, Naylor S, Udseth HR, Smith RD, Clemmer DE (2006) Toward plasma proteome profiling with ion mobility-mass spectrometry. J Proteome Res 5:2977–2984
26. Liu XY, Miller BR, Rebec GV, Clemmer DE (2007) Protein expression in the striatum and cortex regions of the brain for a mouse model of Huntington's disease. J Proteome Res 6:3134–3142
27. Taraszka JA, Kurulugama R, Sowell RA, Valentine SJ, Koeniger SL, Arnold RJ, Miller DF, Kaufman TC, Clemmer DE (2005) Mapping the proteome of Drosophila melanogaster: analysis of embryos and adult heads by LC-IMS-MS methods. J Proteome Res 4:1223–1237
28. Benesch JLP, Ruotolo BT, Simmons DA, Robinson CV (2007) Protein complexes in the gas phase: technology for structural genomics and proteomics. Chem Rev 107:3544–3567
29. Ruotolo BT, Giles K, Campuzano I, Sandercock AM, Bateman RH, Robinson CV (2005) Evidence for macromolecular protein rings in the absence of bulk water. Science 310:1658–1661
30. Ruotolo BT, Hyung SJ, Robinson PM, Giles K, Bateman RH, Robinson CV (2007) Ion mobility-mass spectrometry reveals long-lived, unfolded intermediates in the dissociation of protein complexes. Angew Chem Int Ed Engl 46:8001–8004
31. Jackson SN, Ugarov M, Egan T, Post JD, Langlais D, Schultz JA, Woods AS (2007) MALDI-ion mobility-TOFMS imaging of lipids in rat brain tissue. J Mass Spectrom 42:1093–1098
32. McLean JA, Ridenour WB, Caprioli RM (2007) Profiling and imaging of tissues by imaging ion mobility-mass spectrometry. J Mass Spectrom 42:1099–1105

33. Mason EA, McDaniel EW (1988) Transport properties of ions in gases. Wiley, New York, NY, pp 31–102
34. Dugourd P, Hudgins RR, Clemmer DE, Jarrold MF (1997) High-resolution ion mobility measurements. Rev Sci Instrum 68:1122–1129
35. Merenbloom SI, Glaskin RS, Henson ZB, Clemmer DE (2009) High-resolution ion cyclotron mobility spectrometry. Anal Chem 81:1482–1487
36. Wyttenbach T, Kemper PR, Bowers MT (2001) Design of a new electrospray ion mobility mass spectrometer. Int J Mass Spectrom 212:13–23
37. Dwivedi P, Wu P, Klopsch SJ, Puzon GJ, Xun L, Hill HH Jr (2008) Metabolic profiling by ion mobility mass spectrometry (IMMS). Metabolomics 4:63–80
38. Furche F, Ahlrichs R, Weis P, Jacob C, Gilb S, Bierweiler T, Kappes MM (2002) The structures of small gold cluster anions as determined by a combination of ion mobility measurements and density functional calculations. J Chem Phys 117:6982–6990
39. Ruotolo BT, Verbeck GF, Thomson LM, Woods AS, Gillig KJ, Russell DH (2002) Distinguishing between phosphorylated and nonphosphorylated peptides with ion mobility-mass spectrometry. J Proteome Res 1:303–306
40. Tao L, McLean JR, McLean JA, Russell DH (2007) A collision cross-section database of singly-charged peptide ions. J Am Soc Mass Spectrom 18:1232–1238
41. Mason EA (1984) Ion mobility: its role in plasma chromatography. In: Carr TW (ed) Plasma chromatography. Plenum, New York, NY, pp 43–93
42. Revercomb HE, Mason EA (1975) Theory of plasma chromatography/gaseous electrophoresis. Review. Anal Chem 47:970–983
43. Fenn LS, McLean JA (2008) Biomolecular structural separations by ion mobility-mass spectrometry. Anal Bioanal Chem 391:905–909
44. Giles K, Pringle SD, Worthington KR, Little D, Wildgoose JL, Bateman RH (2004) Applications of a travelling wave-based radio-frequency-only stacked ring ion guide. Rapid Commun Mass Spectrom 18:2401–2414
45. Pringle SD, Giles K, Wildgoose JL, Williams JP, Slade SE, Thalassinos K, Bateman RH, Bowers MT, Scrivens JH (2007) An investigation of the mobility separation of some peptide and protein ions using a new hybrid quadrupole/travelling wave IMS/oa-ToF instrument. Int J Mass Spectrom 261:1–12
46. Ruotolo BT, Benesch JLP, Sandercock AM, Hyung S-J, Robinson CV (2008) Ion mobility-mass spectrometry analysis of large protein complexes. Nat Protoc 3:1139–1152
47. Williams JP, Scrivens JH (2008) Coupling desorption electrospray ionisation and neutral desorption/extractive electrospray ionisation with a travelling-wave based ion mobility mass spectrometer for the analysis of drugs. Rapid Commun Mass Spectrom 22:187–196
48. Vakhrushev SY, Langridge J, Campuzano I, Hughes C, Peter-Katalinic J (2008) Ion mobility mass spectrometry analysis of human glycourinome. Anal Chem 80:2506–2513
49. Riba-Garcia I, Giles K, Bateman RH, Gaskell SJ (2008) Evidence for structural variants of a- and b-type peptide fragment ions using combined ion mobility/mass spectrometry. J Am Soc Mass Spectrom 19:609–613
50. Gidden J, Bowers MT (2003) Gas-phase conformations of deprotonated trinucleotides (dGTT(−), dTGT(−), and dTTG(−)): the question of zwitterion formation. J Am Soc Mass Spectrom 14:161–170
51. Gidden J, Bowers MT (2003) Gas-phase conformations of deprotonated and protonated mononucleotides determined by ion mobility and theoretical modeling. J Phys Chem B 107:12829–12837
52. Shvartsburg AA, Jarrold MF (1996) An exact hard-spheres scattering model for the mobilities of polyatomic ions. Chem Phys Lett 261:86–91
53. Wyttenbach T, Witt M, Bowers MT (2000) On the stability of amino acid zwitterions in the gas phase: the influence of derivatization, proton affinity, and alkali ion addition. J Am Chem Soc 122:3458–3464
54. Fenn LS, McLean JA (2011) Structural resolution of carbohydrate positional and structural isomers based on gas-phase ion mobility-mass spectrometry. Phys Chem Chem Phys 13:2196–2205
55. Williams JP, Grabenauer M, Holland RJ, Carpenter CJ, Wormald MR, Giles K, Harvey DJ, Bateman RH, Scrivens JH, Bowers MT (2010) Characterization of simple isomeric oligosaccharides and the rapid separation of glycan mixtures by ion mobility mass spectrometry. Int J Mass Spectrom 298(1–3): 119–127
56. Fenn LS, Kliman M, Mahsut A, Zhao SR, McLean JA (2009) Characterizing ion mobility-mass spectrometry conformation space for the analysis of complex biological samples. Anal Bioanal Chem 394:235–244
57. Lee S, Wyttenbach T, Bowers MT (1997) Gas phase structures of sodiated oligosaccharides by ion mobility/ion chromatography methods. Int J Mass Spectrom Ion Process 167–168:605–614

58. Hoaglund CS, Valentine SJ, Clemmer DE (1997) An ion trap interface for ESI Ion mobility experiments. Anal Chem 69:4156–4161
59. Liu Y, Clemmer DE (1997) Characterizing oligosaccharides using injected-ion mobility/mass spectrometry. Anal Chem 69:2504–2509
60. Leavell MD, Gaucher SP, Leary JA, Taraszka JA, Clemmer DE (2002) Conformational studies of Zn-ligand-hexose diastereomers using ion mobility measurements and density functional theory calculations. J Am Soc Mass Spectrom 13:284–293
61. Clowers BH, Dwivedi P, Steiner WE, Hill HH, Bendiak B (2005) Separation of sodiated isobaric disaccharides and trisaccharides using electrospray ionization-atmospheric pressure ion mobility-time of flight mass spectrometry. J Am Soc Mass Spectrom 16:660–669
62. Clowers BH, Hill HH Jr (2005) Mass analysis of mobility-selected ion populations using dual gate, ion mobility, quadrupole ion trap mass spectrometry. Anal Chem 77:5877–5885
63. Dwivedi P, Bendiak B, Clowers BH, Hill HH Jr (2007) Rapid resolution of carbohydrate isomers by electrospray ionization ambient pressure ion mobility spectrometry-time-of-flight mass spectrometry (ESI-APIMS-TOFMS). J Am Soc Mass Spectrom 18:1163–1175
64. Zhu M, Bendiak B, Clowers B, Hill H (2009) Ion mobility-mass spectrometry analysis of isomeric carbohydrate precursor ions. Anal Bioanal Chem 394:1853–1867
65. Yamagaki T, Sato A (2009) Peak width-mass correlation in CID MS/MS of isomeric oligosaccharides using traveling-wave ion mobility mass spectrometry. J Mass Spectrom 44:1509–1517
66. Yamagaki T, Sato A (2009) Isomeric oligosaccharides analyses using negative-ion electrospray ionization ion mobility spectrometry combined with collision-induced dissociation MS/MS. Anal Sci 25:985–988
67. Gabryelski W, Froese KL (2003) Rapid and sensitive differentiation of anomers, linkage, and position isomers of disaccharides using High-Field Asymmetric Waveform Ion Mobility Spectrometry (FAIMS). J Am Soc Mass Spectrom 14:265–277
68. Olivova P, Chen W, Chakraborty AB, Gebler JC (2008) Determination of N-glycosylation sites and site heterogeneity in a monoclonal antibody by electrospray quadrupole ion-mobility time-of-flight mass spectrometry. Rapid Commun Mass Spectrom 22:29–40
69. Plasencia MD, Isailovic D, Merenbloom SI, Mechref Y, Clemmer DE (2008) Resolving and assigning N-linked glycan structural isomers from ovalbumin by IMS-MS. J Am Soc Mass Spectrom 19:1706–1715
70. Schenauer MR, Meissen JK, Seo Y, Ames JB, Leary JA (2009) Heparan sulfate separation, sequencing, and isomeric differentiation: ion mobility spectrometry reveals specific iduronic and glucuronic acid-containing hexasaccharides. Anal Chem 81:10179–10185
71. Jin L, Barran PE, Deakin JA, Lyon M, Uhrin D (2005) Conformation of glycosaminoglycans by ion mobility mass spectrometry and molecular modelling. Phys Chem Chem Phys 7:3464–3471
72. McCullough BJ, Kalapothakis JM, Taylor K, Clarke DJ, Eastwood H, Campopiano D, MacMillan D, Dorin J, Barran PE (2010) Binding a heparin derived disaccharide to defensin inspired peptides: insights to antimicrobial inhibition from gas-phase measurements. Phys Chem Chem Phys 12:3589–3596
73. Bagal D, Valliere-Douglass JF, Balland A, Schnier PD (2010) Resolving disulfide structural isoforms of IgG2 monoclonal antibodies by ion mobility mass spectrometry. Anal Chem 82:6751–6755
74. Harvey DJ (2003) Matrix-assisted laser desorption/ionization mass spectrometry of carbohydrates and glycoconjugates. Int J Mass Spectrom 226:1–35
75. Harvey DJ (1999) Matrix-assisted laser desorption/ionization mass spectrometry of carbohydrates. Mass Spectrom Rev 18:349–450
76. Harvey DJ (2001) Identification of protein-bound carbohydrates by mass spectrometry. Proteomics 1:311–328
77. An HJ, Lebrilla CB (2010) Structure elucidation of native N- and O-linked glycans by tandem mass spectrometry (tutorial). Mass Spectrom Rev 30(4):560–578
78. Henderson SC, Valentine SJ, Counterman AE, Clemmer DE (1998) ESI/Ion trap/ion mobility/time-of-flight mass spectrometry for rapid and sensitive analysis of biomolecular mixtures. Anal Chem 71:291–301
79. Hoaglund CS, Valentine SJ, Sporleder CR, Reilly JP, Clemmer DE (1998) Three-dimensional ion mobility/TOFMS analysis of electrosprayed biomolecules. Anal Chem 70:2236–2242
80. Clemmer Cross Section Database. http://www.indiana.edu/~clemmer/Research/cross%20section%20database/cs%20database.htm. Accessed 30 Aug 2010

81. Ridenour WB, Kliman M, McLean JA, Caprioli RM (2010) Structural characterization of phospholipids and peptides directly from tissue sections by MALDI traveling-wave ion mobility-mass spectrometry. Anal Chem 82:1881–1889
82. Park Z-Y, Russell DH (2000) Thermal denaturation: a useful technique in peptide mass mapping. Anal Chem 72:2667–2670
83. Cleland WW (1964) Dithiothreitol, a new protective reagent for SH groups*. Biochemistry 3:480–482
84. Smythe CV (1936) The reaction of iodoacetate and of iodoacetamide with various sulfhydryl groups, with urease, and with yeast preparations. J Biol Chem 114:601–612
85. Kuster B, Harvey DJ (1997) Ammonium containing buffers should be avoided during enzymatic release of glycans from glycoproteins when followed by reducing terminal derivatization. Glycobiology 7:vii–ix
86. Pace CN (1986) Determination and analysis of urea and guanidine hydrochloride denaturation curves. Methods Enzymol 131:266–280
87. Shirley BA (1995) Urea and guanidine hydrochloride denaturation curves. Methods Mol Biol 40:177–190
88. Fenn LS, McLean JA (2009) Simultaneous glycoproteomics on the basis of structure using ion mobility-mass spectrometry. Mol Biosyst 5:1298–1302
89. Lareau NM, Fenn LS, McLean JA (2012) Simultaneous glycomics, proteomics, and lipidomics using ion mobility-mass spectrometry. Anal Chem (in preparation).

Part IV

Quantitation Methods

Chapter 13

Quantitative Analysis of Glycoprotein Glycans

Ron Orlando

Abstract

The ability to quantitatively determine changes in the N- and O-linked glycans is an essential component of comparative glycomics. Multiple strategies are available to by which this can be accomplished, including; both label free approaches and isotopic labeling strategies. The focus of this chapter is to describe each of these approaches while providing insight into their strengths and weaknesses, so that glycomic investigators can make an educated choice of the strategy that is best suited for their particular application.

Key words: Comparative glycomics, Relative quantitation, Isotopic labeling, Label free, Mass spectrometry

1. Introduction

Mass spectrometry is a powerful tool for qualitative glycomics, however several issues arise when this technique is used to obtain quantitative results. Because quantitative glycomics is a relatively new field, it is difficult to evaluate the benefits and limitations of the approaches that have been developed in this area. However, many of these techniques can be directly compared to strategies that have been used for many years in the proteomics field, which also relies heavily upon mass spectrometry. Thus, it seems appropriate to begin with a brief review of the approaches used in quantitative proteomics, including a summary of the lessons learned in proteomics to gain insight into the strategies used for quantitative glycomics. This will be followed by a description of the quantitative approaches developed for glycomics. This later section primarily focuses on the analysis of glycoprotein glycans; however, many of these strategies can be applied to other types of glycans.

Jennifer J. Kohler and Steven M. Patrie (eds.), *Mass Spectrometry of Glycoproteins: Methods and Protocols*,
Methods in Molecular Biology, vol. 951, DOI 10.1007/978-1-62703-146-2_13, © Springer Science+Business Media, LLC 2013

It is important to note at the onset that this chapter focuses on relative quantitation, i.e., how do the levels of individual glycans change between samples. Absolute glycan quantitation is difficult, if not impossible, at the current time because the response for an analyte is going to be dependent on a number of factors, such as the analyte's ionization efficiency, molecular mass, etc., all of which are analyte specific, and thus necessitates the addition of an isotopic variant for each analyte. Unfortunately, standard isotopically labeled glycans of known quantity are not widely available.

Strategies for relative quantitation with MS-based techniques address errors introduced by variability in the sample matrix, the instrument response, instrument to instrument performance, and/or the sample preparation process. Matrix effects, which are often attributed to phenomenon such as ion suppression from the presence of other compounds competing with or interfering with the ionization of the analyte, can alter the response from a particular analyte even when the analyte's concentration does not change. Matrix effects are one of the reasons that ion intensities often do not directly correlate with concentration. Variability in the instrument response further exacerbates the issue of relative quantitation, as it leads to differences in ion abundances between analyses. Of course, these differences are somewhat related to the time between experiments, and thus glycans that are observed in the same spectrum are less prone to this error than glycans that are observed at different time points, such as those that elute at different times in an LC separation. Variability in instrument performance is particularly problematic when long time periods are present between the times when samples are analyzed. Furthermore, different MS systems can produce different ion intensities and different ratios of ion abundances from the same sample. This is particularly true when different MS configurations from different vendors are compared, and thus this type of error can limit cross-laboratory reproducibility. One additional source of quantitative error results from differential analyte losses occurring when samples are treated separately in a parallel manner. The success of the different approaches for relative quantitation depends on how well each of these sources of error is addressed.

The techniques used in proteomics for relative quantitation can be broadly subdivided into two general schemes—those that involve the use of labels and those methods that are label free.

Label-Free Proteomics: In the label free approaches, various aspects of the peptides/proteins such as normalized ion intensities, spectral counts, mass, scan number and signal intensity, and accurate mass plus retention time have been successfully used to assign protein expression levels for comparative investigations (1–4). In some of these label free approaches, the analytes themselves serve as standards. In this way the response of an analyte is compared to the response from one, several, or all other species present in the sample.

With spectral counting, for instance, the number of MS/MS spectra from a particular protein can be expressed as a fraction of all MS/MS spectra acquired for all proteins in the sample. With other methods, one or several of the proteins are assigned as being constant between the samples analyzed, and all results are made relative to these reference species. In this manner, errors associated with changes in instrument performance are decreased since all proteins are analyzed in the same sample, and thus subjected to the same instrument performance. The net result is that this strategy improves the reproducibility of these experiments. However, this is not a perfect solution for variation in instrument performance since there is often sufficient time between acquisition of the analyte and standard signals to allow for altered instrument response, particularly when an LC separation is used. Errors larger than 25% have been reported when the analyte and reference only partially overlap chromatographically (5); larger errors can be expected as the time difference between analyte and reference elution increases. When the analytes serve as their own standards, it can be difficult to deduce if an increased response for a particular protein is due to it being up-regulated or if this increased prevalence is actually caused by a decrease in the abundance of another protein in the sample, since both of these lead to the same result. These approaches do not compensate for issues associated with matrix effects if the analytes are separated chromatographically, variability in instrument to instrument performance, and errors resulting from differential losses from parallel sample preparations. For these reasons, many of these approaches are considered to be only semi-quantitative in nature, and often are attributed to have the inability to reproducibly detect changes in protein expression that are smaller than twofold (6–8). A primary advantage of these approaches is their simplicity, as these do not alter the sample workflow, however, this does add additional data processing steps. Also, these strategies do not need stable isotopes, which elevate the cost of the other internal standard approaches that will be discussed. Label free approaches thus offer a straightforward manner to obtain quantitative results, and are often used as a "screening process" to identify proteins of interest to serve as the subjects of more detailed studies using other methods.

Internal Standards: The use of internal standards is the other general strategy to compensate for the problems associated with quantification by MS. However, the type of internal standard used and when it is introduced into the sample determines if it is capable of compensating for variability in the sample, instrument variability, and/or sample preparation. Consequently, not all internal standard approaches are created equal, which in turn has led to the development of multiple strategies, each of which has its own unique benefits and limitations. In general, the closer the chemical properties of an internal standard to its analyte, the better it compensates for the various sources of error, and thus the optimal internal standard for each analyte is typically an isotopomer of the analyte itself (5, 9). For example, a sample could be mixed with an internal standard consisting of an isotopically labeled (^{13}C, D, ^{15}N, etc.) form of each analyte followed by MS analysis. The mass analyzer resolves the isotopomers, permitting their relative abundances to be determined by comparing the signal intensity of each analyte ion to that from its isotopically labeled form.

Chemical Labeling: It is extremely difficult and/or expensive to obtain isotopically labeled standards for all species observed in a high throughput proteomic analysis of complex samples, hence a variety of isotopic labeling procedures have been developed where one of the samples is modified with a "light" tag while the other is derivatized with a "heavy" tag (10–17). For example, isotope-coded affinity tags (ICAT) chemically target specific amino acids, typically cysteine, in the peptide sequence for differential labeling (10). Additional chemical labeling approaches have been developed to target other functional groups of the polypeptides (11, 13–17). Using these procedures, an isotopically labeled internal standard can easily be created for all components in the mixture. The use of these approaches compensate for ion suppression and variability in instrument to instrument performance when both isotopic variants co-elute. However, since many of the strategies introduce the isotopic label after significant sample processing, variable losses during this sample processing can introduce errors. Typical standard deviations for these labeling procedures have been reported to be 20–25% (5), which is significantly better than the 100% standard deviations often reported for the label free approaches. These strategies can also improve the throughput of the experiment as two or more samples can be analyzed simultaneously; however, these processes alter the discovery workflow as additional steps are needed to label and/or clean-up the sample prior to analysis, and increase the cost due to the isotopic labeling reagent.

Metabolic Labeling: Stable isotopes can also be introduced into biological systems through metabolic labeling. For instance, stable isotope labeling with amino acids in cell culture (SILAC) provides a simple and straightforward method for the incorporation of an isotopic label into proteins prior to MS-based proteomics (12). In a SILAC experiment, two cell populations are grown in culture media that are identical except that one of them contains a "light" and the other a "heavy" form of particular amino acids (^{12}C and ^{13}C labeled lysine and arginine for example). The labeled analogs of amino acids are supplied to cells in culture instead of the natural amino acids, and become incorporated into all newly synthesized proteins. After a number of cell divisions, each instance of the particular amino acids is replaced by its isotope labeled analog. An advantage of this approach over the chemical tagging approaches is that the cells are mixed together immediately after cell lysis. Thus, proteins from both cell types are subjected to the exact same experimental conditions during sample handling, digestion, purification, etc., eliminating the differential losses that can occur when the samples are treated separately in a parallel manner. For this reason, SILAC is often considered the "gold standard" for quantitative proteomic analyses (18). The primarily limitation of these in vivo labeling procedures is that they are typically limited to cells grown in culture.

2. Quantitative Approaches for Glycomics

Many of these quantitative proteomic tools have been adapted for glycomic analysis, and can be broadly divided once again into those that are label free and those where an isotopic label is introduced. The later of these can be subdivided based on how the isotope is incorporated, i.e., metabolically or via chemical derivatization. These three strategies are expected to share similarity with their proteomic counterparts as far as the quantitative issues which each compensate for, and thus the level of reproducibility each can attain. These approaches will be described in the following sections, with the overall purpose of this work being to familiarize glycomics researchers with these quantitative techniques.

2.1. Label Free Approaches

The dominant method for quantitative glycomics currently used is some variation of a label free approach. This process involves releasing the glycans from a sample, often followed by a derivatization step, then acquisition of a MALDI-MS or ESI-MS spectrum, or performing an LC-MS experiment, as shown in Fig. 1. In many of these studies, the response from any one glycan is reported as the percentage of the response for all glycans in the sample. This follows the general procedure utilized for data processing in label free proteomics, and thus compensates for instrument variability and matrix effects to some extent. The Human Proteome Organization (HUPO) recently published the results from a multi-institutional study on the label free profiling of glycoprotein glycans, which highlights the strengths and weaknesses of this methodology (19). In general, this study revealed that MS quantitation was consistent with and comparable to chromatographic analysis of reductively aminated glycans, which is the generally accepted standard method for glycan quantitation (19). A second study reached the same conclusion when ESI-MS was compared with the more traditional LC-based methods with fluorescent detection (20). The HUPO study also noted several issues with quantitation, one of which is prompt, in-source, fragmentation of sialylated glycans ions created by the MALDI process. ESI was found to be gentle enough so that

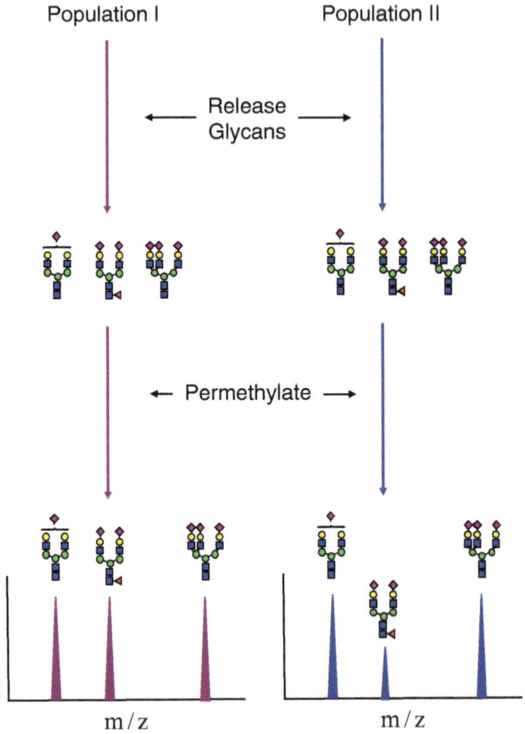

Fig. 1. Flow chart for quantitative glycan analysis using label free approaches. Here, the glycans from two biological populations are released and permethylated in a parallel manner. The derivatized glycans are analyzed by MS and changes in glycan abundance are determined from the glycans' relative peak heights in the two spectra.

the extent of sialylation was not perturbed. The presence of sialic acids on a glycan had previously been reported to decrease the relative ionization efficiencies of sialylated glycans (21). Both of these factors contributed to significant errors in the quantification of sialylated glycans (19). Permethylation of glycans, which is a commonly performed prior to MS analysis of glycans, was found to be critical in order to perform glycan quantitation by MALDI-MS (19), as this process stabilized the sialic acid residues and thus decreased the extent of in-source fragmentation. In addition, this derivatization process leads to more uniform ionization as it convert highly polar –OH and $-COO^-$ groups into nonpolar, chemically homogeneous derivatives, which overcomes the issue associated with decreased ionization efficiency of sialylated glycans. The multitude of steps in the permethylation reaction does however introduce the possibility of differential losses, which in part has led to the development of a solid-phase procedure that is expected to minimize this potential problem (22, 23). These observations led the HUPO study to conclude that permethylation was needed for MALDI-MS glycan quantitation.

The simplicity of label free glycan quantitation makes this an attractive procedure. The HUPO study concluded that MALDI-MS of permethylated glycans was as reliable as chromatographic methods of quantification. The inter laboratory reproducibility of this study on technical replicates (i.e., sample was prepared, then divided into three for three replicate MALDI-MS experiments) was very good, with coefficients of variation ranging between 1.3 and 8.8% for abundant glycans and 12–34% for a glycan of lower abundance. The good reproducibility of these experiments was expected as the response for each glycan was normalized to that from all glycans, all of the glycans were observed in the same MALDI-MS spectrum (eliminating instrument variation), and the technical replicate method negated changes in intensity from matrix effects. The across laboratory portion of this study also indicated very good reproducibility for all the major glycans, with coefficients of variation ranging between 6 and 12%, however the variation was very large (>100%) for low abundance glycans. This observation parallels those from label-free proteomic studies which have observed very good reproducibility for intense species, but significantly worse deviations for proteins of lower abundance (7). A limitation of MALDI-MS is that it cannot be used to quantify structurally distinct isomeric glycans when these are present in isomeric mixtures as this approach provides a measure of the total abundance of the collection of isomers at a particular mass, rather than the abundances of individual species in that collection. This is an advantage of the label free LC-MS of approach for quantitating glycans (24).

The Glycoprotein Research Group (gPRG) of the Association of Biomolecular Resource Facilities (ABRF) performed a study to evaluate the ability to accurately quantitate glycans in 2010 (25). In this double blind study, a mixture of glycoproteins was used, and the ratios of the individual glycoproteins were manipulated to provide three samples with known glycan changes. This allowed the accuracy of the results to be evaluated, unlike the HUPO study which could only evaluate the reproducibility across laboratories. Nineteen laboratories participated in the ABRF study, and used a variety of techniques, including MALDI, ESI, and LC-MS and employed a variety of derivatization strategies. However, all of these laboratories used a label free approach. Results from this study showed an average error of just over 100%, and concluded that none of the approaches used could provide an accurate quantitative value. Results from the ABRF study are in agreement with results from label proteomic experiments, which are typically reported to accuracy error of approximately 100% (i.e., 2×) (6–8).

In general, the label free approaches offer a simple quantitative approach that does not alter the sample workflow. These strategies do not need stable isotopes, which elevate the cost of the other internal standard approaches that will be discussed. Consequently, these approaches thus offer a straightforward manner to screen samples for major glycan changes, which can serve as the subjects of more detailed studies using more accurate approaches, which often involve an isotopic labeling strategy.

2.2. Isotopic Labeling Strategies

Several approaches have been developed that introduce a stable isotope to a glycan via chemical derivatization. Glycans are typically derivatized prior to analysis either by tagging the reducing terminus with a chromophore when subsequent analyses are chromatographic or by permethylation when the sample is to be analyzed by MS. Methods that introduce an isotopic label have also followed these two strategies, and in essence have utilized the commercial availability of isotopically labeled reagents for these procedures. A typical workflow for these in vitro labeling approaches involves the parallel release of glycans from the sample populations under investigation, derivatization with an isotopic label after which the samples are mixed, followed by MS analysis, as shown in Fig. 2, where permethylation with $^{12}C/^{13}C$ methyliodide was used as an example. Each of these procedures will be summarized in the following sections.

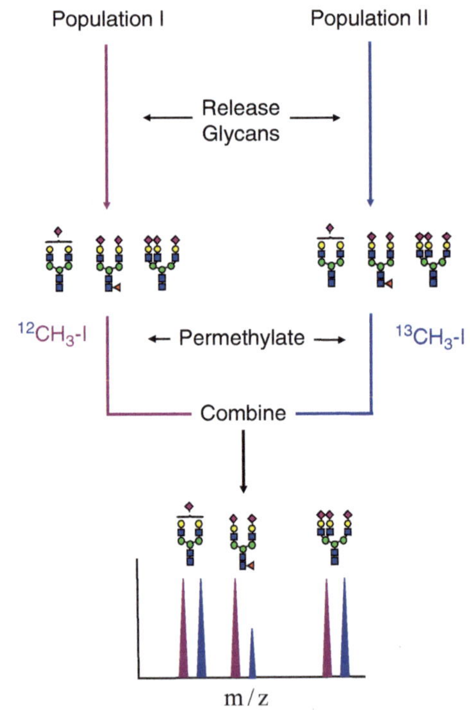

Fig. 2. Flow chart for quantitative glycan analysis using isotopic labeling during the permethylation step. Glycans from two biological samples are released followed by permethylation with $^{13}CH_3I$ or $^{12}CH_3I$ and mixed together prior to MS analysis. Changes in glycan abundance are determined by comparing the peak heights of the light and heavy labeled glycans.

2.2.1. Labeling on the Reducing Terminus

Several of the reagents typically used to label the reducing terminus of glycans are available in their deuterated form, and thus provide a straightforward method to incorporate an isotopic label through reductive amination using standard protocols (26–28). Tags have also been synthesized that can modify oligosaccharides with four isotope-enriched variants allowing up to four samples to be analyzed simultaneously (29). One application of these tetraplex tags is to designate a reference glycan mixture, which is labeled with one isotopic variant of the tag, and used as an internal standard for all analyses. This approach enables the comparison of a large number of samples. In all of these cases, tags were added to the reducing terminus of the glycan, which is problematic for O-linked glycans since the reducing termini of these glycans are usually reduced as a result of β-elimination. The use of sodium borodeuteride in the place of sodium borohydride during β-elimination has been reported as a method for incorporation of an isotope for the quantification of O-linked glycans (30). The 1 Da shift introduced by this procedure is insufficient to shift the deuterated species away from the naturally occurring ^{13}C isotope peak of the species labeled with sodium borohydride, which in turn can lead to challenging quantitation. Overall, the commercial availability of these reagents makes this approach relatively easy to implement in labs that typically label the reducing terminus.

Each of these isotopic tags uses hydrogen/deuterium as the light/heavy isotopic species. This pair introduces the largest isotopic effect as the atomic mass of deuterium is twice that hydrogen, and thus this issue should be addressed. For the reductive amination tags, the kinetic isotope effect is expected to be fairly small as the substituted isotope is not directly involved in the bond that is breaking or forming and thus this is classified as a secondary effect. This is not the case when sodium borodeuteride is used during β-elimination as the deuterium is directly involved in the reaction, and thus, the relative change in rate associated with the incorporation of these isotopes is a function of the inverse square root of the ratios of the reduced masses of the atoms involved in the bonds based on the calculation of the vibrational energy required to break or form a bond. All of these labeling processes are intended to proceed to completion, and thus a kinetic isotope effect may decrease the rate of reaction, but if care is taken to drive the reaction to completion this kinetic effect is not expected to be problematic. One area where the isotope effect introduced by hydrogen/deuterium substitution is often noticed involves LC separations. This phenomenon is widely reported in the proteomics literature, where the peptide labeled with the deuterated tag elutes at a later time than the same peptide with the non-deuterated tag and has been reported to cause errors in excess of 25% (5). Glycans labeled with deuterated forms of standard reductive amination reagents have also been reported to be partially resolved from their hydrogen containing counterparts (26). As discussed above, the rationale for introducing an isotopically labeled version of the analyte is to ensure that the analyte and standard are analyzed at exactly the same time to compensate for differential instrument response and matrix effects; this criterion is not met when the analyte does not co-elute with standard, and thus care should be taken when using hydrogen/deuterium isotopic tags.

The introduction of an isotopic label with a reductive amination tag is an attractive procedure because this procedure does not alter existing workflows, i.e., the researcher simply derivatizes with the deuterated form of the tag. This labeling strategy also introduces a fixed shift in mass between the light and heavy pair, thus simplifying the identification of matched glycans. When this type of approach has been utilized for relative quantitation, coefficients of variation been reported as 20% (29), which makes this approach significantly more precise than the label free procedures discussed above. One limitation for these procedures is that they cannot be used to quantify structurally distinct isomeric glycans when these are present in isomeric mixtures as this approach provides a measure of the total abundance of the collection of isomers at a particular mass, rather than the abundances of individual species in that collection. One possible solution in these instances is to chromatographically separate these isomers, however this may be difficult due to the chromatographic shifts associated with the use of deuterium as the heavy isotope.

2.2.2. Isotopic Incorporation During Permethylation

Because of the advantages associated with permethylation (discussed above) many glycans are subjected to this procedure prior to MS analysis. This process provides the opportunity to isotopically label oligosaccharides with the use of heavy methyl iodide ($^{13}CH_3$ or $^{12}CD_3$) and light methyl iodide ($^{12}CH_3$) in the standard permethylation procedures (20, 31, 32), as shown in Fig. 3 with a permethylated milk oligosaccharide. The use of methyl iodide with varying degrees of deuterium content (i.e., $^{12}CH_3$, $^{12}CH_2D$, $^{12}CHD_2$, and $^{12}CD_3$) allows for the simultaneous analysis of four different samples (32). The kinetic isotope effect resulting from the use of CH_2DI has been estimated to be less than 2%, and thus this is not expected to significantly alter the permethylation process (33). However, the use of deuterated methyl iodide is expected to lead to a chromatographic shift due to isotope effects, which because of the large number of methyl groups introduced is expected to be more pronounced than that observed with the reductive amination tags where only a few deuteriums are added to the glycan (26). This chromatographic shift is not expected when $^{13}CH_3$ and $^{12}CH_3$ are used, and was one of the reasons this pair was selected (31).

A potential issue unique to the permethylation labeling strategy is that large errors can be introduced by small variation in labeling efficiency since the number of modified sites is so large.

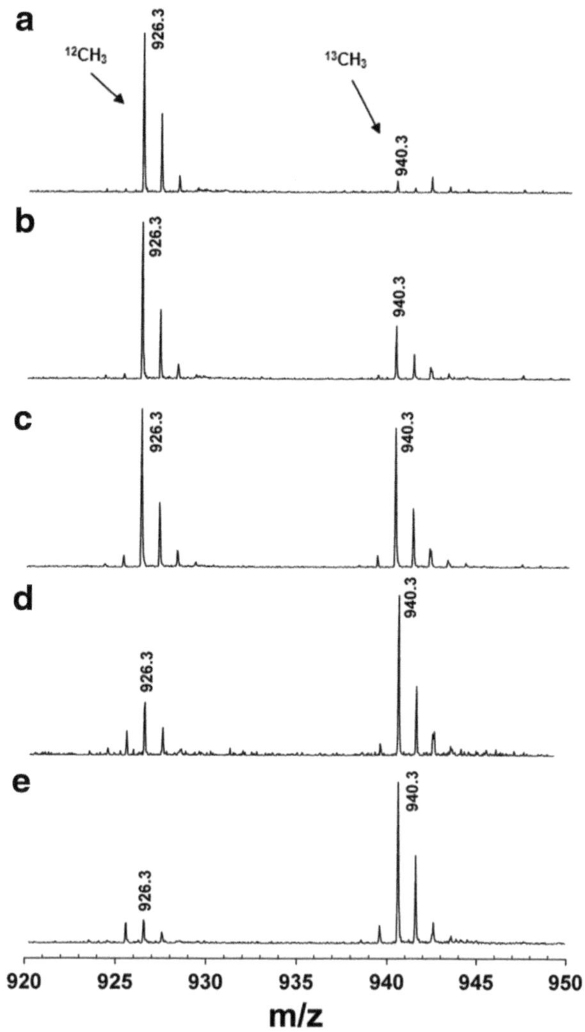

Fig. 3. MALDI-TOF MS spectra showing the isotopic pairs for the tetrasaccharide mixture (Gal2-GlcNAc-Glc) labeled with different proportions of ^{12}C and ^{13}C methyl iodide. For this experiment, two aliquots with similar amounts of oligosaccharides from human milk were separately labeled with either ^{12}C or ^{13}C methyl iodide. Before MS, these aliquots were mixed in the following ^{12}C:^{13}C proportions: 9:1 (**a**), 7.5:2.5 (**b**), 1:1 (**c**), 2.5:7.5 (**d**), and 1:9 (**e**). The monoisotopic *m/z* for the ^{12}C- and ^{13}C-labeled glycans are 926.3 and 940.3, respectively.

For instance, changing the labeling efficiency from 99.1 to 99.0% leads to a 0.1% error when the isotope is introduced via reductive amination since this label is only added to a single site on the glycan. A similar 0.1% decrease during permethylation has a much more profound effect on the error level since glycans have many sites that become methylated. A fully sialylated triantennary complex glycan, for example, has 50 sites of permethylation, and thus the 0.1% change in labeling efficiency results in a 5.0% error. Consequently, with the permethylation approach reproducible quantitation is much more dependent on reproducible labeling than with the other methods of introducing an isotopic label.

An important aspect of incorporating the isotopic label during permethylation is that it does not alter the typical glycan workflow and can be performed on any glycan that is amenable to permethylation. One limitation of these isotopic labeling approaches, however, is that the mass difference (Δm) between the heavy and light forms of each glycan is variable and can be very large, as Δm is proportional to the number of methylation sites on the glycan (31). This variability can confound the analysis of complex mixtures, as it can be difficult to match the differentially labeled forms of the same chemical species. In addition, this approach cannot be used to quantify the structurally distinct isomeric glycans that are often encountered in glycomic analyses, as was the case with the labeled reductive amination tags. Despite these limitations, permethylation tagging procedures have been reported to provide linear quantification over two orders of magnitude and yielded coefficients of variation in the range of 15–30% (31, 32).

An isobaric labeling approach based upon the incorporation of an isobaric label during pemethylation, called quantitation by isobaric labeling (QUIBL), has recently been introduced (33, 34). In this approach, glycans are permethylated with either $^{13}CH_3I$ or $^{12}CH_2DI$. This pair of reagents has the same nominal mass but differ in their exact mass by 0.002922 Da/label. This mass difference is difficult to resolve with current mass spectrometers in cases where only a single label is attached to the analyte. However, glycans typically contain multiple hydroxyl groups, which increase the delta mass between the two samples. Since the number of hydroxyl groups increases with the mass of the glycan, the difference between these isobaric species also increases and thus the resolution needed is approximately 25,000 $\Delta M/M$ and is independent of the glycan's size for typical N- and O-linked species. This level of mass resolution is easily attained with FT-MS and Orbitrap MS systems.

The advantages of QUIBL are numerous, and result primarily from the isobaric ions appearing at the same nominal mass to charge ratio. This characteristic leads to increased ion intensity as ions from both samples are not distributed between isotopic species having different m/z values. The small mass difference between these isobars allows the two species to be simultaneously selected for MS^n analysis, permitting the relative quantitation of isomeric glycans, as shown in Fig. 4. This later characteristic was used to demonstrate a decrease in the level of N-linked glycans containing the Lewis X structure glycans when mouse embryonic stem cells were allowed to spontaneously differentiate. In addition, the light and heavy analyte ions resulting from reagents not having 100% isotope incorporation can still be resolved, unlike the typical isotope labeling strategies where the ion produced by under incorporation of the heavy isotope results in an ion that is indistinguishable from an ion

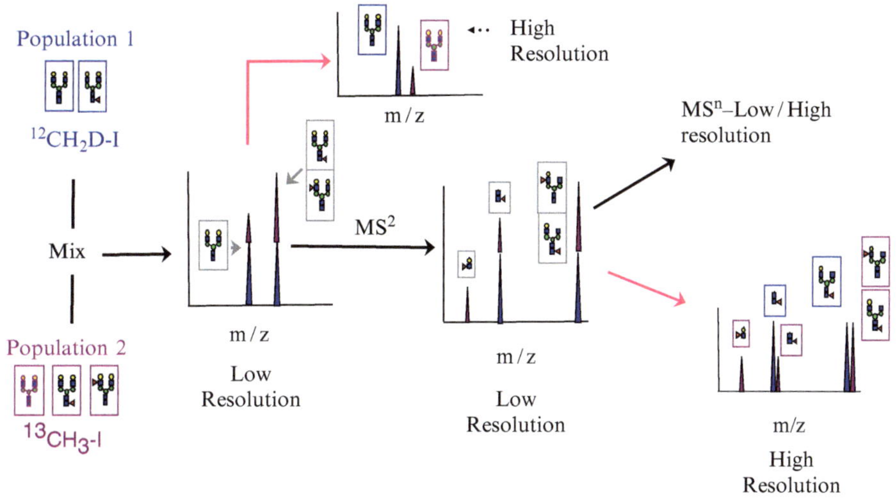

Fig. 4. Flow chart for quantitative glycan analysis using isobaric labeling. Glycans from two biological samples are permethylated in either $^{13}CH_3I$ or $^{12}CH_2DI$ and mixed together prior to analysis. At low mass resolution, the two labeled species appear at the same *m/z* value thereby increasing their abundance and decreasing sample complexity. Analysis of the glycans by high-resolution MS separates the differentially labeled glycan precursor ions permitting their relative quantitation by comparing the peak intensities from the $^{13}CH_3$ to the $^{12}CH_2D$ labeled glycans. Structural information on the glycan is provided by low resolution MS^n, which does not alter the ratio of isobaric labels. High-resolution analysis of the MS^n fragment ions permits the isomeric glycans to be quantified.

produced by the light species. The characteristic of QUIBL results from the presence of multiple labeling sights on the glycan, and thus replacing one of the many ^{13}C atoms with a ^{12}C atom or replacing one of the many D (or ^2H) atoms with an ^1H atom decreases the analyte's mass by approximately 1 Da, however, the resulting ion is detected in the appropriate ($^{13}CH_3$-labeled or $^{12}CH_2D$-labeled) ion series because it still contains a large number of isotopic labels. This greatly simplifies quantitation, which is accomplished by summing the ion abundances for the $^{13}CH_3$-labeled and $^{12}CH_2D$-labeled series and comparing these two values.

QUIBL offers a relatively straightforward approach to compensate for quantitative errors resulting from instrument performance and matrix effects, provided that the researcher has access to a high-resolution MS system. This procedure does not alter the workflow for labs that routinely permethylate glycans prior to analysis. As with the other glycan tagging procedures discussed here, QUIBL does not compensate for differential losses resulting from multiple samples processed in parallel. A chromatographic shift is also expected between the light- and heavy-labeled species as a result of the CH_2D label. Despite these limitations, QUIBL has been reported to provide linear quantification over at least two orders of magnitude and yielded coefficients of variation in the range of 10–26% (33, 34), as demonstrated in Fig. 5. This level of accuracy and dynamic range is approximately the same as that seen using the other glycan labeling strategies. The one advantage of

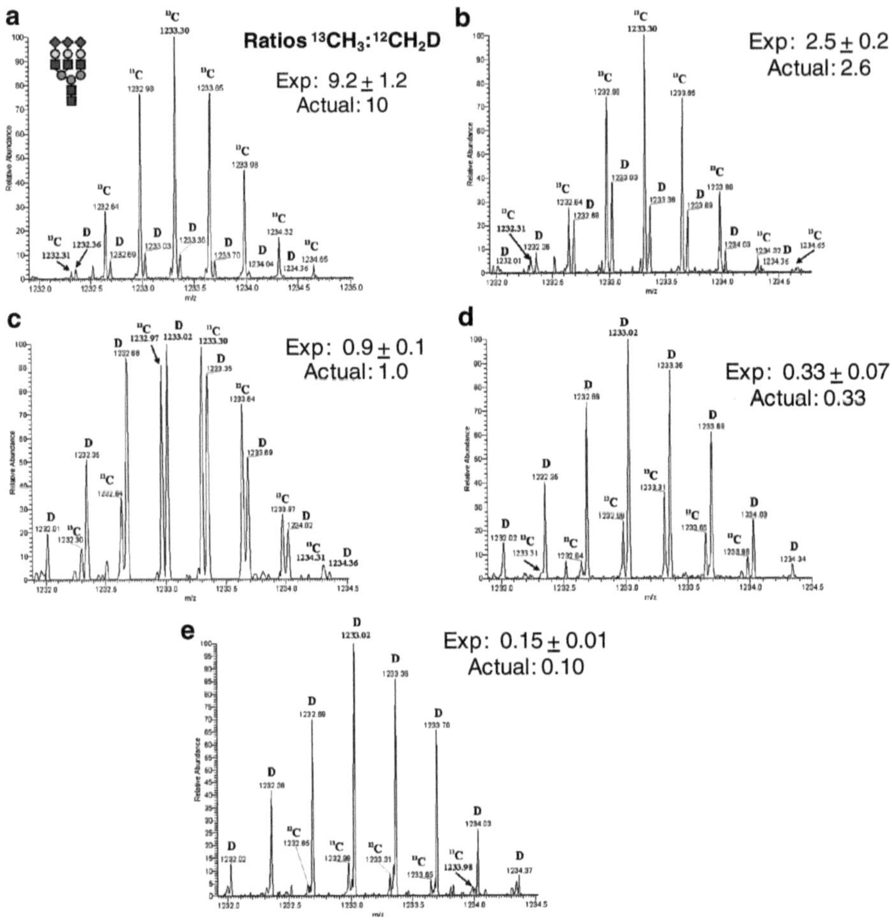

Fig. 5. Quantitation by isobaric labeling (QUIBL) analysis of a differently labeled fetuin glycan mixed at five different ratios. Two fetuin glycan mixtures were permethylated in either $^{13}CH_3I$ or $^{12}CH_2DI$. The two differentially labeled glycan mixtures were then mixed together at the ratios 10:1, 8:3, 1:1, 3:8, and 1:10 ($^{13}CH_3$:$^{12}CH_2D$) and analyzed by FT-MS (**a**, **b**, **c**, **d**, **e**). Accurate quantitation was achieved at all ratios over two orders of magnitude.

QUIBL over these other methods is that it is capable of quantitating individual glycan present in isomeric mixtures (Fig. 6).

2.2.3. In Vivo Labeling

An in vivo labeling strategy has recently been described for glycomic studies (35). This methodology termed IDAWG, Isotopic Detection of Aminosugars With Glutamine, relies on the side-chain of glutamine being the sole donor source of nitrogen for aminosugars in the production of sugar nucleotides according to the hexosamine biosynthetic pathway, as shown in Fig. 7. Thus, introduction of glutamine with a ^{15}N labeled side-chain (amide-^{15}N-Gln) into Gln-free media allows for the incorporation of one ^{15}N into all aminosugars, including GlcNAc, GalNAc, and sialic acids. Consequently, the mass of all N- and O-linked glycans, glycolipids, and extracellular matrix polysaccharides, is increased by +1 Da/aminosugar. This approach was demonstrated by the

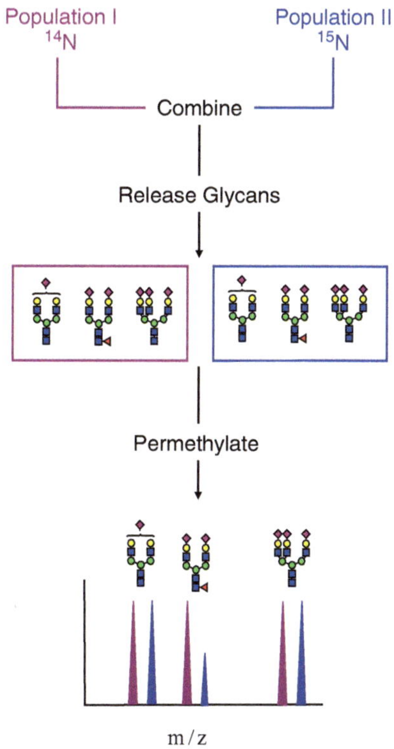

Fig. 6. Flow chart for quantitative glycan analysis using in vivo isotopic labeling. Here, glycans in one of the populations are labeled with ^{14}N while glycans in the other population are labeled with ^{15}N. The two populations are mixed, followed by glycan release, permethylation, and MS analysis. Changes in glycan abundance are determined by comparing the peak heights of the light- and heavy-labeled glycans.

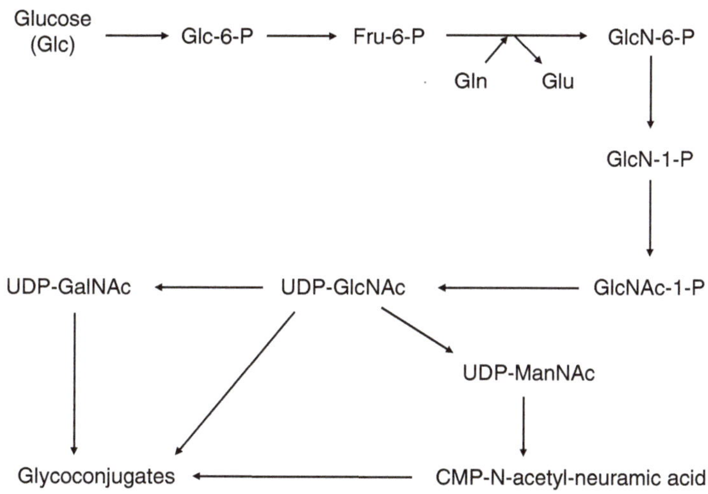

Fig. 7. The hexosamine biosynthetic pathway demonstrating that the side-chain of glutamine is the sole donor source of nitrogen for aminosugars in the production of sugar nucleotides, which allows the introduction of an ^{15}N isotopic tag into all aminosugars, including GlcNAc, GalNAc, and sialic acids. Species containing ^{15}N are indicated.

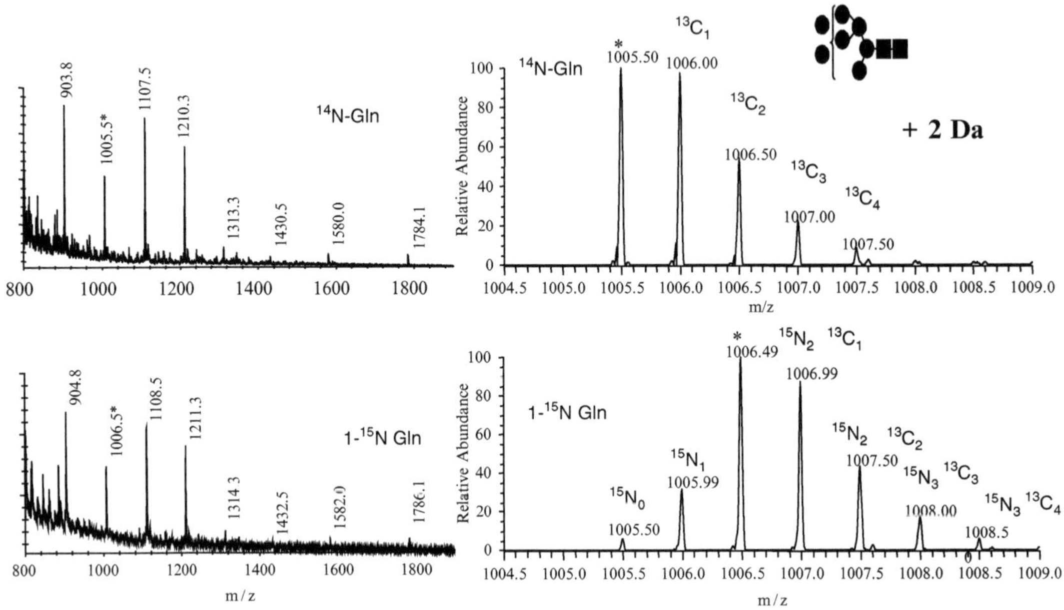

Fig. 8. Isotopically labeled N-linked glycans. Full spectra from 850 to 2,000 *m/z* of the permethylated N-linked glycans released from cells grown in either ^{14}N-Gln or amide-^{15}N-Gln. Expanded region of the spectra showing the expected 2 Da mass shift for ^{15}N incorporation into the two core GlcNAc residues.

analysis of both N-linked and O-linked glycans released from proteins of murine embryonic stem cells grown in both the light and amide-^{15}N-Gln, as shown in Fig. 8. By incorporating the isotopic label into glycans as they are being synthesized, IDAWG is similar to the SILAC procedure used in proteomics. Both of these approaches share the advantage that the differentially labeled cells can be mixed together at the beginning of the analytic procedure, minimizing the contribution of handling and work-up to overall variability. Although there are no published reports using IDAWG for quantitation, this technique is expected to compare with SILAC and thus have a linear response for at least two orders of magnitude and yielded coefficients of variation in the range of 10–20%.

IDAWG is a relatively straightforward procedure. Cell culture media is typically supplied glutamine free since this amino acid rapidly decomposes in aqueous solutions, which leads to glutamine being one of the most common supplements to cell culture media. This phenomenon greatly simplifies the IDAWG approach, since the only change in standard operating procedures it to use amide-^{15}N-Gln when preparing fresh media. The concentration of amide-^{15}N-Gln used for IDAWG is the same as that used for normal cell growth with ^{14}N-Gln. Using these conditions, 96% incorporation of ^{15}N into N- and O-linked glycans have been reported after labeling mouse embryonic stem cells for 3 days, as shown in Fig. 8 (35). Consequently, this in vivo labeling strategy provides a strategy to isotopically label a glycan population without significantly altering the experimental procedures.

Metabolic labeling of glycans provides new opportunities for assessing the dynamics of glycan turnover during the course of any cellular behavior that can be induced or sustained in culture. By completely labeling cells with heavy Gln and then replacing the media supplement with light Gln, the half-life of any aminosugar-containing glycans can be determined. Previously, glycan turnover studies required incorporation of radioactive monosaccharide and extensive subsequent fractionation to identify specific changes in glycan expression. Generally, these radiotracer techniques allowed for very sensitive detection of glycan classes, but lacked the resolution to follow individual glycan structures or subsets of biosynthetically related species. The stable isotope incorporation method reported here merges the analytic advantages of high-resolution mass spectrometry to the biological necessity of understanding the dynamics of glycan turnover. Thus, IDAWG appears to be a powerful quantitative tool for exploring the biological role of glycans, glycoproteins, and glycolipids in cell culture systems.

3. Conclusions

A range of procedures have been developed for quantitative glycomics, each of which has its own strengths and weaknesses. In general, the label free approaches are the easiest to implement, however these also show the lowest accuracy and reproducibility, particularly in the analysis of glycan with low abundance. The label free approaches, thus, provide a method for quickly surveying samples for glycan changes that can be targeted by the more accurate, but also more expensive, isotopic labeling strategies. The in vivo labeling approach is expected to offer the highest level of accuracy, but this appears to be limited to the study of cell culture systems. The presence of isomeric glycans causes problems with quantitation, and only the QUIBL approach has demonstrated an ability to quantitate individual glycans present in isomeric mixtures. Despite these limitations, all of these quantitative approaches offer powerful new tools in the glycomics toolbox.

References

1. Liu H, Sadygov RG, Yates JR III (2004) A model for random sampling and estimation of relative protein abundance in shotgun proteomics. Anal Chem 76:4193–4201
2. Radulovic D, Jelveh S, Ryu S, Hamilton TG, Foss E, Mao Y, Emili A (2004) Informatics platform for global proteomic profiling and biomarker discovery using liquid chromatography-tandem mass spectrometry. Mol Cell Proteomics 3:984–997
3. Silva JC, Denny R, Dorschel CA, Gorenstein M, Kass IJ, Li GZ, McKenna T, Nold MJ, Richardson K, Young P, Geromanos S (2005) Quantitative proteomic analysis by accurate mass retention time pairs. Anal Chem 77:2187–2200

4. Wang W, Zhou H, Lin H, Roy S, Shaler TA, Hill LR, Norton S, Kumar P, Anderle M, Becker CH (2003) Quantification of proteins and metabolites by mass spectrometry without isotopic labeling or spiked standards. Anal Chem 75:4818–4826
5. Goshe MB, Smith RD (2003) Stable isotope-coded proteomic mass spectrometry. Curr Opin Biotechnol 14:101–109
6. Zybailov B, Mosley AL, Sardiu ME, Coleman MK, Florens L, Washburn MP (2006) Statistical analysis of membrane proteome expression changes in *Saccharomyces cerevisiae*. J Proteome Res 5:2339–2347
7. Old WM, Meyer-Arendt K, Aveline-Wolf L, Pierce KG, Mendoza A, Sevinsky JR, Resing KA, Ahn NG (2005) Comparison of label-free methods for quantifying human proteins by shotgun proteomics. Mol Cell Proteomics 4:1487–1502
8. Hendrickson EL, Xia Q, Wang T, Leigh JA, Hackett M (2006) Comparison of spectral counting and metabolic stable isotope labeling for use with quantitative microbial proteomics. Analyst 131:1335–1341
9. Tao WA, Aebersold R (2003) Advances in quantitative proteomics via stable isotope tagging and mass spectrometry. Curr Opin Biotechnol 14:110–118
10. Gygi SP, Rist B, Gerber SA, Turecek F, Gelb MH, Aebersold R (1999) Quantitative analysis of complex protein mixtures using isotope-coded affinity tags. Nat Biotechnol 17:994–999
11. Liu P, Regnier FE (2002) An isotope coding strategy for proteomics involving both amine and carboxyl group labeling. J Proteome Res 1:443–450
12. Ong SE, Blagoev B, Kratchmarova I, Kristensen DB, Steen H, Pandey A, Mann M (2002) Stable isotope labeling by amino acids in cell culture, SILAC, as a simple and accurate approach to expression proteomics. Mol Cell Proteomics 1:376–386
13. Rao KC, Carruth RT, Miyagi M (2005) Proteolytic 18O labeling by peptidyl-Lys metalloendopeptidase for comparative proteomics. J Proteome Res 4:507–514
14. Schnolzer M, Jedrzejewski P, Lehmann WD (1996) Protease-catalyzed incorporation of 18O into peptide fragments and its application for protein sequencing by electrospray and matrix-assisted laser desorption/ionization mass spectrometry. Electrophoresis 17:945–953
15. Vosseller K, Hansen KC, Chalkley RJ, Trinidad JC, Wells L, Hart GW, Burlingame AL (2005) Quantitative analysis of both protein expression and serine/threonine post-translational modifications through stable isotope labeling with dithiothreitol. Proteomics 5:388–398
16. Wells L, Vosseller K, Cole RN, Cronshaw JM, Matunis MJ, Hart GW (2002) Mapping sites of O-GlcNAc modification using affinity tags for serine and threonine post-translational modifications. Mol Cell Proteomics 1:791–804
17. Yao X, Freas A, Ramirez J, Demirev PA, Fenselau C (2001) Proteolytic 18O labeling for comparative proteomics: model studies with two serotypes of adenovirus. Anal Chem 73:2836–2842
18. Ong SE, Foster LJ, Mann M (2003) Mass spectrometric-based approaches in quantitative proteomics. Methods 29:124–130
19. Wada Y, Azadi P, Costello CE, Dell A, Dwek RA et al (2007) Comparison of the methods for profiling glycoprotein glycans—HUPO Human Disease Glycomics/Proteome Initiative multi-institutional study. Glycobiology 17:411–422
20. Aoki K, Perlman M, Lim JM, Cantu R, Wells L, Tiemeyer M (2007) Dynamic developmental elaboration of N-linked glycan complexity in the *Drosophila melanogaster* embryo. J Biol Chem 282:9127–9142
21. Sutton CW, O'Neill JA, Cottrell JS (1994) Site-specific characterization of glycoprotein carbohydrates by exoglycosidase digestion and laser-desorption mass spectrometry. Anal Biochem 218:34–46
22. Kang P, Mechref Y, Klouckova I, Novotny MV (2005) Solid-phase permethylation of glycans for mass spectrometric analysis. Rapid Commun Mass Spectrom 19:3421–3428
23. Kang P, Mechre Y, Novotny MV (2008) High-throughput solid-phase permethylation of glycans prior to mass spectrometry. Rapid Commun Mass Spectrom 22:721–734
24. Karlsson NG, Wilson NL, Wirth HJ, Dawes P, Joshi H, Packer NH (2004) Negative ion graphitised carbon nano-liquid chromatography/mass spectrometry increases sensitivity for glycoprotein oligosaccharide analysis. Rapid Commun Mass Spectrom 18:2282–2292
25. Orlando R, Leymarie N, Keck R (2010) Results of the 2010 Glycoprotein Research Group's (gPRG) study on quantitative glycoprotein analysis. J Biomol Tech 21(3 suppl):S17
26. Yuan J, Hashii N, Kawasaki N, Itoh S, Kawanishi T, Hayakawa T (2005) Isotope tag method for quantitative analysis of carbohydrates by liquid chromatography–mass spectrometry. J Chromatogr A 1067:145–152

27. Hitchcock AM, Costello CE, Zaia J (2006) Glycoform quantification of chondroitin/dermatan sulfate using an LC/MS/MS platform. Biochemistry 45:2350–2361
28. Hsu J, Chang SJ, Franz AH (2006) MALDI-TOF and ESI-MS analysis of oligosaccharides labeled with a new multifunctional oligosaccharide tag. J Am Soc Mass Spectrom 17:194–204
29. Bowman MJ, Zaia J (2007) Tags for the stable isotopic labeling of carbohydrates and quantitative analysis by mass spectrometry. Anal Chem 79:5777–5784
30. Xie Y, Liu J, Zhang J, Hedrick JL, Lebrilla CB (2004) Method for the comparative glycomic analyses of O-linked, mucin-type oligosaccharides. Anal Chem 76:5186–5197
31. Alvarez-Manilla G, Warren NL, Abney T, Atwood J III, Azadi P, York WS, Pierce M, Orlando R (2007) Tools for glycomics: relative quantitation of glycans by isotopic permethylation using 13CH3I. Glycobiology 17:677–687
32. Kang P, Mechref Y, Kyselova Z, Goetz JA, Novotny MV (2007) Comparative glycomic mapping through quantitative permethylation and stable-isotope labeling. Anal Chem 79:6064–6073
33. Botelho JC, Atwood JA III, Cheng L, Alvarez-Manilla G, York WS, Orlando R (2008) Quantification by isobaric labeling (QUIBL) for the comparative glycomic study of O-linked glycans. Int J Mass Spectrom 278(2–3): 137–142
34. Atwood JA III, Cheng L, Alvarez-Manilla G, Warren NL, York WS, Orlando R (2008) Quantitation by isobaric labeling: applications to glycomics. J Proteome Res 7:367–374
35. Orlando R, Lim J, Atwood JA III, Angel PM, Fang M, Alvarez-Manilla G, Moremen KW, York WS, Tiemeyer M, Pierce M, Dalton S, Wells L (2009) IDAWG: metabolic incorporation of stable isotope labels for quantitative glycomics of cultured cells. J Proteome Res 8:3816–3823

Chapter 14

Stable Isotope Labeling of N-Glycosylated Peptides by Enzymatic Deglycosylation for Mass Spectrometry-Based Glycoproteomics

Hiroyuki Kaji and Toshiaki Isobe

Abstract

Protein glycosylation is one of the most common and crucial post-translational modifications that regulates many biological processes. Because abnormal glycosylation is also associated with various pathologies, including cancer, and inflammatory and degenerative diseases, technology for comprehensive analysis of glycoproteins, or glycoproteomics, is important not only for biological studies but also for biomedical and clinical research, including the discovery of biomarkers for disease diagnosis, prognosis, and therapeutic response to drugs. Here, we describe a protocol for peptide-*N*-glycanase-mediated ^{18}O labeling of N-glycosylated peptides, termed "isotope-coded glycosylation site-specific tagging." Coupled with advanced mass spectrometry-based proteomics technology, this method facilitates the identification of hundreds to thousands of N-glycoproteins, coupled with their sites of glycosylation, from a complex biological mixture.

Key words: *N*-glycosylation, Glycoproteomics, Mass spectrometry, Stable isotope labeling, Peptide-*N*-glycanase, Lectin, Oxygen-18 (^{18}O)

1. Introduction

Recent advances in mass spectrometry-based proteomics technology, coupled with biochemical procedures to capture specific functional groups attached to polypeptide chains, have facilitated comprehensive analysis of many post-translational modifications (PTMs) in proteins, such as phosphorylation, ubiquitination, and acetylation. For the analysis of N-linked protein glycosylation, one of the most common and important PTMs, two different approaches, either independent or dependent on the glycan structure attached

Jennifer J. Kohler and Steven M. Patrie (eds.), *Mass Spectrometry of Glycoproteins: Methods and Protocols*, Methods in Molecular Biology, vol. 951, DOI 10.1007/978-1-62703-146-2_14,

Table 1
Methods of glycopeptide capture

Method	References
Glycan structure-independent capture	
Periodate oxidation–hydrazide coupling	(1)
Hydrophilic interaction chromatography	(2–4)
Borate column chromatography	(19)
Multiple (mixed) lectin affinity chromatography	(20, 21)
Size-exclusion chromatography	(22)
Glycan structure-specific capture	
Lectin affinity chromatography on column	(5–8, 13, 15)
Lectin affinity chromatography on ultrafiltration membrane (FASP)	(14)
Metabolic incorporation of functional group into a specific monosaccharide	(9–12)
Chemoenzymatic modification of glycan (O-GlcNAc)	(23)
BEMAD (β-elimination/Michael addition with DTT, O-GlcNAc)	(24)

to the polypeptide chain, have been utilized to capture N-glycosylated peptides (Table 1). The glycan structure-independent approaches include periodate oxidation–hydrazide coupling (1), and some chromatographic procedures such as hydrophilic interaction chromatography (HILIC; (2–4)). These methods capture or concentrate glycosylated peptides regardless of their glycan structure, and are thereby suited for comprehensive analysis of N-linked glycosylation. The glycan structure-dependent approaches, on the other hand, include affinity capture via lectins (5–8), or via azide or alkyne groups introduced metabolically onto distinct glycan residues such as GlcNAc, sialic acid, GalNAc, and fucose (9–12). These methods are rather specific to a particular type of glycan structure, and are thereby used for the systematic analysis of N-glycoproteins, as well as for the analysis of a particular subset of N-glycoproteins of biological interest, such as those carrying glycan motifs including α1,6-core fucose, Lewis antigens, and branched or extended polylactosamine.

The method termed "isotope-coded glycosylation site-specific tagging (IGOT)" described herein is designed for the incorporation of a stable isotope, ^{18}O, specifically into N-glycosylated sites of proteins that are affinity-captured from a complex biological mixture, such that the ^{18}O label can serve as a specific tag of the N-glycosylated site during the MS-based identification of N-glycoproteins (13). The steps of this approach are: (1) lectin column-mediated affinity capture of glycopeptides generated by

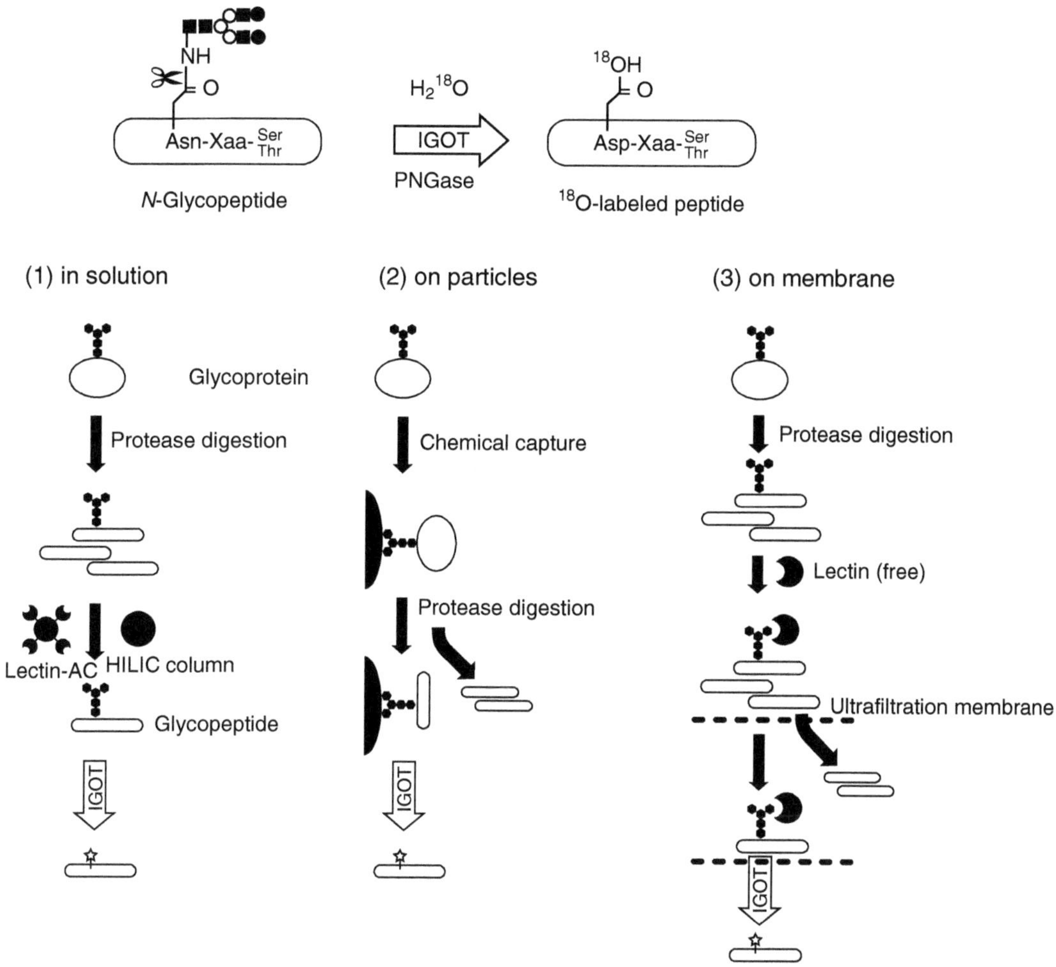

Fig. 1. Stable isotope labeling of glycopeptides by enzymatic deglycosylation (isotope-coded glycosylation site-specific tagging (IGOT)). (**a**) Schematic presentation of the IGOT reaction. (**b**) Variations of the IGOT reaction. The IGOT reaction can be performed (1) in solution, (2) on a solid support in the suspension, and (3) on an ultrafiltration membrane (see ref. (14) for details of the procedure (3)).

protease digestion of protein mixtures; (2) purification of the enriched glycopeptides by HILIC; (3) peptide-*N*-glycanase (PNGase)-mediated incorporation of a stable isotope tag, ^{18}O, specifically at the N-glycosylation site; and (4) identification of ^{18}O-tagged peptides by LC–MS-based proteomics technology. The IGOT approach can be performed in solution, on a solid support to capture glycopeptides via a covalent bond, or on an ultrafiltration membrane (14) (Fig. 1). The protocol is applicable to N-linked glycopeptides captured by a variety of lectin columns (Table 2) and to those captured by other methods such as HILIC. It is also applicable to protein mixtures from various sources including cells, cell-culture medium, organelles, tissues, body fluids, and whole organisms such as the nematode *Caenorhabditis elegans* (15).

Table 2
Representative lectins

Lectins (abbreviation)	Source	Specificity (monosaccharide)	Typical elution sugar
Con A	Jack bean (*Canavalia ensiformis*)	Man, Glc	0.2 M αMM
WGA	Wheat germ (*Triticum aestivum*)	GlcNAc	Chitin hydrolysate (GlcNAc oligo)
AAL	Orange peel mushroom (*Aleuria aurantia*)	Fuc	5 mM fucose
RCA120	Castor oil bean (*Ricinus communis*)	Gal	50 mM lactose
SSA	Red-berried elder (*Sambucus sieboldiana*)	NeuAc	50 mM lactose

Man mannose; *Glc* glucose; *GlcNAc* *N*-acetylglucosamine; *Fuc* fucose; *Gal* galactose; *NeuAc* *N*-acetylneuraminic acid; α*MM* methyl α-D-mannopyranoside

2. Materials

Prepare all solutions using ultrapure water and analytical grade reagents, unless stated otherwise.

2.1. Peptide Mixture Preparation and Lectin-Mediated Glycopeptide Capture

1. Ethylenediaminetetraacetic acid (EDTA): 0.5 M solution in water (pH 8.0). Weigh 18.6 g of EDTA disodium salt dihydrate and transfer to a 100-mL glass beaker containing about 60 mL of water. Stirring gently with a magnetic stir bar, add 1 M NaOH to the solution to adjust pH to 8.0. Make up to 100 mL with water in a 100-mL graduated cylinder. Store at 4°C.
2. Protein extraction buffer: 0.5 M Tris–HCl, pH 8.5, 7 M guanidine-HCl, 10 mM EDTA. Weigh 6.06 g of Tris and 66.9 g of guanidine-HCl. Transfer to a 100-mL glass beaker containing about 20 mL of water. Add 2 mL of 0.5 M EDTA. Adjust to pH 8.5 with 1 M HCl inside the fume hood. Make up to 100 mL with water in a 100-mL graduated cylinder. Store at 4°C (see Note 1).
3. Disulfide reduction reagent: Dithiothreitol (DTT).
4. Thiol alkylation reagent: Iodoacetamide (Wako Pure Chemical Industries, Osaka, Japan). Store at 4°C. Light sensitive. Use in the dark.
5. Tris buffer (TB): 0.1 M Tris–HCl (pH 7.4) (for lectin chromatography). Store at 4°C. Weigh 12.1 g of Tris and transfer to a 1-L glass beaker containing about 800 mL of water. Adjust to pH 7.4 with 0.5 M HCl. Make up to 1 L with water in a 1-L graduated cylinder.

6. TPCK (Tosyl phenylalanyl chloromethyl ketone)-treated trypsin (Thermo Fisher Scientific, Mass spectrometry grade). Store at −20°C. TPCK inhibits the chymotryptic activity in a trypsin preparation and thereby increases the efficiency of peptide assignment by MS/MS-ion searching.
7. *Achromobacter* protease I: Lysyl endopeptidase (Wako Pure Chemical Industries, Sequence-grade).
8. Phenylmethanesulfonyl fluoride (PMSF) (Sigma): 100 mM solution in MeOH. Store at 4°C.
9. Elution buffer for ConA affinity chromatography: 0.2 M Methyl α-D-mannopyranoside (αMM) in TB. Stable for several weeks at 4°C. Watch out for molding.
10. ConA affinity column (LA-ConA; 4.6 mm ID × 150 mm, Seikagaku Corporation, Tokyo): Concanavalin A from jack bean, *Canavalia ensiformis.*
11. Protein assay reagents (based on Lowry method, Bio-Rad Laboratories).

2.2. Glycopeptide Purification by Hydrophilic Interaction Chromatography

1. 75% (v/v) Acetonitrile (MeCN) in 0.1% (v/v) trifluoroacetic acid (TFA). Stable at room temperature (see Note 2).
2. 40% (v/v) MeCN in 0.1% (v/v) TFA. Stable at room temperature (see Note 3).
3. 10% (v/v) TFA in water (see Note 2).
4. HILIC column (Amide-80, 2 mm ID × 50 mm, TOSOH, Tokyo).

2.3. PNGase-Mediated ^{18}O Labeling of N-Glycosylated Peptides

1. ^{18}O-labeled water ($H_2{}^{18}O$, >99 atom % ^{18}O) (Taiyo Nippon Sanso Corp., Tokyo).
2. PNGase buffer: 1 M Tris–AcOH, pH 7.5. Prepare with ^{18}O-labeled water.
3. PNGase F (Takara Bio Inc., Shiga, Japan): Dissolve the lyophilyzed enzyme powder with $H_2{}^{18}O$ at 5 mU/μL. Store at −20°C as 5-μL aliquots.

3. Methods

3.1. Peptide Mixture Preparation and Lectin-Mediated Glycopeptide Capture

1. Homogenize a sample tissue (typically 0.2–1 g) in a tenfold volume of protein extraction buffer using Polytron or other appropriate equipment (see Note 4). Remove insoluble and floating materials by centrifugation at 10,000 × *g* for 30 min at 4°C.
2. Quantify protein concentration by Lowry assay.
3. Insert capillary tubing connected to an N_2 gas line and introduce N_2 gas bubbles for 15 min at 1–2 bubbles/s. Add DTT

at a weight equal to the amount of protein and continue N_2 gas bubbling for 2 h (see Note 5).

4. Add iodoacetamide (at a 2.5-fold weight of protein), dissolve the reagent by vortex mixing, and leave the solution in the dark for 2 h.
5. Dialyze the reaction mixture at 4°C against 100 volumes of TB. Change the buffer solution twice, each time after 2 h of dialysis, and then leave overnight (see Note 6).
6. Transfer the dialyzed solution to a polypropylene (PP) tube. Add 1/50 (w/w) trypsin or 1/100 (w/w) lysyl endopeptidase, and leave the mixture overnight at 37°C (see Note 7).
7. Add 1/100 volume of 100 mM PMSF solution (see Note 8).
8. Load the peptide mixture onto a lectin affinity column, e.g., a ConA column (4.6 mm ID × 150 mm), equilibrated with TB. Other lectin columns with different substrate specificities can be used for capturing glycopeptides with other types of sugar moieties attached to the polypeptide chain (Table 2) (see Note 9).
9. Wash the column with ~5-column volumes of TB until the effluent's absorbance at 280 nm becomes <0.1.
10. Reload the flow-through fraction onto the same column and repeat steps 8 and 9 to maximize glycopeptide recovery.
11. Elute the column with 20 mL of elution buffer (0.2 M αMM in TB) (see Notes 10 and 11).

3.2. Glycopeptide Purification by HILIC

1. To the eluate, add MeCN and 10% TFA to a final concentration of 75% (v/v) and 0.1% (v/v), respectively. Remove precipitate, if any, by centrifugation.
2. Load the solution onto the Amide 80 HILIC column (2 mm ID × 50 mm) equilibrated with 75% MeCN in 0.1% TFA.
3. Wash the column with the same solvent until the effluent's absorbance at 220 nm becomes <0.05.
4. Elute the column with 0.8 mL of 40% MeCN in 0.1% TFA to recover the glycopeptides.
5. Transfer 0.1 mL of the eluate into 0.6-mL polypropylene (PP) tubes. Add 2 μL of glycerol and evaporate the solvent by a centrifugal vacuum concentrator. Add another 0.1-mL aliquot of the eluate into the same tube and evaporate. Repeat the process until the total eluate is dry (see Note 12).

3.3. PNGase-Mediated ^{18}O-Labeling of *N*-Glycosylated Peptides

1. Add 5 μL of 1 M Tris–AcOH, pH 8.0, prepared with $H_2{}^{18}O$, into the PP tube.
2. Add 40 μL of $H_2{}^{18}O$ and dissolve the glycopeptide by vortex mixing. Ensure that the pH of the solution is ~8 with a test paper. If the pH is lower than ~8, add 1 μL of 1 M Tris–AcOH, pH 8.0.

3. Add 5 μL of 1 mU/μL PNGase in $H_2{}^{18}O$. Incubate the mixture at 37°C for 5 h for simultaneous deglycosylation and ^{18}O labeling.

3.4. LC/MS Analysis for Peptide Identification

1. Acidify the sample solution by adding an aliquot of 1% formic acid. Remove precipitates, if any, by centrifugation.
2. Load the sample solution to an LC-coupled ESI-tandem MS system equipped with a Q-TOF, ion-trap, or Orbitrap mass spectrometer for the identification of ^{18}O-labeled peptides.
3. Search the spectral data by an MS/MS ion-searching program, such Mascot® or/and Sequest®, against protein sequence database(s), including NCBI-RefSeq (ftp://ftp.ncbi.nih.gov/refseq/) and EBI-IPI (ftp://ftp.ebi.ac.uk/pub/databases/IPI/current/). Parameters for identifying the ^{18}O-labeled, former glycopeptides are as follows. Fixed modifications: carbamidomethylation (carbamoylmethylation) of Cys. Variable modifications: deamination of Gln at the peptide N-terminus (pyroglutamination) and of carbamidomethylated Cys at the peptide N-terminus; oxidation of Met; and deamidation plus incorporation of ^{18}O in Asn (IGOT).
4. Extract and evaluate the search results by a software program such as Scaffold (16) or STEM (17). Our criteria for the glycopeptide identification are:
   (a) The "expect" value of the Mascot result is <0.05; (b) the identified peptide sequence is ranked top of the candidates; (c) the sequence contains one or more consensus sequences for N-linked glycosylation, NX(S/T), where X is any amino acid except P; (d) the identified peptide contains one or more IGOT-labeled Asn residues; and (e) the number of labeled Asn residues is less than or equal to the number of consensus sequence (see Note 13).

4. Notes

1. Guanidine-HCl; harmful if swallowed. Avoid contact with eyes, skin, and clothing.
2. TFA; Danger! Corrosive. Causes burns. Harmful if swallowed, inhaled, or absorbed through the skin.
3. MeCN is toxic. Harmful if swallowed, inhaled, or absorbed through the skin. Wear gloves. Flammable liquid and vapor.
4. For the analysis of body fluids such as serum or cerebrospinal fluids, dilute 100 μL of the sample with 900 μL of the protein extraction buffer. Remove floating materials by centrifugation at 10,000 × *g* for 30 min at 4°C.

5. Addition of an aliquot of EtOH into the solution reduces the froth of the protein. For a small-volume sample, perform the reduction reaction in a sealed tube flushed with N_2 gas.
6. For a small-volume sample, typically less than 0.2 mL, add ice-cold acetone at a final concentration of 80%. Leave the mixture at −20°C for 30 min, and recover the precipitated proteins by centrifugation. Dissolve the precipitate in 30 μL of 0.1 M Tris–HCl, pH 8.0, containing 6 M urea by vortex mixing. Do not heat the sample to avoid *N*-cyanoethylation of lysine. Dilute the sample with 60 μL of 0.1 M Tris–HCl, pH 8.0, to make a 2-M urea solution before protease digestion.
7. If precipitation occurs during the dialysis, add protease to the suspension without centrifugation. The precipitated proteins might be digested by protease to give a clear solution. Any precipitate in the final mixture should be removed by centrifugation. We recommend SDS-PAGE to examine the extent of digestion. If protein bands are detected after the overnight digestion, add another aliquot (1/100–1/50 w/w) of protease and extend the digestion for >5 h.
8. PMSF stops the digestion and protects the lectin column used in the subsequent step. The sample can be stored frozen at −20°C for several weeks.
9. Use a column with sufficient capacity to capture target glycopeptides in the sample mixture. The capacity of column depends on the amount of immobilized lectin; a commercial ConA column has a relatively high ConA density (e.g., 20 mg/mL support), whereas others might have lower density (e.g., 2 mg/mL). See suppliers' instruction for the characteristics of each lectin column.
10. The sample can be stored frozen at −20°C for several weeks.
11. Typical elution sugars to recover glycopeptides bound to several other lectin columns are given in Table 2.
12. This step is time-consuming but is crucial to minimize loss of glycopeptides due to nonspecific adsorption to the PP tube wall. Addition of glycerol into the effluent avoids complete dryness that may reduce the recovery of glycopeptides.
13. The number of glycopeptides and glycosylation sites identified by this protocol depends largely on the performance of the LC–MS system. The LC–MS system that we use for the analysis of ^{18}O-labeled peptides is described in refs. (13, 15, 18). In brief, the system consists of a direct nano-flow two-dimensional LC apparatus equipped with a cation exchange column (Bioassist-S, 7 μm particles, TOSOH, 0.75 mm ID × 40 mm; first dimension) and a reversed-phase tip column (Mightysil-C18, 3 μm particle, 0.15 mm ID × 50 mm, Kanto Chemical Co., Tokyo; second dimension), which is coupled to a Q-TOF

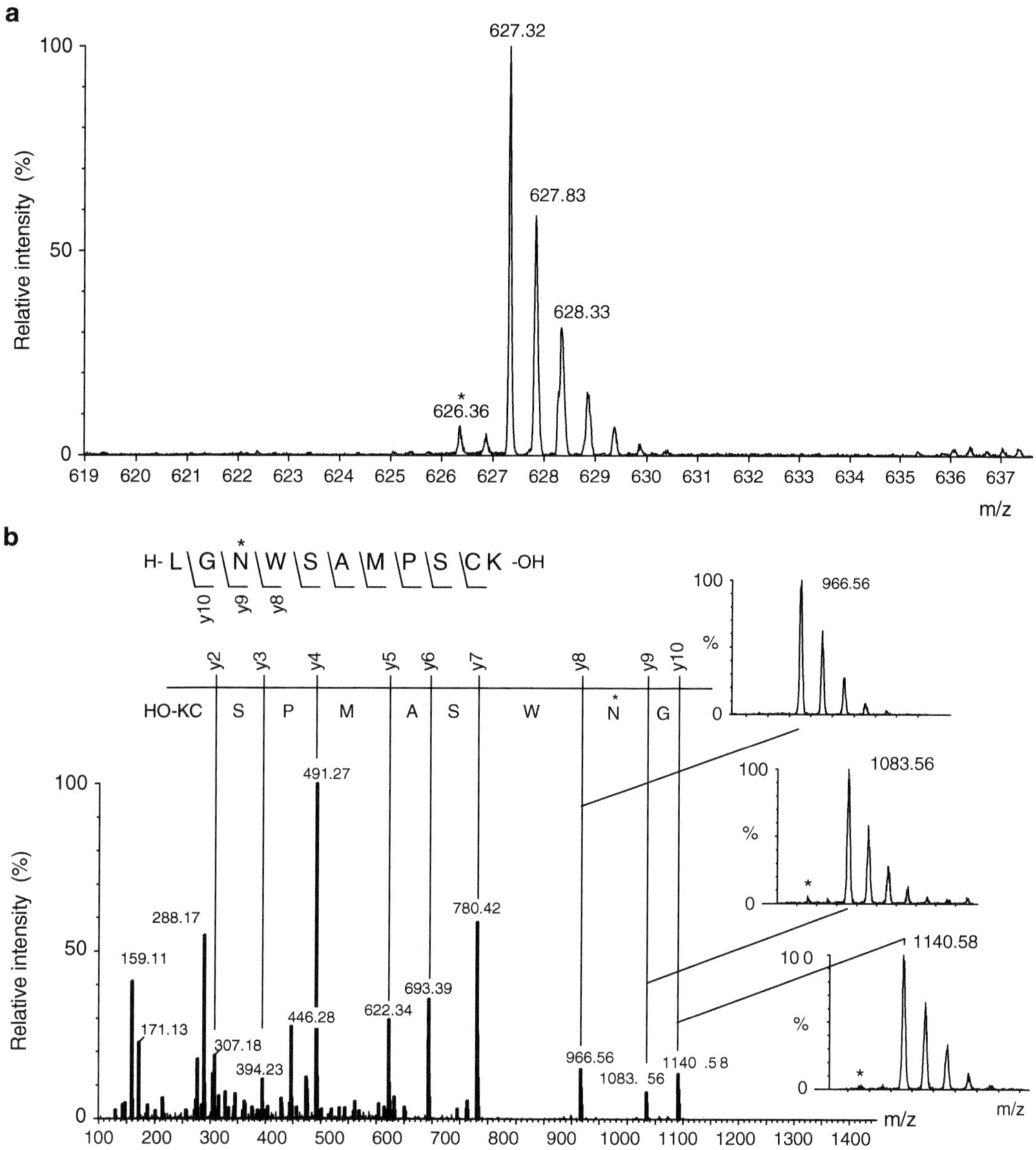

Fig. 2. Typical MS and MS/MS spectra of an IGOT-peptide derived from human apolipoprotein H, spanning residues 251–261 with the sequence LGNWSAMPSCK, where N-253 was previously glycosylated. (**a**) Magnified MS spectrum selected for collision-induced dissociation analysis. The IGOT peptide shows characteristic signals (*) on the low mass side of the monoisotopic signal of the labeled peptide, which results from the incorporation of ^{16}O. (**b**) MS/MS spectrum of the labeled peptide and the signal annotation (y2–y10). The fragment signals y9 and y10, which include deamidated Asn (Asp), show the characteristic signal (*) indicating the incorporation of $^{16}O/^{18}O$.

Ultima API mass spectrometer (Waters-Micromass) through an ESI interface. In a typical analysis using ~10 μg of peptides, this system facilitates the identification of hundreds up to a thousand glycopeptides and glycosylation sites in a single analytical run. Typical MS and MS/MS spectra of IGOT-peptide are shown in Fig. 2.

References

1. Zhang H, Li XJ, Martin DB, Aebersold R (2003) Identification and quantification of N-linked glycoproteins using hydrazide chemistry, stable isotope labeling and mass spectrometry. Nat Biotechnol 21:660–666
2. Mysling S, Palmisano G, Højrup P, Thaysen-Andersen M (2010) Utilizing ion-pairing hydrophilic interaction chromatography solid phase extraction for efficient glycopeptide enrichment in glycoproteomics. Anal Chem 82:5598–5609
3. Calvano CD, Zambonin CG, Jensen ON (2008) Assessment of lectin and HILIC based enrichment protocols for characterization of serum glycoproteins by mass spectrometry. J Proteomics 3:304–317
4. Wohlgemuth J, Karas M, Eichhorn T, Hendriks R, Andrecht S (2009) Quantitative site-specific analysis of protein glycosylation by LC-MS using different glycopeptide-enrichment strategies. Anal Biochem 395:178–188
5. Jung K, Cho W, Regnier FE (2009) Glycoproteomics of plasma based on narrow selectivity lectin affinity chromatography. J Proteome Res 8:643–650
6. Comunale MA, Wang M, Hafner J, Krakover J, Rodemich L, Kopenhaver B, Long RE, Junaidi O, Bisceglie AM, Block TM, Mehta AS (2009) Identification and development of fucosylated glycoproteins as biomarkers of primary hepatocellular carcinoma. J Proteome Res 8:595–602
7. Kaji H, Yamauchi Y, Takahashi N, Isobe T (2006) Mass spectrometric identification of N-linked glycopeptides using lectin-mediated affinity capture and glycosylation site-specific stable isotope tagging. Nat Protoc 1:3019–3027
8. Wei X, Dulberger C, Li L (2010) Characterization of murine brain membrane glycoproteins by detergent assisted lectin affinity chromatography. Anal Chem 82:6329–6333
9. Vocadlo DJ, Hang HC, Kim EJ, Hanover JA, Bertozzi CR (2003) A chemical approach for identifying O-GlcNAc modified proteins in cells. Proc Natl Acad Sci USA 100: 9116–9121
10. Sampathkumar SG, Li AV, Jones MB, Sun Z, Yarema KJ (2006) Metabolic installation of thiols into sialic acid modulates adhesion and stem cell biology. Nat Chem Biol 2:149–152
11. Prescher JA, Bertozzi CR (2006) Chemical technologies for probing glycans. Cell 126:851–854
12. Dube DH, Prescher JA, Quang CN, Bertozzi CR (2006) Probing mucin-type O-linked glycosylation in living animals. Proc Natl Acad Sci USA 103:4819–4824
13. Kaji H, Saito H, Yamauchi Y, Shinkawa T, Taoka M, Hirabayashi J, Kasai K, Takahashi N, Isobe T (2003) Lectin affinity capture, isotope-coded tagging and mass spectrometry to identify N-linked glycoproteins. Nat Biotechnol 21:667–672
14. Zielinska DF, Gnad F, Wiśniewski JR, Mann M (2010) Precision mapping of an in vivo N-glycoproteome reveals rigid topological and sequence constraints. Cell 141:897–907
15. Kaji H, Kamiie J, Kawakami H, Kido K, Yamauchi Y, Shinkawa T, Taoka M, Takahashi N, Isobe T (2007) Proteomics reveals N-linked glycoprotein diversity in Caenorhabditis elegans and suggests an atypical translocation mechanism for integral membrane proteins. Mol Cell Proteomics 6:2100–2109
16. http://www.proteomesoftware.com/Proteome_software_prod_Scaffold.html
17. Shinkawa T, Taoka M, Yamauchi Y, Ichimura T, Kaji H, Takahashi N, Isobe T (2005) STEM: a software tool for large-scale proteomic data analyses. J Proteome Res 4:1826–1831
18. Natsume T, Yamauchi Y, Nakayama H, Shinkawa T, Yanagida M, Takahashi N, Isobe T (2002) A direct nanoflow liquid chromatography-tandem mass spectrometry system for interaction proteomics. Anal Chem 74:4725–4733
19. Xu Y, Wu Z, Zhang L, Lu H, Yang P, Webley PA, Zhao D (2009) Highly specific enrichment of glycopeptides using boronic acid-functionalized mesoporous silica. Anal Chem 81:503–508
20. Yang Z, Hancock WS (2004) Approach to the comprehensive analysis of glycoproteins isolated from human serum using a multi-lectin affinity column. J Chromatogr A 1053:79–88
21. Heo SH, Lee SJ, Ryoo HM, Park JY, Cho JY (2007) Identification of putative serum glycoprotein biomarkers for human lung adenocarcinoma by multilectin affinity chromatography and LC-MS/MS. Proteomics 7:4292–4302
22. Alvarez-Manilla G, Atwood J III, Guo Y, Warren NL, Orlando R, Pierce M (2006)

Tools for glycoproteomic analysis: size exclusion chromatography facilitates identification of tryptic glycopeptides with N-linked glycosylation sites. J Proteome Res 5:701–708

23. Khidekel N, Ficarro SB, Peters EC, Hsieh-Wilson LC (2004) Exploring the O-GlcNAc proteome: direct identification of O-GlcNAc-modified proteins from the brain. Proc Natl Acad Sci U S A 101:13132–13137

24. Wells L, Vosseller K, Cole RN, Cronshaw JM, Matunis MJ, Hart GW (2002) Mapping sites of O-GlcNAc modification using affinity tags for serine and threonine post-translational modifications. Mol Cell Proteomics 1:791–804

Chapter 15

Approaches for Site Mapping and Quantification of O-Linked Glycopeptides

Peng Zhao, Stephanie H. Stalnaker, and Lance Wells

Abstract

As a complex post-translational event, the biosynthesis, structures, and functions of O-linked glycans have attracted research interests in various aspects. The recent development of novel technologies for the analysis of glycans and glycoproteins sheds new insights with regard to determining site occupancy, structure–function relationships, and the contributions of O-linked glycosylation to physiological and pathological processes. In this chapter, we refer to several approaches for the structural characterization and quantification of O-linked glycopeptides, with a focus on O-GlcNAc and O-Mannose modified glycoproteins.

Key words: O-linked glycosylation, Isolation, Antibody, Lectin, Quantification, O-GlcNAc, O-Mannose, Site mapping, HCD/ETD

1. Introduction

Glycosylation is one of the most frequently occurring post-translational modifications in eukaryotic cells. Oligosaccharide moieties are attached to proteins mainly through the amine group of asparagine (N-linked) or the hydroxyl group of serine and threonine (O-linked). N-linked glycosylation is initiated in the endoplasmic reticulum and further processed in the Golgi apparatus. Three main classes of N-linked structures exist: high-mannose, complex, and hybrid type glycans. Classical mucin-like O-linked glycosylation begins in the Golgi apparatus with the addition of an *N*-acetylgalactosamine (GalNAc) to serine or threonine residues catalyzed by a polypeptide GalNAc transferase using UDP-GalNAc as the sugar donor. A stepwise enzymatic elongation is subsequently carried out by specific transferases and yields several core

Jennifer J. Kohler and Steven M. Patrie (eds.), *Mass Spectrometry of Glycoproteins: Methods and Protocols*,
Methods in Molecular Biology, vol. 951, DOI 10.1007/978-1-62703-146-2_15,

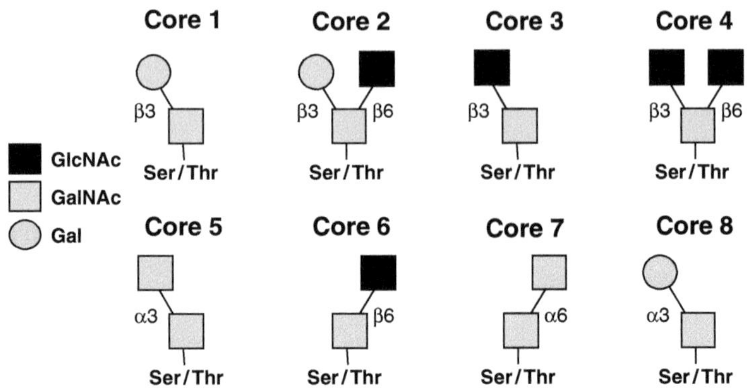

Fig. 1. Core structures of O-GalNAc-initiated glycans.

structures, which are further extended or modified by sialylation, sulfatation, acetylation, fucosylation, and polylactosamine extension. Eight core structures have been identified to date (1, 2) and are shown in Fig. 1. In addition to O-GalNAc-initiated structures, several other types of O-linked glycosylation have been observed. For example, O-Fucose and O-Glucose were found on a specific consensus-sequence in the epidermal growth factor (EGF) protein domains (3). Reversible O-linked glycosylation by *N*-acetylglucosamine (GlcNAc) occurs on many cytoplasmic and nuclear proteins (4). O-Mannose structures are commonly found in yeast where O-linked glycosylation is initiated in endoplasmic reticulum instead of Golgi (5), and have recently been characterized on a single mammalian protein α-Dystroglycan (6). Thus, O-glycosylation can be initiated by a number of monosaccharides and then further extended to more complex structures.

Studies have revealed that the presence of oligosaccharide moieties in soluble and membrane bound proteins can modulate protein function by improving their solubility in water, contributing to molecular folding, and influencing biological lifetime (7, 8). Some glycan structures are cell specific and regulated to allow the phenotypes of cells to change in response to environmental conditions, stage of development, etc. (9), which necessitates the structural analysis of glycosylation with respect to both the occurrences and localizations within a protein. Compared to N-linked structures, the analysis of O-linked glycosylation has proven to be more difficult. First, contrary to N-glycosylation, where the sequence motif N-X-S/T (X being any amino acid except proline) is known to form a specific glycosylation target, no particular sequon has been described for most types of O-linked glycosylation, leaving the site of modification unpredictable. Second, the intact N-linked glycans can be released from the protein with peptide *N*-glycosidase F or A (PNGase F/A); however, no equivalent enzyme is available for O-linked glycans, which hampers the separate analysis of O-linked glycans and proteins. Furthermore, because of the diversity of O-linked glycosylation, such as O-Fucose, O-Mannose, O-GlcNAc, and mucin-type O-glycans, each of those modifications brings along a variety of specific methodological difficulties. For instance, because of their regulatory function, O-GlcNAc modifications are generally substoichiometric and hence of very low abundance, which requires selective enrichment of O-GlcNAc peptides prior to analysis by mass spectrometry. With regard to O-Mannose and O-Fucose initiated structures, isolation of proteins modified by these rare glycans from the bulk of mucin-type O-glycosylated or N-glycosylated proteins represents a major challenge.

In this chapter, we refer to several mass spectrometry-based approaches for the structural characterization and quantification of O-linked glycopeptides. Specifically, the isotopic labeling of O-linked glycopeptides using β-elimination followed by Michael addition with dithiothreitol (BEMAD), and the subsequent characterization and quantification by mass spectrometry (MS) will be discussed; furthermore, label-free techniques for the enrichment of O-linked glycopeptides, such as immunoprecipitation and lectin affinity chromatography, will also be reviewed. The isolation and detection of O-linked glycopeptides using site assignment and quantification of O-GlcNAc and O-Mannose modified glycoproteins will be discussed.

2. Materials

2.1. Base-Catalyzed β-Elimination of O-Linked Glycopeptides Followed by Isotopic Labeling with Dithiothreitol (BEMAD)

1. Performic acid oxidation buffer (made fresh): 45% (v/v) formic acid, 5% (v/v) hydrogen peroxide, in Milli-Q water.
2. 1 mM $MgCl_2$.
3. Alkaline phosphatase (Promega).
4. Dithiothreitol (DTT): light (d0) and heavy (d6) (C/D/N Isotopes).
5. BEMAD solution (made fresh): 1.5% (v/v) triethylamine (TEA), 0.15% (v/v) NaOH, 20 mM DTT (either d0 or d6).
6. C18 Reversed-phase macro-spin or micro-spin columns (The Nest Group).
7. 1% (v/v) Trifluoroacetic acid (TFA).
8. 1% (v/v) TFA in 75% (v/v) acetonitrile.
9. Thiol column buffer (made fresh), degassed: 20 mM Tris, pH 7.6, 150 mM NaCl, 1 mM EDTA (see Note 1).
10. Thiol column elution buffer (made fresh), degassed: 20 mM Tris, pH 7.6, 150 mM NaCl, 1 mM EDTA, 20 mM DTT (see Note 1).

11. Thiopropyl Sepharose 6B (Amersham Biosciences).
12. 0.1% (v/v) Formic acid.
13. Savant Speed-Vac concentrator.
14. Liquid chromatography (LC) coupled with tandem MS equipped with appropriate ion source (see Note 2).
15. Control peptides (see Note 3).
16. Approximately 1–100 pmol of protein sample in 40 mM NH_4HCO_3, pH 8.0 (see Note 4).
17. Seal-Rite™ Natural microcentrifuge tubes (USA Scientific) (see Note 5).

2.2. Antibody Enrichment of O-GlcNAc Modified Glycoproteins from Cell Lysate (see Note 6)

1. Protein A/G PLUS agarose.
2. Normal mouse IgG AC (Santa Cruz).
3. O-GlcNAc-specific monoclonal antibodies: 18B10.C7(3), 9D1.E4(10), and 1F5.D6(14) (see Note 7).
4. Phosphate buffered saline (PBS) or Tris buffered saline (TBS), pH 7.5.
5. Dimethyl sulfoxide (DMSO) (Sigma).
6. Disuccinimidyl suberate (DSS) (Thermo Fisher Scientific).
7. 100 mM Glycine (pH 2.5).
8. 1 M Tris (pH 8.8).
9. 1% (v/v) NP40 (IGEPAL630).
10. 0.1% (w/v) Sodium dodecyl sulfate (SDS).
11. 0.1% (v/v) Formic acid.
12. 0.1% (v/v) Formic acid in 80% (v/v) acetonitrile.
13. C18 Reversed-phase macro-spin or micro-spin columns (The Nest Group).
14. Savant Speed-Vac concentrator.
15. Seal-Rite™ Natural microcentrifuge tubes (USA Scientific) (see Note 5).
16. LTQ Orbitrap XL™ ETD mass spectrometer (Thermo Fisher Scientific).

2.3. Wisteria Floribunda Agglutinin (WFA) Lectin Enrichment of O-Glycosylated (O-Mannosylated) Peptides That Terminate with a GalNAc Residue (see Note 8)

1. *N*-acetylgalactosamine (Toronto Research Chemicals).
2. Agarose bound WFA Lectin (Vector Labs).
3. Sequence grade trypsin (Promega).
4. 40 mM NH_4HCO_3, pH 8.0.
5. Iodoacetamide (Sigma).
6. 2-mL Chromatography column (Bio-Rad).

7. Lectin Column buffer: 10 mM Tris–HCl, pH 7.4; 0.15 M NaCl; 1 mM $CaCl_2$; 1 mM $MnCl_2$.
8. Elution Buffer: 200 mM *N*-acetylgalactosamine (GalNAc, Toronto Research Chemicals Inc.) prepared in lectin column buffer.
9. 0.1% (v/v) Formic acid.
10. 0.1% (v/v) Formic acid in 80% (v/v) acetonitrile.
11. C18 Reversed-phase macro-spin or micro-spin columns (The Nest Group).
12. Savant Speed-Vac concentrator.
13. Seal-Rite™ Natural microcentrifuge tubes (USA Scientific) (see Note 5).
14. LTQ Orbitrap XL™ mass spectrometer (Thermo Fisher Scientific).

3. Methods

3.1. Base-Catalyzed β-Elimination to Isotopically Label and Quantify O-Linked Glycoproteins (BEMAD) (Fig. 2)

3.1.1. Performic Acid Oxidation

1. Suspend protein sample in 300 μL performic acid oxidation buffer (see Note 9).
2. Spike with 1–10 pmol of control peptides (see Note 10).
3. Incubate on ice for 1 h.
4. Dry down in Speed-Vac.

3.1.2. Trypsin Digestion

1. Resuspend protein sample in 40 mM NH_4HCO_3.
2. Digest by addition of 1:10–1:100 (w/w) sequencing-grade trypsin overnight (12–16 h) at 37°C.
3. Acidify the digest by the addition of TFA to 1% (v/v) final concentration.
4. Clean up over a C18 reversed-phase column (see manufacturer's instructions).
5. Dry down peptides in Speed-Vac.

3.1.3. Phosphatase Treatment

1. Resuspend peptides in 40 mM NH_4HCO_3, 1 mM $MgCl_2$.
2. Add alkaline phosphatase (1 U/10 μL) and incubate at 37°C for 4 h (see Note 11).
3. Dry down peptides in Speed-Vac.

3.1.4. BEMAD Treatment

1. Resuspend peptides in 500 μL BEMAD solution (either the d0 or d6 version) and adjust pH to 12.5–13 with TEA if necessary (see Note 12).
2. Incubate reaction at 52°C for 1.5 h.

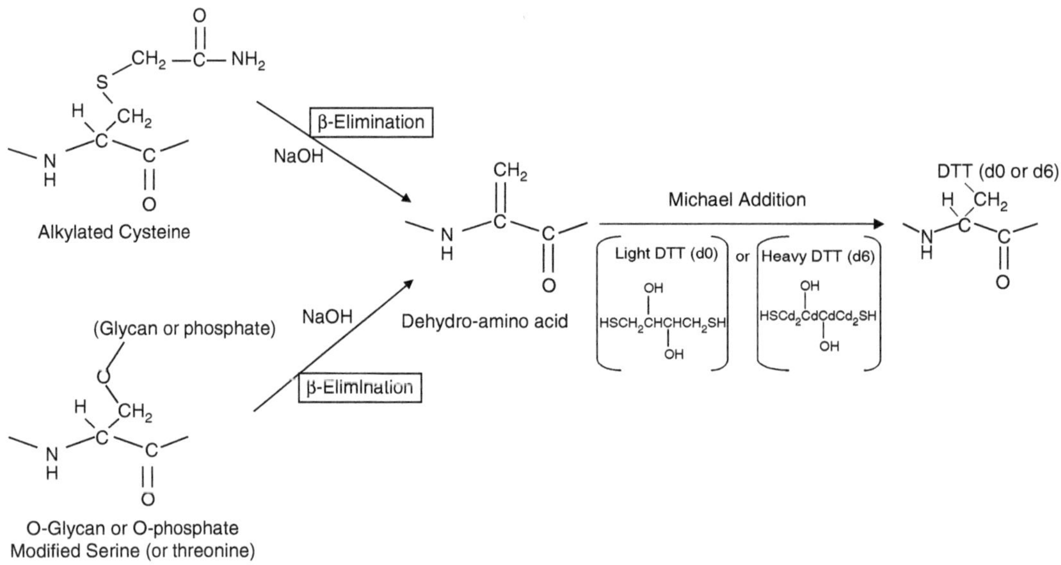

Fig. 2. Schematic of expanded BEMAD approach β-elimination of O-GlcNAc and replacement with DTT (BEMAD) through Michael addition chemistry. Alkylated cysteines as well as modified serines/threonine can be β-eliminated and for quantification purposes either standard, "light," d0-DTT or deuterated, "heavy," d6-DTT can be added so that a mass difference of 6 Da is generated for the otherwise identical peptides.

3. Stop reaction by adding TFA to 2% (v/v) final concentration (below pH 5).
4. Clean up over C18 reversed-phase column (see manufacturer's instructions).
5. Dry down peptides in Speed-Vac.

3.1.5. Thiol-Affinity Chromatography

1. Swell and wash thiopropyl sepharose resin several times in degassed thiol column buffer.
2. Resuspend peptides in thiol column buffer.
3. Bind peptides to thiol column at room temperature for 1 h (see Note 13).
4. Wash column with 20 mL thiol column buffer.
5. Elute peptides three times sequentially with 150 μL thiol column elution buffer.
6. Acidify peptides by adding TFA to 0.5% (v/v) final concentration.
7. Clean up over C18 reversed-phase column to remove free DTT.
8. Dry down peptides in Speed-Vac concentrator.

3.1.6. Liquid Chromatography Coupled with Tandem Mass Spectrometry (LC-MS/MS) Analysis

1. Resuspend peptides in 0.1% (v/v) formic acid.
2. Load sample onto capillary 75 mm i.d. column packed with C18.

3. Separate sample over an 80 min linear gradient (see Note 14) of increasing acetonitrile at a flow rate of 200–300 nL/min into the MS source. Throughout the LC gradient, spectral data may be recorded continuously with an MS scan followed by MS/MS scans of the most intense ions (top two to eight). Dynamic exclusion should also be applied to prevent repetitive selection of the same ions within a preset time. Collision energies may be adjusted automatically according to the charge state and mass value of the precursor ion (see Note 15).

3.1.7. Quantification of BEMAD-Labeled O-Linked Glycopeptides Using Mass Spectrometry Data

1. The acquired raw spectral data may be processed and searched against protein sequence databases using available and compatible proteomic search algorithms, such as MASCOT (Matrix Science), SEQUEST (Thermo Fisher Scientific), or any in-house programs. When specifying database search parameters, the following differential modification should be accounted for: DTT (d0) labeled serine and threonine (+136.1 Da); DTT (d6) labeled serine and threonine (+142.1 Da); oxidized cysteine (+48.0 Da); oxidized tryptophan (+48.0 Da); and oxidized methionine (+32.0 Da) (see Note 16).
2. Quantification of ion pair ratios may be performed manually by averaging isotope envelope area over the time of elution of a given ion pair using Xcalibur (Thermo Fisher Scientific) (Fig. 3).

3.2. Site Mapping and Sequence Characterization of O-Linked Glycopeptides by Nano Liquid Chromatography Coupled with Electron Transfer Dissociation (ETD) and High-Energy Collisional Dissociation (HCD) Tandem Mass Spectrometry (nLC-MS/MS)

3.2.1. Enrichment of O-GlcNAc Modified Glycoproteins from Cell Lysates Using Antibodies (see Notes 6 and 17)

1. Weigh out 1 g Protein A/G PLUS agarose into 20 mL disposable column and add 15 mL of PBS into stopped column, cap and let rotate for at least 30 min (do not exceed 2 h) at RT.
2. Wash with 3 × 10 mL of PBS, taking care not to let column dry out completely, and add PBS to reach an approximately 50/50 slurry of the Protein A/G PLUS agarose.
3. Add 900 μL of Protein A/G PLUS agarose slurry (~450 μL bed volume) into a clean tube, centrifuge at 2,000 rpm for 5 min, and discard supernatant (see Note 18).
4. Wash the slurry with PBS, 2 × 400 μL, and discard supernatant.
5. Add 450 μg of antibody mab-14 (or mab-3, mab-10) to the slurry and rotate at 4°C for 30 min (see Note 19).
6. Wash the coupled antibody slurry with PBS, 3 × 500 μL; add 400 μL of PBS at the last time.
7. Dissolve 2 mg of DSS in 80 μL of DMSO.
8. Add 25 μL of the dissolved DSS to the bound antibody, mix for 30–60 min with rotation at RT.
9. Centrifuge it and discard supernatant.

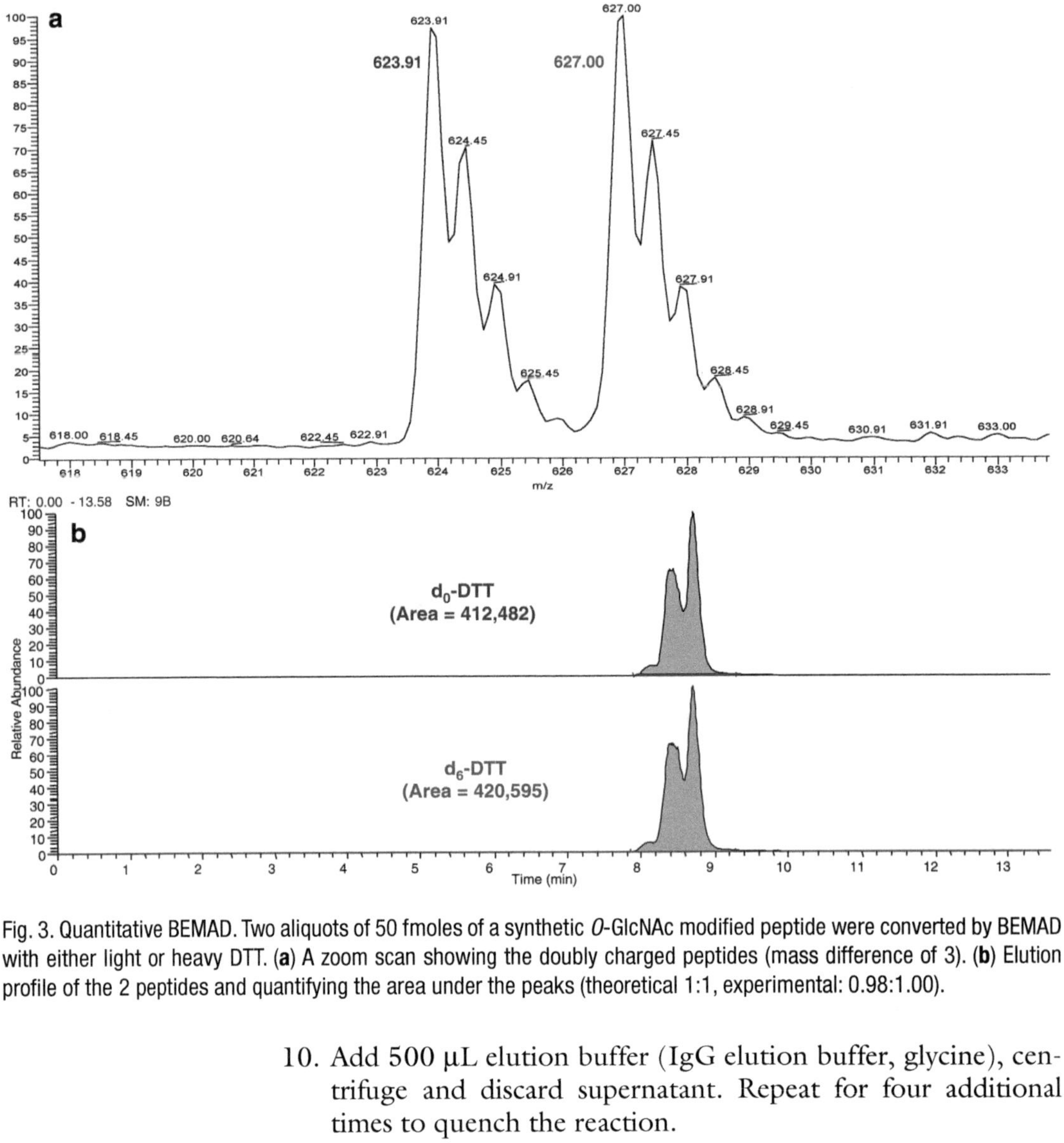

Fig. 3. Quantitative BEMAD. Two aliquots of 50 fmoles of a synthetic *O*-GlcNAc modified peptide were converted by BEMAD with either light or heavy DTT. (**a**) A zoom scan showing the doubly charged peptides (mass difference of 3). (**b**) Elution profile of the 2 peptides and quantifying the area under the peaks (theoretical 1:1, experimental: 0.98:1.00).

10. Add 500 μL elution buffer (IgG elution buffer, glycine), centrifuge and discard supernatant. Repeat for four additional times to quench the reaction.
11. Wash the cross-linked antibody with PBS, 3 × 500 μL.
12. Store at 4°C as a 50/50 slurry in PBS (0.9 mL).
13. Preclear cell lysates with normal mouse IgG-agarose conjugate and protein A/G PLUS agarose slurry for 30 min at 4°C with rotation.
14. Pass the precleared sample through an empty 10-mL disposable column.
15. Incubate precleared sample with mab-14 (or mab-3 or mab-10) Ab-agarose covalent coupled beads at 4°C overnight with rotation.
16. Allow column to flow-through, collect flow-through and pass back through column 1×, collect and save flow-through.

17. Wash column 3× with 1% NP40 (IGEPAL630), 0.1% SDS in PBS (9 mL each), then 3× with 5 mL PBS.
18. Stopper the column then add 600 μL of 100 mM glycine pH 2.5, let sit 3 min, drain into a clean tube containing 150 μL 1 M Tris pH 8.8. Repeat elution with another 600 μL of glycine and collect into the same tube.
19. Immediately regenerate the pH in antibody column by washing with PBS 3 × 3 mL, store the antibody column in ~1 mL PBS at 4°C.

3.2.2. Trypsin Digestion of Enriched Glycoproteins and Reverse-Phase Desalting of Glycopeptides

1. Add 1 M DTT in 1:100 ratio to sample.
2. Incubate at 56°C for 1 h.
3. Add 10 mg/mL iodoacetamide (55 mM) to sample for a final concentration of 5 mg/mL.
4. Incubate in dark at RT for 45 min with vortexing every 15 min.
5. Add trypsin to sample in the ratio of 1:50 or 1:100 (w/w), incubate at 37°C overnight.
6. Quench reaction with 1% (v/v) TFA to a final concentration of 0.1%.
7. Clean peptides with C18 reversed-phase macro-spin or micro-spin columns according to manufacturer's instruction using 0.1% (v/v) formic acid.
8. Elute peptides with 0.1% (v/v) formic acid in 80% (v/v) acetonitrile and collect them into clean 1.5-mL tubes.
9. Dry peptides using Speed-Vac.

3.2.3. LC-MS/MS Analysis of Enriched O-GlcNAc Modified Glycopeptides

1. Resuspend dry peptides in 0.1% formic acid and 0.1% formic acid in 80% acetonitrile (ratio 38:1 v/v).
2. Load sample onto a nanobore capillary column (75 μm i.d.) packed with C18.
3. Sample may be separated over a 160 min linear gradient (see Note 14) of increasing acetonitrile and eluted at a flow rate of ~250 nL/min into the nanospray ion source of an LTQ Orbitrap XL ETD mass spectrometer (Thermo Fisher Scientific) (see Note 20). Full MS scans may be acquired from *m/z* 150 to 2,000 at a resolution of 60,000 (FWHM at *m/z* 400), followed by data-dependent HCD MS/MS scans of the 5 most abundant precursors. When a HexNAc oxonium ion (*m/z* 204.09) is observed in the HCD scans (Fig. 4), a subsequent ETD MS/MS scan (Fig. 5) will be triggered to analyze the precursor ion (see Note 21). A similar strategy can be applied for Hex oxonium ions. The HCD normalized collision energy

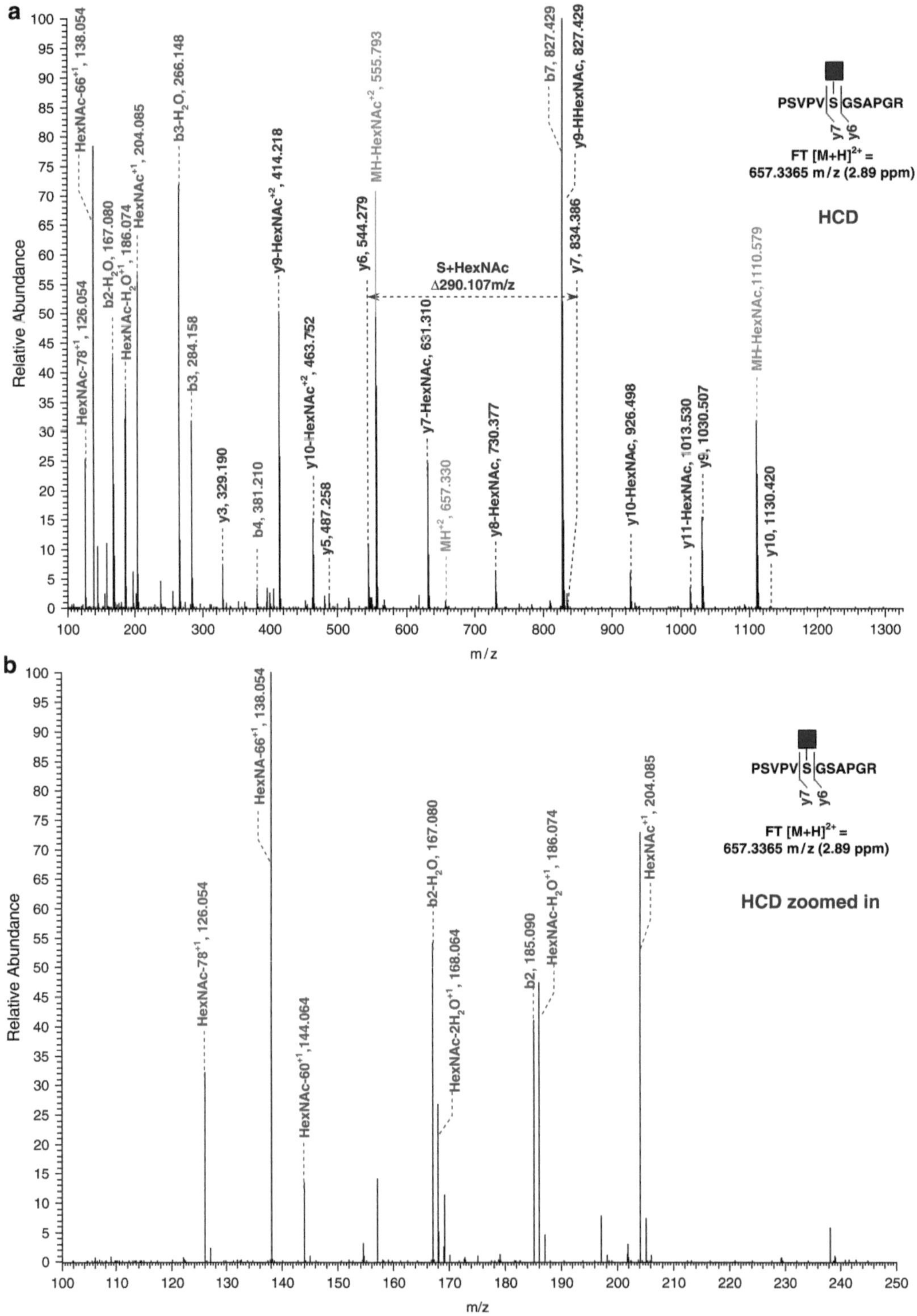

Fig. 4. (**a**) HCD spectrum of O-GlcNAc modified standard peptide BPP. (**a**) Most ions exhibit the loss of HexNAc (indicated as—HexNAc) and the HexNAc oxonium ion (*m/z* 204.09) shows high intensity. (**b**) A zoomed in HCD spectrum of O-GlcNAc modified standard peptide BPP. Besides HexNAc oxonium ion (*m/z* 204.09), a series of HexNAc fragments are observed at the low mass range of HCD scan.

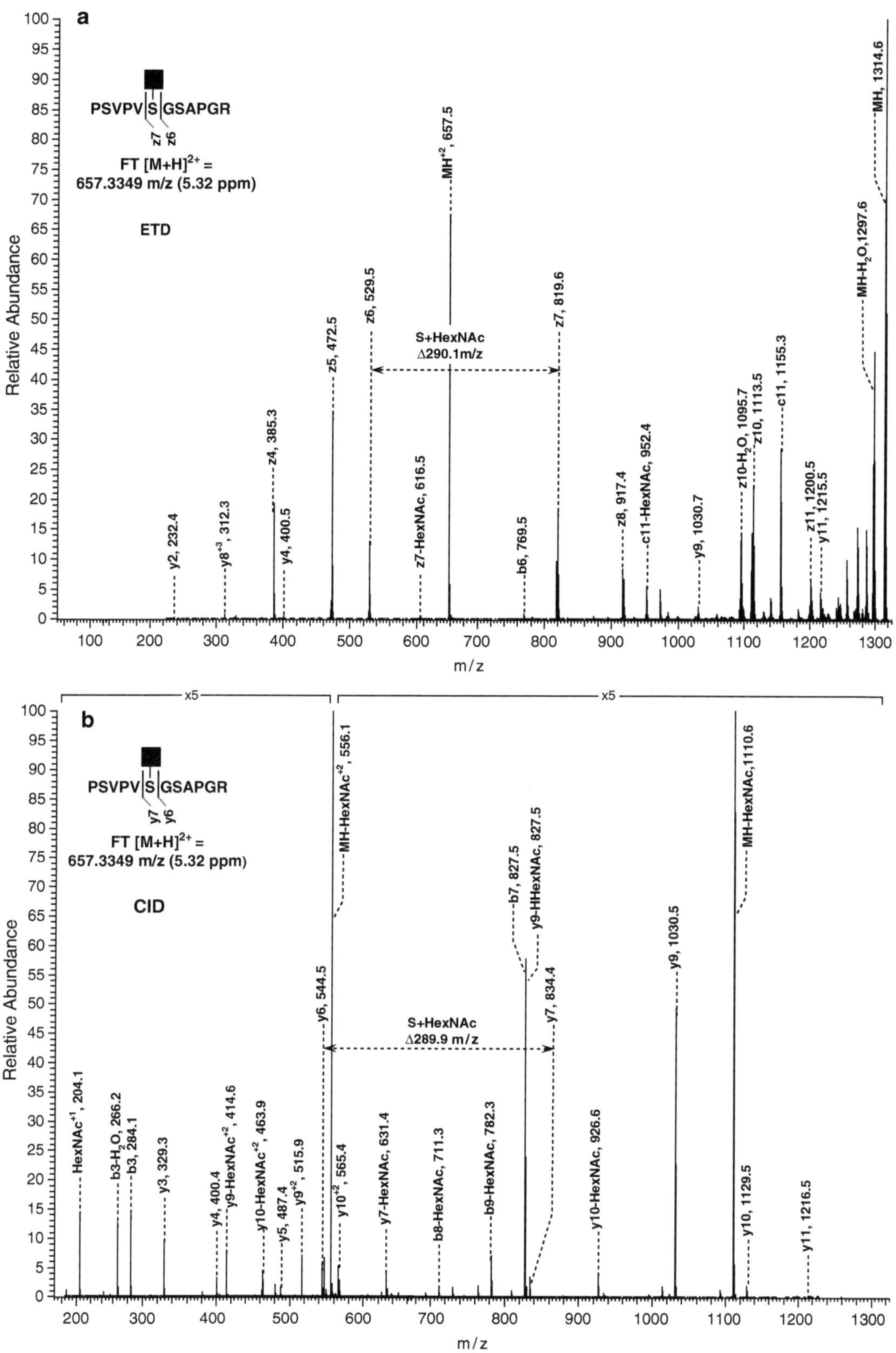

Fig. 5. (**a**) ETD spectrum of O-GlcNAc modified standard peptide BPP. No HexNAc loss is observed in ETD spectra, and the modification site can be unambiguously assigned. (**b**) CID spectrum of O-GlcNAc modified standard peptide BPP. The most dominant ions are created by the loss of HexNAc from the precursor ion, which suppresses the production of other ions, and the information on site assignment cannot be obtained with high confidence

may be set to 45%, and fragment ions may be detected in the Orbitrap at a resolution of 7,500 (FWHM at *m/z* 400).

4. Acquired spectra may be searched against a protein sequence database (NCBI nonredundant database, or Uniprot database, etc.) using search algorithms such as MASCOT (Matrix Science) or SEQUEST (Thermo Fisher Scientific) (see Note 15). Dynamic modifications of HexNAc on serine and threonine (+203.08 Da), alkylated cysteine (+57.02 Da), and oxidized methionine (+15.99 Da) should be allowed.

3.3. Site Mapping of O-Glycosylated Peptides Terminating with a GalNAc Residue (see Note 8)

3.3.1. WFA Enrichment of O-Linked Glycopeptides (see Note 8)

1. Dry down protein sample in Speed-Vac.
2. Reconstitute protein sample with 40 mM NH_4HCO_3.
3. Add trypsin to sample to a final concentration of 1:50–1:100 (w/w) and incubate overnight at 37°C.
4. Quench the reaction by adding 1% TFA to a final concentration of 0.1%.
5. Dry down peptides in Speed-Vac.
6. Reconstitute peptides with 500 μL WFA lectin column buffer (freshly prepared).
7. Add 500 μL of WFA resin slurry to a 2-mL chromatography column (see Note 22).
8. Wash resin with 5 column volumes of lectin column buffer.
9. Incubate peptides with WFA column at 4°C for 2 h with end-over-end rotation.
10. Wash unbound peptides off with 5 column volumes of lectin column buffer.
11. Elute bound glycopeptides with 5 column volumes of 200 mM GalNAc (prepared in lectin column buffer), and collect eluates.
12. Dry down eluted glycopeptides in Speed-Vac.
13. Clean up glycopeptides using C18 macro-spin or micro-spin columns.
14. Dry peptides in Speed-Vac.

3.3.2. Assignment of Modification Sites on O-Glycopeptides by LC-MS/MS/MS

1. Resuspend peptides in 48.5 μL of 0.1% formic acid and 1.5 μL of 0.1% formic acid/80% acetonitrile.
2. Load sample onto a nanobore capillary column (75 μm i.d.) packed with C18.
3. Sample may be separated over a 160 min linear gradient (see Note 14) of increasing acetonitrile and eluted at a flow rate of ~200 nL/min into the nanospray ion source of an LTQ Orbitrap XL mass spectrometer (Thermo Fisher Scientific) (see Note 20). A full MS at 60,000 resolution may be acquired

from *m/z* 400–2,000, followed by three data-dependent MS/MS spectra of the most intense ions. When a neutral loss corresponding to a monosaccharide (see Note 23) is detected as one of the three most intense ions in the MS/MS spectra, this fragment ion is fragmented again by collision-induced dissociation (CID) to yield an MS/MS/MS spectrum.

4. Acquired spectra may be searched against a protein sequence database (NCBI nonredundant database, or Uniprot database, etc.) using search algorithms such as MASCOT (Matrix Science) or SEQUEST (Thermo Fisher Scientific) (see Note 24). Dynamic modifications of Hex, HexNAc, HexHexNAc, and HexHexNAcHexNAc (see Note 25) on serine and threonine should be allowed.

4. Notes

1. Both buffers for thiol affinity chromatography can be made with either Tris-based (TBS) (10) or phosphate-based (PBS) (11).
2. The use of certain LC-MS/MS instruments such as Finnigan LCQ™ (Thermo Fisher Scientific) and QSTAR Pulsar (MDS Sciex) has been reported (10, 11); however, other mass spectrometers that are compatible with liquid chromatography, such as LTQ™, LTQ XL™ series (Thermo Fisher Scientific), may also be considered.
3. As an internal control for monitoring BEMAD and quantification, sample should be spiked with 1–20 pmol of known O-glycosylated peptides. Glycosylated peptides can be synthesized as described in (12).
4. The amount of starting material will vary depending on the sensitivity of the LC-MS/MS instrument and the purity of the sample. With the Finnigan LCQ, which is able to reach sensitivities in the final range, pico-molar amounts of starting protein may be enough, given that its purity is approximately 90% and assuming that the stoichiometry of the O-GlcNAc modification is approximately 10%. Additional details on protein sample preparation for BEMAD may be found in (10, 11).
5. In order to lessen plastic contamination, we recommend the use of these tubes. All plastic tubes and columns should be rinsed with 50% acetonitrile prior to use and never autoclaved. Also, clean pipet tips should be used.
6. This protocol can be modified to enrich for other types of O-linked glycopeptides from cell lysates.

7. The three O-GlcNAc-specific antibodies used in this protocol were characterized in (13).
8. This protocol may be applied to enriching other types of O-linked glycopeptides. Different lectin and elution buffer will need to be modified accordingly, for example, using succinylated wheat germ agglutinin (sWGA) to enrich *O*-GlcNAc modified glycoproteins or glycopeptides and eluting with GlcNAc. For more details regarding lectin enrichment of glycoproteins/glycopeptides, please refer to Chapter 2.
9. The purpose of performing performic acid oxidation instead of alkylation for denaturing cysteine-rich proteins is to increase the specificity of quantitative BEMAD labeling of O-linked glycopeptides, since alkylated cysteine will also undergo β-elimination in basic environment and therefore become labeled with DTT and enriched by thiol affinity chromatography. Performic acid oxidation can be performed before or after trypsin digestion.
10. The amount of control peptides is dependent on the sensitivity of the LC-MS/MS instrument being used.
11. Phosphatase treatment is also performed to increase the specificity of quantifying O-linked glycopeptides by avoiding the interference from DTT-labeled and enriched O-phosphorylated peptides.
12. This method may be adapted for serine and threonine phosphorylation sites as follows: Instead of phosphatase treatment, the sample should be acidified to pH 4.5 with TFA and treated with (1 U/20 μL) β-hexosaminidase (e.g., O-GlcNAcase) (New England Biolabs) at 37°C for 16 h. Also, the BEMAD solution should be modified to 2% (v/v) TEA, 0.2% (v/v) NaOH, 10 mM DTT, and the reaction allowed to proceed for 5 h at 50°C.
13. This is a minimum incubation time. Peptides may be bound to the thiol column for longer than 1 h.
14. The gradient may be adjusted based on the complexity of the samples, and/or to achieve different depths of dynamic range.
15. These MS and MS/MS methods should be used as a general guide only. Parameters, such as dynamic exclusion, MS/MS fragmentation, and collision energy, etc., should be optimized according to the specific instrument being used.
16. When analyzing oxidized samples, variable mass additions due to incomplete oxidation (e.g., +16, +32, or +48 for cysteine and tryptophan and +16 or +32 for methionine and histidine) may also be considered.
17. This section is modified from Harlow and Lane's protocol in (14).

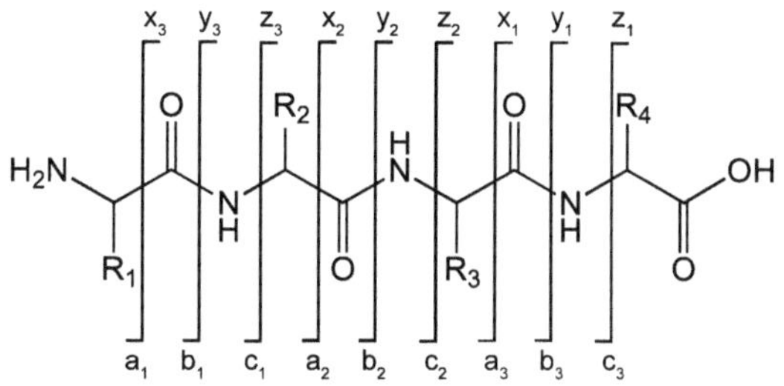

Fig. 6. The nomenclature for fragment ions observed in an MS/MS spectrum (15).

18. The amount of Protein A/G PLUS agarose slurry may be adjusted to different antibodies.
19. The amount of antibodies used may be modified by preference.
20. Other mass spectrometer equipped with nano ESI ion source may also be used. Instrument methods should be modified accordingly.
21. In this experiment, HCD is used to produce signature ions of O-GlcNAc peptides, e.g., oxonium ions of the cleaved GlcNAc residue, and triggers a subsequent ETD scan on the same precursor to produce c- and z- ions with GlcNAc attached to them. As a result, the product ions of an HCD scan will direct ETD scan to target O-GlcNAc peptides, which will decrease analysis time. Moreover, unlike when undergoing HCD, O-GlcNAc modification stays attached to peptide backbone during ETD, therefore the information of site localization is preserved. The types of fragment ions observed in an MS/MS spectrum are depicted in Fig. 6 (15).
22. The amount of WFA resin may be adjusted based on the protein amount of a particular sample.
23. Frequently used neutral losses for monosaccharides: Hexose, 162 Da; HexNAc, 203 Da; Sialic acid, 291 Da (Neu5Ac) or 307 Da (Neu5Gc).
24. Manual interpretation of spectra especially neutral-loss-triggered MS/MS/MS may be required.
25. The search for dynamic modifications may be adjusted to account for O-GlcNAc (HexNAc), O-GalNAc (HexNAc), O-Mannose (Hex, HexHexNAc, HexHexNAcHexNAc) modified glycopeptides.

Acknowledgements

This work was supported in part by NIH grants 1R01DK075069 (to LW) and 5P41RR018502 (LW, senior investigator).

References

1. Hounsell EF, Davies MJ, Renouf DV (1996) O-linked protein glycosylation structure and function. Glycoconj J 13:19–26
2. Yamashita Y, Chung YS, Horie R, Kannagi R, Sowa M (1995) Alterations in gastric mucin with malignant transformation: novel pathway for mucin synthesis. J Natl Cancer Inst 87:441–446
3. Harris RJ, Spellman MW (1993) O-linked fucose and other post-translational modifications unique to EGF modules. Glycobiology 3:219–224
4. Hart GW (1997) Dynamic O-linked glycosylation of nuclear and cytoskeletal proteins. Annu Rev Biochem 66:315–335
5. Herscovics A, Orlean P (1993) Glycoprotein biosynthesis in yeast. FASEB J 7:540–550
6. Stalnaker SH, Hashmi S, Lim JM, Aoki K, Porterfield M, Gutierrez-Sanchez G, Wheeler J, Ervasti JM, Bergmann C, Tiemeyer M, Wells L (2010) Site mapping and characterization of O-glycan structures on alpha-dystroglycan isolated from rabbit skeletal muscle. J Biol Chem 285:24882–24891
7. Varki NM, Varki A (2007) Diversity in cell surface sialic acid presentations: implications for biology and disease. Lab Invest 87:851–857
8. Drickamer K, Taylor ME (1998) Evolving views of protein glycosylation. Trends Biochem Sci 23:321–324
9. Varki A (2006) Nothing in glycobiology makes sense, except in the light of evolution. Cell 126:841–845
10. Vosseller K, Hansen KC, Chalkley RJ, Trinidad JC, Wells L, Hart GW, Burlingame AL (2005) Quantitative analysis of both protein expression and serine/threonine post-translational modifications through stable isotope labeling with dithiothreitol. Proteomics 5:388–398
11. Wells L, Vosseller K, Cole RN, Cronshaw JM, Matunis MJ, Hart GW (2002) Mapping sites of O-GlcNAc modification using affinity tags for serine and threonine post-translational modifications. Mol Cell Proteomics 1:791–804
12. Greis KD, Hayes BK, Comer FI, Kirk M, Barnes S, Lowary TL, Hart GW (1996) Selective detection and site-analysis of O-GlcNAc-modified glycopeptides by beta-elimination and tandem electrospray mass spectrometry. Anal Biochem 234:38–49
13. Teo CF, Ingale S, Wolfert MA, Elsayed GA, Not LG, Chatham JC, Wells L, Boons GJ (2010) Glycopeptide-specific monoclonal antibodies suggest new roles for O-GlcNAc. Nat Chem Biol 6:338–343
14. Harlow E, Lane D (1999) Using antibodies: a laboratory manual. Cold Spring Harbor Laboratory Press, New York
15. Roepstorff P, Fohlman J (1984) Proposal for a common nomenclature for sequence ions in mass spectra of peptides. Biomed Mass Spectrom 11:601

Chapter 16

Glycan Profiling: Label-Free Analysis of Glycoproteins

Yoshinao Wada

Abstract

Profiling of glycans requires both characterization of structure and determination of the relative abundance of each glycan. Label-free approaches enable facile and efficient profiling, while detailed structures and precise quantitation require derivatization. For glycan profiling by mass spectrometry, correlating the ion abundance in the mass spectrum to the content of each glycoform in the sample is acceptable, when one has adequate knowledge of ionization mode and ionization efficiency in mass spectrometry. Glycopeptide is a suitable analyte for this label-free approach.

Key words: Mass spectrometry, Glycopeptides, Ionization, Biomarker

1. Introduction

"Label-free quantitation" of glycans or glycoproteins is a term describing mass spectrometric measurement of relative abundance of different glycoforms without chemical derivatization or stable isotope labeling. Relative quantitation of each glycoform among various glycans attached to a specific glycosylation site or among total glycans of a glycoprotein, whole cells, or an organism is called "glycan profiling," which is an essential part of glycobiological analysis as well as structural characterization of glycans. The label-free method requires no specific sample preparation but is only based on the reading of mass spectrum, thus allowing facile and rapid analysis. On the other hand, derivatization of reducing end of glycans or of hydroxyl groups improves sensitivity in mass spectrometry or chromatography and also stabilizes glycosidic bonds during ionization. However, most derivatization methods require a procedure for removal of chemicals, resulting in a loss of samples, which is especially problematic for small amounts of starting materials.

Jennifer J. Kohler and Steven M. Patrie (eds.), *Mass Spectrometry of Glycoproteins: Methods and Protocols*,
Methods in Molecular Biology, vol. 951, DOI 10.1007/978-1-62703-146-2_16, © Springer Science+Business Media, LLC 2013

The analyte of label-free quantitation by mass spectrometry is either glycans released from glycoproteins or glycopeptides obtained by enzymatic proteolysis. A series of multi-institutional studies of the HUPO Human Disease Glycomics/Proteome Initiative (HGPI) assessed the validity of these strategies as well as various derivatization and found that the label-free quantitation is reliable especially for the glycans without charged groups (1, 2). This conclusion was reasonable, because protonation occurs on the peptide backbone of glycopeptides which bear neutral glycans. Even in the cases of negatively charged glycans, which are underestimated in positive ion mode and vice versa in negative ion mode, quantitation is reproducible if instrumental parameters are constant. In this section, glycan profiling by MS and data analysis to give the content of sugar unit are described (3).

2. Materials

Prepare all solutions using resin-filtered and deionized water such as "Milli-Q water" and analytical grade reagents. Prepare and store all reagents at room temperature (unless indicated otherwise).

2.1. Affinity Gel Components

1. HiTrap NHS-activated HP (GE Healthcare, Piscataway, NJ): 1 mL column volume.
2. *Coupling buffer*: 0.2 M $NaHCO_3$, 0.5 M NaCl, pH 8.3. Weigh 1.7 g $NaHCO_3$ and 2.9 g NaCl and transfer to a beaker. Add water to a volume of 90 mL. Mix and adjust pH with NaOH. Make up to 100 mL with water. Store at room temperature.
3. *Washing and deactivation buffer A*: 0.5 M ethanolamine (2-amino-1-ethanol), 0.5 M NaCl, pH 8.3. Add about 80 mL water and 3 mL ethanolamine to a 100 mL beaker. Weigh 2.9 g NaCl and transfer to the beaker. Mix and adjust pH with HCl. Make up to 100 mL with water. Store at room temperature.
4. *Washing and deactivation buffer B*: 0.1 M acetate, 0.5 M NaCl, pH 4.0. Add about 80 mL water and 0.57 mL acetic acid to a 100 mL beaker. Weigh 2.9 g NaCl and transfer to the beaker. Mix and adjust pH with NaOH. Make up to 100 mL with water. Store at room temperature.

2.2. Immunoaffinity Purification Components

1. Rabbit IgG against human IgA.
2. *Binding buffer*: phosphate-buffered saline (PBS). 0.14 M NaCl, 2.7 mM KCl, 1.5 mM KH_2PO_4, 8.1 mM Na_2HPO_4. Store at room temperature.
3. *Elution buffer*: 0.1 M glycine-HCl, pH 2.7. Weigh 0.75 g glycine and transfer to a 100 mL beaker. Add about 90 mL water.

Mix and adjust pH with HCl. Make up to 100 mL with water. Store at room temperature.

4. *Neutralization buffer*: 1 M Tris–HCl, pH 9.0. Weigh 12.1 g Tris(hydroxymethyl)aminomethane (2-amino-2-hydroxymethyl-1,3-propanediol) and transfer to 100 mL beaker. Add about 90 mL water. Mix and adjust pH with HCl. Make up to 100 mL with water. Store at room temperature.

2.3. Carbamidomethylation and In-Solution Digestion Components

1. *Stock solution A* for reduction and alkylation: 1 M Tris–HCl, 4 mM EDTA, pH 8.5 guanidine. Weigh 12.1 g Tris and 0.16 g EDTA-3Na and transfer to a 100 mL beaker. Add about 90 mL water. Mix and adjust pH with HCl. Make up to 100 mL with water. Store at room temperature.
2. *Stock solution B* for reduction and alkylation: 8 M guanidine hydrochloride (GndCl). Weigh 76.4 g GndCl and transfer to a 100 mL beaker. Dissolve and make up to 100 mL.
3. *Reduction/alkylation buffer*: Mix Stock Solution A and B at a 1:3 ratio to make a solution of 6 M GndCl, 0.25 M Tris–HCl, and 1 mM EDTA prior to use.
4. 1 M dithiothreitol (DTT) in reduction/alkylation buffer. Weigh approximately 10 mg DTT (Sigma-Aldrich, St. Louis, MO) in a microcentrifuge tube. Add reduction/alkylation buffer to make 1 M solution.
5. 1 M iodoacetamide (IA) in reduction/alkylation buffer. Weigh approximately 10 mg IA (Sigma-Aldrich) in a microcentrifuge tube. Add reduction/alkylation buffer to make 1 M solution.
6. 0.05 N HCl. Add about 100 mL water to a 100 mL beaker. Add 0.43 mL HCl to the beaker. Make up to 100 mL with water.
7. NAP-5 gel filtration column (GE Healthcare). The column is equilibrated with 10 mL of 0.05 N HCl by gravity flow just prior to use.
8. 1.5 M Tris. Weigh 1.81 g Tris and transfer to a 15 mL plastic conical tube. Add 10 mL water. No pH adjustment required.
9. *Trypsin stock solution*. Dissolve 20 μg trypsin (TPCK-treated, methylated; Sequencing Grade Modified Trypsin) (Promega, Madison, WI) in a vial with 40 μL of 50 mM NH_4HCO_3 (pH 7.8 without pH adjustment). Store at −80°C.

2.4. Glycopeptide Enrichment Components

1. *Glycopeptide binding solution*: 1-butanol/ethanol/H_2O (4:1:1, v/v).
2. *Glycopeptide elution solution*: 50% ethanol.
3. Sepharose CL-4B (GE Healthcare). Add 0.2 mL gel volume of Sepharose CL-4B in a microcentrifuge tube. Wash the gel with glycopeptide binding solution twice.

2.5. Desialylation Components

1. 2 M acetic acid. Add 11.5 mL acetic acid to 80 mL water in a beaker. Make up to 100 mL.

2.6. Desalting Components

1. ZipTip C18 pipette tips (Millipore, Bedford, MA).
2. *Wetting solution*: 100% acetonitrile.
3. *Equilibration/wash solution*: 0.1% trifluoroacetic acid (TFA).
4. *Elution solution*: 0.1% TFA/50% acetonitrile.

2.7. Mass Spectrometry Components

1. *MALDI Matrix solution*: 10 mg/mL 2,5-dihydroxybenzoic acid (DHB) (Proteomics grade, Wako, Osaka, Japan). Weigh a few mg DHB in another microcentrifuge tube. Add appropriated volume of 0.1% TFA/50% acetonitrile to make 10 mg/mL (see Note 1).

3. Methods

In this section, an example of IgA is described. Carry out all procedures at room temperature unless otherwise specified.

3.1. Purification of Glycoproteins with Affinity Gel [6 h]

1. A polyclonal antibody (rabbit IgG against human IgA) is dissolved in the coupling buffer at a concentration of 1–10 mg/mL. Affinity gel preparation using a HiTrap NHS with coupling, washing, and deactivation buffers is carried out according to the manufacturer's instruction.
2. Open to the packed column by a cutter, and recover the gel. The gel can be stored in 20% (v) ethanol at 4°C.
3. Wash the gel in binding buffer three times.
4. Add 20 μL gel, 0.5 mL coupling buffer and then 10 μL serum in a microcentrifuge tube.
5. Gently agitate the tube at room temperature for 3 h or at 4°C overnight.
6. Centrifuge the tube briefly at 3,000–5,000 ×*g*.
7. Add 1 mL binding buffer to the pellet for washing.
8. Repeat centrifugation and wash cycle three times.
9. Add 50 μL elution buffer to the pellet.
10. Gently agitate the tube for 3 min.
11. Centrifuge the tube briefly at 3,000–5,000 ×*g*.
12. Remove the supernatant to a new microcentrifuge tube.
13. Add 5 μL (or one part to ten) of neutralization buffer to the immunopurified IgA sample solution.

3.2. Carbamidomethylation and In-Solution Digestion [4 h]

1. Add 300 μL reduction/alkylation buffer to 50 μL immunopurified IgA solution in a microcentrifuge tube.
2. Add 20 μL of 1 M DTT solution (dissolved in reduction/alkylation buffer), and then make up to 0.4 mL with reduction/alkylation buffer.
3. Incubate at 55°C for 2 h.
4. Add 100 μL of 1 M iodoacetamide solution (dissolved in reduction/alkylation buffer).
5. Incubate in the dark at room temperature for 30 min.
6. Add the 0.5 mL sample to a NAP-5 column equilibrated with 0.05 N HCl. Allow the sample to enter the gel bed completely.
7. Add 1.0 mL of 0.05 N HCl to the column and elute the purified sample in a microcentrifuge tube.
8. Add 0.1 mL of 1.5 M Tris to the eluent for digestion to raise pH above 7.8.
9. Add 1 μL (0.5 μg) of trypsin stock solution (see Note 2).
10. Incubate overnight at 37°C for digestion (see Note 3).

3.3. Glycopeptide Enrichment (4) [3 h]

1. Concentrate the trypsinized IgA solution (approximately 1.1 mL) to 0.2 mL by SpeedVac.
2. Add 0.2 mL ethanol and 0.8 mL 1-butanol.
3. Add 20 μL gel volume of Sepharose CL-4B (see Note 4).
4. Gently agitate the tube for 45 min.
5. Centrifuge the tube briefly at 3,000–5,000 × *g*.
6. Discard the supernatant.
7. Add 1 mL glycopeptide binding solution to the pellet for washing.
8. Repeat the centrifugation and wash cycle three times.
9. Add 100 μL glycopeptide elution solution to the pellet.
10. Gently agitate the tube for 10 min.
11. Centrifuge the tube briefly at 3,000–5,000 × *g*.
12. Remove the supernatant to a new microcentrifuge tube.
13. Dry up the tube by SpeedVac.

3.4. Desialylation of Glycopeptides [2 h]

1. Add 50 μL of 2 M acetic acid to the dried glycopeptides in the microcentrifuge tube.
2. Incubate the tube at 80°C for 2 h.
3. Add 100 μL water to the tube and mix.

3.5. Desalting of Glycopeptides [1 h]

1. Pre-wet ZipTip by aspirating 100% acetonitrile (wetting solution) into tip. Dispense to waste and repeat.
2. Equilibrate the tip for binding by washing with the equilibration/wash solution (0.1% TFA) three times.
3. Bind glycopeptides to ZipTip by fully depressing the pipettor plunger to a dead stop. Aspirate and dispense sample 3–7 cycles.
4. Wash tip and dispense to waste using at least 5 cycles of equilibration/wash solution.
5. Dispense 5–10 μL of elution solution into a new microcentrifuge tube. Carefully, aspirate and dispense eluate through ZipTip at least three times without introducing air.

3.6. Mass Spectrometry [0.5 h]

1. Mix 1 μL sample solution with 1 μL matrix solution on the MALDI sample target.
2. Operate the mass spectrometer in a linear time-of-flight mode. This is essential for accurate measurements (see Note 5).
3. Acquire a mass spectrum after accumulation of the data from 100 laser shots or more.

3.7. Calculation (3, 5)

Relative quantitation is based on the intensities (heights) of the signals in the mass spectrum. Calculation for label-free quantitation is carried out according to the following equations, and is valid for glycopeptides bearing the same peptide backbone sequences with neutral glycans (see Note 6). However, the results of calculation on a mixture of neutral and acidic glycans such as sialylated ones are reproducible and useful as well (1).

Figure 1 shows the mass spectrum of glycopeptides from immunoglobulin A (IgA). IgA1 bears mucin-type core-1 O-glycans, with Galβ1-3GlcNAc-(Ser/Thr) as the core structure but different patterns of glycan distribution, in the hinge region of the Fc chain, and tryptic cleavage yields the 38 amino acid hinge O-glycopeptide: $H^{208}YTNPSQDVTVPC^{220}PVPSTPPTPSPSTPPTPSPS^{240}CCHPR^{245}$. The mucin-type O-glycosylation is initiated by linking *N*-acetylgalactosamine (GalNAc) to the Ser or Thr of the protein backbone and this reaction is catalyzed by UDP-*N*-acetyl-α-d-galactosamine: polypeptide *N*-acetylgalactosaminyltransferases. Galactosylation to form core-1 O-glycan is catalyzed by a core-1 β1-3 galactosyltransferase (C1 Gal-T1). However, this reaction is incomplete in this region, leaving substantial amounts of the GalNAc monosaccharide (Tn antigen).

3.7.1. Content of Each Glycopeptide Species (Eq. 1)

The percent content of each glycopeptide species is calculated using Eq. 1 (see Note 7 for ESI mass spectrum).

$$\text{Glycopeptide peak\%} = [\text{Glycopeptide peak intensity}] / [\text{Total glycopeptide intensity}] \times 10^2 \quad (1)$$

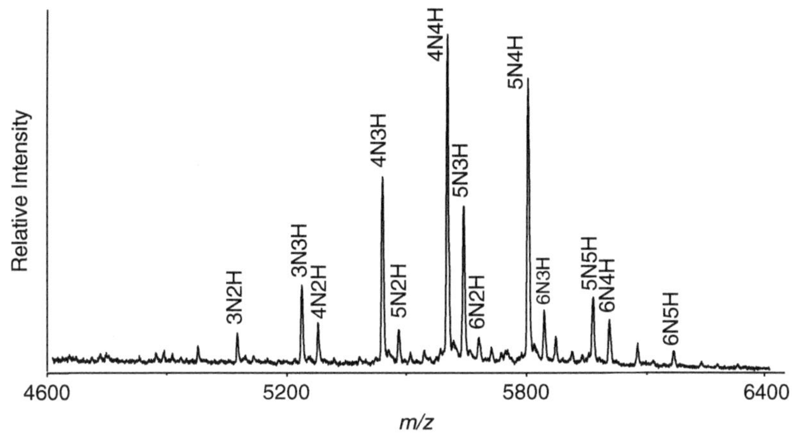

Fig. 1. MALDI linear TOF mass spectrum of glycopeptides from IgA. Glycopeptides were enriched and desialylated before measurement. Relative molecular mass of the unglycosylated peptide is 4138.6, and that of sugar residues are 203.2 and 162.1 for GalNAc (N) and Gal (H), respectively. The mass spectrum was acquired with Voyager DE Pro (Applied Biosystems, Foster City, CA) equipped with a nitrogen laser (20 Hz repetition rate and 10 ns duration). The peak intensities are used for calculation.

For example, 25% for the major glycopeptide species with four GalNAcs and four Gals, i.e., 4N4H in the figure.

3.7.2. Content of Glycans (Eq. 2)

The molar content of O-glycans attached to this region is calculated using Eq. 2

$$\text{Glycan content(mol/glycopeptide)} = \sum \left\{ \begin{matrix} \text{[Glycopeptide peak\%]} \times \\ \text{[Number of glycans attached to glycopeptide]} \end{matrix} \right\} \times 10^{-2} \quad (2)$$

where the number of glycans attached to a glycopeptide equals to that of GalNAc residues in the molecule, in the cases of mucin type core-1 O-glycans. Therefore, the glycan content is the same with GalNAc content, and calculated to be 4.51 mol/peptide from the mass spectrum in Fig. 1.

3.7.3. Content of Saccharide Component (Eq. 3)

The following equation is analogous to Eq. 2, but is useful to *N*-glycans (Fig. 2).

$$\text{GalNAc, Gal(mol/glycopeptide)} = \sum \left\{ \begin{matrix} \text{(Glycopeptide peak\%)} \times \\ \text{(Number of GalNAc or Gal in the glycopeptide)} \end{matrix} \right\} \times 10^{-2} \quad (3)$$

GalNAc and Gal are 4.51 and 3.51 mol/peptide, respectively, from the mass spectrum in Fig. 1.

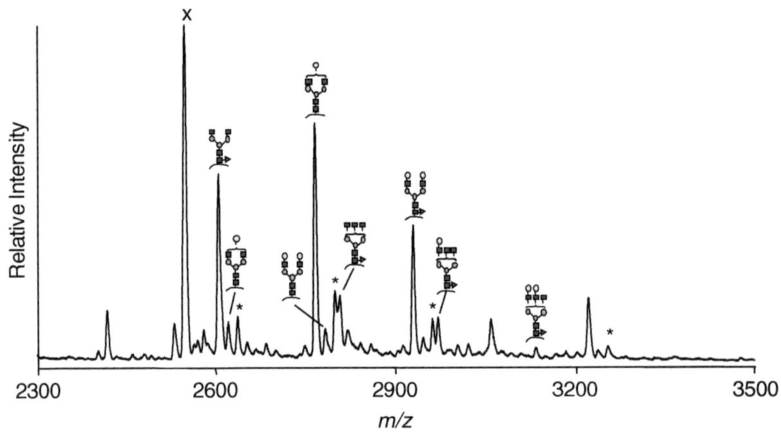

Fig. 2. MALDI linear TOF mass spectrum of glycopeptides from IgG. Human IgG is purified by a Protein G affinity column. Subsequent sample preparation and measurements are carried out according to the protocol described in this section. Serum IgG is polyclonal and is thus a mixture composed of different primary protein structures. The amino acid sequence of the tryptic peptide involving the N-glycosylation site Asn-297 is heterogeneous, and EEQYNSTYR and EEQFNSTFR representing two subclass molecules, IgG1 and IgG2, respectively, are abundant. In this mass spectrum, the peaks indicated by illustrated glycoforms are derived from IgG2, and those marked by *asterisks* are from IgG1. Calculation based on Eqs. 1 and 2 yields the galactosylation content of IgG2 glycopeptides to be 0.90 mol/peptide. X, contamination signal. See Fig. 1 for instrument and setting.

4. Notes

1. For example, α-cyano-4-hydroxycinnamic acid (CHCA) is used in the majority of proteomics application for the analysis of peptides, but CHCA is a "hot" matrix producing considerable in-source decay of the oligosaccharide portion of glycopeptides. Positive mode analysis with DHB is most widely used for MS of glycopeptides.
2. Combined use of lysylendopeptidase (Wako, Osaka, Japan) with trypsin enhances cleavage of Lys-Xxx bonds.
3. Digestion with microwave oven allows very short incubation time. (Method) Add 300 mL water to a 500 mL beaker. A microcentrifuge tube containing protein and trypsin in digestion buffer is floated on the water and put the beaker in a microwave oven. Set a low power level of microwave or defrost mode. The incubation time is determined by a rehearsal run without samples. At the end of incubation, water should be heated to 55°C, and preferably it takes 5–10 min.
4. Approximately 20 μL gel volume is enough for a few mg glycoprotein digests. A larger gel volume causes contamination of unglycosylated peptides.

5. Linear time-of-flight mode, but not the reflector mode, is the choice for quantitative measurements. MALDI MS of glycopeptides is accompanied by post-source decay (PSD) as well as in-source fragmentation. Broad and strange signals derived from PSD are found in the reflector mode mass spectrum.
6. This assumption is based on the ionization of glycopeptides. Protonation occurs on the peptide backbone but not on the glycans. Therefore, efficiency of ionization of the glycopeptides with same peptide backbone sequences is mostly independent of the glycan structure. Dissociation of glycosidic bonds is minimal during ionization or a mass analysis process such as ion trapping.
7. In the case of the ESI mass spectrum, the relative abundances of the peaks for the same glycopeptide composition but differing in the charge state are combined for calculation. Interestingly, the calculation results from MALDI and ESI mass spectra are consistent with each other (3).

References

1. Wada Y, Azadi P, Costello CE et al (2007) Comparison of the methods for profiling glycoprotein glycans—HUPO Human Disease Glycomics/Proteome Initiative multi-institutional study. Glycobiology 17:411–422
2. Wada Y, Dell A, Haslam SM et al (2010) Comparison of methods for profiling *O*-glycosylation: Human Proteome Organisation Human Disease Glycomics/Proteome Initiative multi-institutional study of IgA1. Mol Cell Proteomics 9:719–727
3. Wada Y, Tajiri M, Ohshima S (2010) Quantitation of saccharide compositions of *O*-glycans by mass spectrometry of glycopeptides and its application to rheumatoid arthritis. J Proteome Res 9:1367–1373
4. Wada Y, Tajiri M, Yoshida S (2004) Hydrophilic affinity isolation and MALDI multiple-stage tandem mass spectrometry of glycopeptides for glycoproteomics. Anal Chem 76: 6560–6565
5. Rebecchi KR, Wenke JL, Go EP et al (2009) Label-free quantitation: a new glycoproteomics approach. J Am Soc Mass Spectrom 20: 1048–1059

Part V

Computational Tools

Chapter 17

Introduction to Informatics in Glycoprotein Analysis

Kiyoko F. Aoki-Kinoshita

Abstract

Although the field of glycome informatics has established several methods, standards and technologies for carbohydrate analysis, the analysis of glycoproteins and other glycoconjugates is still in its infancy. However, from even before the term "glycome informatics" emerged, several groups have developed methods and tools on the analysis of glycosylation sites. In particular, the Expasy server has provided such tools to aid in the prediction of glycosylation sites of N- and O-glycans, while glycosciences.de has provided tools for the analysis of the amino acid distribution around glycosylation sites in 3D space, based on data from the Protein Data Bank (PDB). In addition to these tools, databases of glycoprotein information are available that may aid in glycoprotein prediction; GlycoProtDB is a database of glycoprotein information characterized by the Japanese Consortium for Glycobiology and Glycotechnology, and UniProt includes glycosylation site information along with its protein sequence data. Furthermore, the providers of the glycosylation tools on Expasy, the Center for Biological Sequence Analysis, also provide a database of O-glycosylation called O-GlycBase. Such databases may eventually aid in the development of glycoprotein-analysis tools as more consistent data is accumulated, and some prospects on this area of research will be given.

Key words: Glycome informatics, Glycoprotein analysis, Glycosylation, Databases, Glycobiology

1. Introduction

This chapter will introduce various tools and methods that are available for the analysis of glycoproteins in general. To date, the majority of these tools pertain to glycosylation site prediction. A few tools provided by the glycosciences.de web portal provide statistical tools for the analysis of amino acids surrounding glycans as found in the data of the Protein Data Bank (PDB). A description of potentially useful databases pertaining to glycoproteins will also be introduced. The URLs for these resources are listed in Table 1. Each resource will be described in different subsections, and in summary, perspectives on future glycoproteomic research will be given.

Jennifer J. Kohler and Steven M. Patrie (eds.), *Mass Spectrometry of Glycoproteins: Methods and Protocols*,
Methods in Molecular Biology, vol. 951, DOI 10.1007/978-1-62703-146-2_17,

Table 1
The URLs for each of the resources described in this chapter are listed

Resource name	URL
Expasy	http://expasy.org/tools/#ptm
Big-PI Predictor	http://mendel.imp.ac.at/sat/gpi/gpi_server.html
Center for Biological Sequence Analysis (CBS)	http://www.cbs.dtu.dk/services/
GPI-SOM	http://gpi.unibe.ch/
OGPet	http://ogpet.utep.edu/OGPET/
EnsembleGly	http://turing.cs.iastate.edu/EnsembleGly/
Glycosciences.de	http://www.glycosciences.de/tools/index.php
JCGGDB	http://jcggdb.jp
UniProt	http://www.uniprot.org/

2. Materials

2.1. Expasy Proteomics Server

The “Expasy Proteomics Server” provides a number of useful tools for glycoprotein analysis, and in particular, glycosylation site analysis, listed under “Post-translational modification prediction.” These include DictyOGlyc, NetCGlyc, NetOGlyc, NetGlycate, NetNGlyc, OGPet, YinOYang, Big-PI Predictor, and GPI-SOM. A brief description of these tools is described in Table 2. In order to use these tools, protein sequences are generally required as input. An explanation of their usages is given in Sect. 3.

2.2. Glycosciences.de

Tools for the statistical analysis of carbohydrate structures derived from the PDB are GlyTorsion, GlyVicinity, and GlySeq. The latter two pertain to the statistical analysis of amino acids surrounding amino acid residues and glycosylation sites. The data of PDB are used in the analysis, and the only input data required are parameters for which to search the statistical data.

2.3. Glycoprotein Databases and Analyses

The use of glycoprotein databases normally only requires few inputs, such as keywords for sequences or lectins that may be used to recognize glycans on glycoproteins. Their analyses, on the other hand, would require an all-inclusive data set of not only sequence information, but also spatial conformational information and binding affinity data. Since these data resources have just recently become available, their integration with one another in the near future should expect to provide an invaluable resource for the development of tools to give a true picture of glycosylation and glycan binding mechanisms.

Table 2
A table of the tools available for glycoprotein analysis at Expasy, along with their descriptions

Tool name	Description
Big-PI Predictor	GPI-modification site prediction
DictyOGlyc	Prediction of GlcNAc O-glycosylation sites in *Dictyostelium*
EnsembleGly	Prediction of N-linked, O-linked, and C-linked glycosylation sites using an ensemble of Support Vector Machines (SVMs)
GlySeq	Statistical analysis of amino acids in the sequential context of N- and O-glycosylation sites
GlyTorsion	Statistical analysis of torsion angles of carbohydrate structures
GlyVicinity	Statistical analysis of amino acids found in the spatial vicinity of carbohydrate residues
GPI-SOM	Identification of GPI-anchor signals by a Kohonen Self Organizing Map
NetCGlyc	Prediction of C-mannosylation sites in mammalian proteins
NetGlycate	Prediction of glycation of epsilon amino groups of lysines in mammalian proteins
NetNGlyc	Prediction of N-glycosylation sites in human proteins
NetOGlyc	Prediction of O-GalNAc (mucin type) glycosylation sites in mammalian proteins
OGPet	Prediction of O-GalNAc (mucin type) glycosylation sites in eukaryotic (non-protozoan) proteins
YinOYang	Prediction of O-beta-GlcNAc attachment sites in eukaryotic protein sequences

3. Methods

3.1. Expasy Proteomics Server

Here we describe each of the glycosylation prediction tools available at Expasy, which are originally developed at the Center for Biological Sequences (CBS). Most of these tools by the CBS are based on neural networks trained on known glycosylated peptide sequences as well as other necessary information such as cell-surface accessibility. As these tools utilize similar methodology, which will be described in the following chapter, only a brief description of their usage will be presented here. The specific methods utilized by the other tools will be described in detail.

3.1.1. Big-PI Predictor

In developing this tool, a meta-analysis of proprotein sequences in protein sequence databases (1) had revealed the following four sequence signal elements for GPI-modification, called the ω-site:

1. An unstructured linker region of about 11 residues (ω−11 … ω−1);
2. A region of small residues (ω−1 … ω+2), including the ω-site for propeptide cleavage and GPI-attachment;

3. A spacer region (ω+3 … ω+9) of moderately polar residues; and
4. A hydrophobic tail beginning with ω+9 or ω+10 up to the C-terminal end.

Each of these signal elements was incorporated into the prediction software, whereby a scoring function was computed to indicate the probability of a GPI-modification given a particular sequence. The scoring function in part describes the conservation of physical properties in the GPI-modification signal arising from the interaction of few or many sequence positions:

1. Side-chain volume limitations and mutual volume compensation effects for residues ω−1 … ω+2 expected to be located within the catalytic cleft of the putative GPI-modification transamidase;
2. Backbone flexibility requirements within the segment ω−1 … ω+2;
3. Propeptide length ranges (from ω−1 to the C end);
4. Spacer region (ω+3 … ω+8) hydrophilicity and sequence volume per residue;
5. Hydrophobicity limits averaged over the C-terminal hydrophobic region and conditions for even distribution of hydrophobic residues;
6. The presence of aliphatic hydrophobic residues (LVI-contents in the tail) and the absence of long stretches of residues with a flexible backbone (GS-content in a window) in the C-terminal hydrophobic tail.

Because of the limited availability of GPI-anchored sequences at the time of the tool's development, and because of taxon-specific characteristics of GPI-modifications, the scoring function has only been developed for and tested on metazoan and protozoan sequences.

This tool takes as input an amino acid sequence in FASTA format. Assuming that the end of the sequence is the C-terminus, it computes a score based on a combination of 20 features including the properties listed above, together with penalties for various missing elements. The results page includes a detailed explanation of the predicted GPI-modification sites and the derivation of the scores.

3.1.2. DictyOGlyc

This tool, provided by the CBS, utilizes artificial neural networks in an attempt at predicting O-GlcNAc glycosylation sites in membrane and secreted proteins of *Dictyostelium discoideum* (2). The knowledge of glycosylation sites and the context of the sequence within which they are found were extracted from in vivo experimental data (3) combined with surface accessibility prediction to develop a method for predicting putative acceptor sites in other *D. discoideum* amino acid sequences (4). The predicted outputs from the glycosylation network and the surface-accessibility network are combined to produce the final prediction results. A value for "glycosylation threshold" is computed depending on whether a site is predicted to be on the surface or buried. Thus the averaged glycosylation potential is compared with the modulated surface threshold, and any site with a potential greater than the threshold is determined as being glycosylated.

This tool takes as input at most 50 protein sequences (70,000 amino acids), with each sequence having no more than 4,000 amino acids each. In the results, the input sequences and their potentially glycosylated sites will be displayed, along with detailed tables and a graph (if requested) will be listed to indicate the glycosylation potential compared to the threshold that resulted in the predicted assignment.

3.1.3. GPI-SOM

GPI-SOM is a tool for the prediction of GPI-anchor sites of one or more protein sequences (5). This tool takes as input one or more protein sequences, which may be in any of the following formats: fasta, embl, genbank, SWISS-PROT, gcg, gcgdata, pir, or raw format. In most cases, the fasta format is the simplest to provide (an example is provided next to the input form). If the input sequences conform to the expected signal sequences and are of sufficient length, a table will be returned as the results, indicating the number of input sequences, the number that were ignored due to insufficient length, the number that had the expected C-terminal GPI-anchor signal sequence, the number of sequences that could not be determined as having or not having a GPI-anchor, and the number of GPI-anchored sequences (those conforming to the assumption that GPI-anchored sequences have a particular signal sequence at the N- and C-termini). Clicking on the link for GPI-anchored sequences, a listing of all of the sequences and a comment indicating which residue is GPI-anchored will be displayed.

3.1.4. NetCGlyc, NetNGlyc, NetOGlyc, NetGlycate, YinOYang

These tools are also developed by the CBS (http://www.cbs.dtu.dk/services/) and use neural networks to learn from the available glycosylation data in order to make predictions on new data. For example, the sequence motif for C-mannosylation is WXXW, where the first tryptophan is modified. However, it is difficult to make accurate predictions based on such a short motif. Thus, similarly to GPI-SOM, NetCGlyc uses currently available glycosylated sequences to train neural networks to learn from the more inherent properties of the sequences which may not be directly obvious from the data (6). As a result, these tools are able to predict glycosylation sites with a much higher accuracy than by identifying sequence motifs alone. NetNGlyc (7) and NetOGlyc (8) have long been around since the late 1990s to prediction N- and

O-glycosylation sites, respectively using a similar methodology. NetOGlyc further incorporates secondary structure and surface accessibility in addition to sequence context in their neural networks in order to predict mucin type O-glycosylation sites. NetGlycate predicts the glycation of epsilon amino groups of lysines in mammalian proteins using neural networks trained on sequence context (9), and YinOYang predicts O-beta-GlcNAc attachment sites in eukaryotic protein sequences (7).

To use these tools, one only needs to enter a sequence for which to predict glycosylation. Few parameters, if any, are required as input. The results of these tools are also quite detailed, describing the justification for the predictions by presenting a breakdown of the scoring function used.

3.2. OGPet

OGPet is a tool for the prediction of O-glycosylation sites in proteins. Input is taken in FASTA format, and an option to select the relative amino acid positions around potential O-glycosylation sites is given. These positions correspond to various peptide substrate specificities that have been observed for different ppGalNAcT family members. Thus, the user must select the appropriate positions for the sequence at hand. The suggested positions to try are the following in order:

- Default (considering all five positions).
- All positions, but −3.
- All positions, but +1.
- All positions, but +3.
- All positions, but −1.
- All positions, but +4.

In the results, a listing of the input sequences and an indication of the GPI-anchor sites is displayed, along with a table of the positions, the profile that matched the GPI-anchor site and its sequential context, and a score.

3.3. EnsembleGly

EnsembleGly is a web site for the prediction of various glycosylation types, including N-linked, O-linked, and C-linked glycosylation. Compared to the neural network models of the CBS, this web server uses ensembles of Support Vector Machine (SVM) classifiers (10). From the main web page, clicking on the "Predict" menu item on the left will lead the user to the input form for the tool. There are several options in this form that will need to be specified based on the prediction method selected. For the Ensemble of SVMs method, an option for the String Kernel is given to use either straight identity comparisons of amino acids (1 if same, 0 if different), or the BLOSUM 62 scoring matrix, which weights biochemically similar amino acids more highly compared to dissimilar amino acids. If Decision Tree or Naïve Bayes is selected for the prediction method, then similar options are provided as "Amino Acid Identity" and "Amino Acid Identity plus physical properties." The resulting output is similar to other glycosylation prediction tools, consisting of the input sequence annotated with glycosylation sites, the scores for the predictions together with threshold values as evidence for the prediction of positive (+) or negative (−) for glycosylation.

3.4. Glycosciences.de

In order to achieve the goal of analyzing glycoprotein structures as a whole, there needs to be an understanding of the spatial environment in which glycosylation takes place. Thus, the glycosciences.de web portal provides a number of tools to analyze these environmental variables, based on available glycoprotein information from the PDB, called the Carbohydrate Structure Suite, or CSS (11). The three major glycoprotein-analysis tools in the CSS are described here.

3.4.1. GlyTorsion(DB)

GlyTorsionDB is a database of torsion angles derived from the carbohydrate structures found in the PDB. In addition, it also contains ring torsions, omega torsions, *N*-acetyl group torsions as well as side-chain torsions of asparagine residues involved in glycosidic bonds. This database can be queried using the tool GlyTorsion, which takes as input either a PDB ID or PDB file of a glycoprotein. It then computes the angles for all glycosidic bonds it finds in the input data. The results are displayed as a Ramachandran plot, produced by the software called carp, which is also available in glycosciences.de as a separate tool.

3.4.2. GlyVicinity(DB)

GlyVicinity computes the amino acids found in the spatial vicinity of carbohydrate residues, to determine the characteristics of glycoproteins, which in turn may be used for the examination of carbohydrate-binding proteins. GlyVicinity performs statistical analyses of the amino acid types surrounding carbohydrate structures and of the atoms that are found to form the closest contacts between protein and carbohydrate residues. All the analyzed data from the PDB are stored in the database GlyVicinityDB.

There are couple analyses that can be targeted with this tool; a global analysis of a particular residue or chain among all the PDB data, or a targeted analysis of particular proteins from the PDB. As a result of a global analysis on a particular monosaccharide, a bar chart of the number of amino acids surrounding the selected residue will be displayed. From these results, a more detailed analysis of the closest atoms among those of the carbohydrate and of the amino acid residues can be performed. Similar results and detailed analyses can also be obtained from a targeted analysis of particular proteins. Either a list of PDB IDs for the input data or a selection can be made of PDB IDs based on their classification (i.e., membrane protein, transferase, etc.), biological source, method of structural determination or resolution.

3.4.3. GlySeq(DB)

GlySeqDB is a database of the glycoprotein sequences from PDB and Swiss-Prot, and GlySeq is a tool for querying and graphically displaying the statistical distribution of amino acids surrounding N- and O-glycosylation sites. It takes as input the following parameters:

- Selection of the data set for which analyses are available: Swiss-Prot, PDB or a combination of the two, and whether or not to include redundant sequences.
- The number of amino acid positions to include surrounding the glycosylation site.
- The amino acid of the glycosylation site (Asn, Ser, Thr, Ser/Thr, Asp, Glu, or All).
- The specification of the types of amino acids to analyze, based on biochemical properties, or a user-defined list, or a particular position may be analyzed.
- The output format by which to display the results (e.g., Absolute numbers or percentages, etc.), as well as whether or not to include a figure of the results as a bar chart. In the case that all amino acids are to be analyzed and listed individually, a bar chart will not be able to be produced.
- Additionally, filters may be specified such that particular positions surrounding a glycosylation site may be required to satisfy a certain condition in order to be included in the results.

The results will list each amino acid and their distribution at each position surrounding the glycosylation site. A bar char will also be displayed of the same information.

3.5. JCGGDB, UniProt, O-GlycBase

In this section, three databases containing glycoprotein information will be introduced.

3.5.1. JCGGDB

The Japan Consortium for Glycobiology and Glycotechnology Data Base (JCGGDB) provides the GlycoProtDB database, which contains characterized glycoproteins and the lectins that have been used to determine to determine the glycan structures on them. The main entry of a glycoprotein contains basic information such as the biological source, amino acid length, and amino acid sequence, with glycosylation motifs underlined in red and glycosylation sites highlighted. A table indicating confirmed glycosylation sites using the IGOT method, which is a proteomics analysis technique based on LC/MS followed by MS/MS (12). This method uses a lectin column to determine the glycan motifs found on the peptides of the glycoprotein, so the information on the lectin found to bind to the glycoprotein is also listed in the GlycoProtDB entry. This lectin information is linked to the Lectin frontier DataBase (LfDB) of JCGGDB, which provides detailed structural and sequential information on the lectin, as well as its binding specificity.

3.5.2. UniProt

UniProt has long been known to contain information on the glycosylation sites on proteins (13). Its detailed protein sequence information consists of a section called "Amino acid modifications" which includes glycosylation sites. It must be noted, however, that many of these sites may be predicted sites based on sequence motifs and other statistics (i.e., the use of tools such as NetNGlyc and NetOGlyc), and not experimentally confirmed. On the other hand, C-linked glycosylation and glycation sites are always annotated based on experimental evidence. Detailed information regarding the annotation policies of glycosylation are described in the online manual available at http://www.uniprot.org/manual/carbohyd.

3.5.3. O-GlycBase

O-GlycBase is a database provided by the CBS, containing O- and C-glycosylated protein information (14). At the time of this writing, version 6.0 of O-GlycBase contains 242 glycoprotein entries that have been experimentally verified to contain at least one O- or C-glycosylation site. Each entry of this database contains the following information:

- Biological source,
- Links to other major databases such as Swiss-Prot and PIR,
- The glycan structure(s) on the glycoprotein,
- The positions of the glycosylation sites, and indication of whether these were determined experimentally or computationally predicted,
- Literature references,
- Amino acid sequence and their corresponding glycosylation sites,
- And any comments.

4. Notes

In this section, perspectives on how the described tools and resources may be used for further bioinformatics analysis will be discussed, in the hopes that they may provide ideas for the development of future analyses and analytical tools.

The development of various tools for the prediction of glycosylation sites illustrates the feasibility of such methods using neural networks. Some sites provide the same types of predictions and may be confusing to the user. For example, CBS provides a similar tool to OGPet to predict O-glycosylation sites of protein sequences, called NetOGlyc. Both of these tools may give completely different results for the same input, illustrating perhaps the lack of O-glycosylation data and analytical models to fully characterize O-glycosylation. However, these tools are both useful in that if results are obtained that are common to both tools may be deemed to have more validity, and users are encouraged to try both in their own analyses. The same may be said for the GPI-anchor prediction tools GPI-SOM and Big-PI Predictor.

On the other hand, statistical analyses of continually accumulating spatial information of glycosylation sites will provide invaluable information by which further predictions and improved analytical tools may be developed. Since these tools take as input characterized protein structures in the PDB, analyses can be separately performed on various groups of proteins based on their folds or biological properties. The PDB data are also updated weekly such that the latest information is included in the analyses, thus ensuring that the maximal amount of information available can be used for analysis. By understanding the environmental conditions under which glycosylation can take place, similar analyses may be applied to glycan binding predictions of lectins, for example. Furthermore, the integration of existing data that have been experimentally confirmed by such advanced technologies as the IGOT method would aid in the confirmation of such predictions as well.

Overall, it is hoped that the comprehensive analysis of glycoprotein glycosylation would also aid in the more difficult task of understanding glycan recognition mechanisms of glycan binding proteins. In general, the development of tools for such analyses would require a number of various data sets, not limited to protein sequence, but also domain information, spatial information, and binding affinity. With the increasing integration and collaborations of major carbohydrate databases such as the databases of the Consortium for Functional Glycomics (CFG) and EuroCarbDB, these kinds of all-inclusive analyses can be expected to be realized in the near future.

References

1. Eisenhaber B, Bork P, Eisenhaber F (1998) Sequence properties of GPI-anchored proteins near the omega-site: constraints for the polypeptide binding site of the putative transamidase. Protein Eng 11:1155–1161
2. Gupta R, Jung E, Gooley AA, Williams KL, Brunak S, Hansen J (1999) Scanning the available Dictyostelium discoideum proteome for O-linked GlcNAc glycosylation sites using neural networks. Glycobiology 9:1009–1022
3. Jung E, Gooley AA, Packer NH, Slade MB, Williams KL, Dittrich W (1997) An *in vivo* approach for the identification of acceptor sites for O-glycosyltransferases: motifs for the addition of O-GlcNAc in *Dictyostelium discoideum*. Biochemistry 36:4034–4040
4. Jung E, Gooley AA, Packer NH, Karuso P, Williams KL (1998) Rules for the addition of O-linked N-acetylglucosamine to secreted proteins in Dictyostelium discoideum. Studies on glycosylation of mucin MUC1 and MUC2 repeats. Eur J Biochem 253:517–524
5. Fankhauser N, Maser P (2005) Identification of GPI anchor attachment signals by a Kohonen self-organizing map. Bioinformatics 21:1846–1852
6. Julenius K (2007) NetCGlyc 1.0: prediction of mammalian C-mannosylation sites. Glycobiology 17(8):868–876
7. Gupta R, Brunak S (2002) Prediction of glycosylation across the human proteome and the correlation to protein function. Pac Symp Biocomput 7:310–322
8. Hansen JE, Lund O, Tolstrup N, Gooley AA, Williams KL et al (1998) NetOGlyc: prediction of mucin type O-glycosylation sites based

on sequence context and surface accessibility. Glycoconj J 15:115–130

9. Johansen MB, Kiemer L, Brunak S (2006) Analysis and prediction of mammalian protein glycation. Glycobiology 16:844–853
10. Caragea C, Sinapov J, Silvescu A, Dobbs D, Honavar V (2007) Glycosylation site prediction using ensembles of support vector machine classifiers. BMC Bioinformatics 8:438
11. Lutteke T, Frank M, von der Lieth CW (2005) Carbohydrate Structure Suite (CSS): analysis of carbohydrate 3D structures derived from the PDB. Nucleic Acids Res 33:D242–D246
12. Kaji H, Saito H, Yamauchi Y, Shinkawa T, Taoka M, Hirabayashi J, Kasai K, Takahashi N, Isobe T (2003) Lectin affinity capture, isotope-coded tagging and mass spectrometry to identify N-linked glycoproteins. Nat Biotechnol 21:667–672
13. Jung E, Veuthey AL, Gasteiger E, Bairoch A (2001) Annotation of glycoproteins in the SWISS-PROT database. Proteomics 1:262–268
14. Gupta R, Birch H, Rapacki K, Brunak S, Hansen J (1999) O-GLYCBASE version 4.0: a revised database of O-glycosylated proteins. Nucleic Acids Res 27:370–372

Chapter 18

Software Tools for Glycan Profiling

Chuan-Yih Yu, Anoop Mayampurath, and Haixu Tang

Abstract

We introduce three software tools, Cartoonist, GlycoWorkbench, and MultiGlycan, for N-glycan profiling of complex biological samples. Detailed instructions for using these tools are provided, and their performances are demonstrated by using real glycan profiling data.

Key words: N-linked glycans, O-linked glycans, Linkage, Fragmentation, Annotation

1. Introduction

Glycosylation is a common post-translational modification that affects the protein function through the attachment of glycans. Alterations of protein glycosylation are indicative of diseases (1–3), which may occur through changes of the glycans (alterations in monosaccharide composition, glycan structure, or linkage), aberrant glycosylation, and dynamic microheterogeneities. The aim of mass spectrometry-based glycomics is to detect these changes through glycan profiling by first characterizing the glycans and then looking for changes across conditions. Software tools in glycomics aim to first detect the glycans from MS platforms such as MALDI-TOF and then annotating them. Annotation can be done at three levels—composition level, sequence level (through glycan *cartoons*), and sequence plus linkage level as shown in Fig. 1. Tools for platforms such as MALDI typically allow annotation through composition and cartoons. Usually, a single spectrum that contains peaks indicating the presence of released glycans from glycoprotein is used for detection and annotation of all putative glycan species within the sample. The putative glycans can be detected through

Jennifer J. Kohler and Steven M. Patrie (eds.), *Mass Spectrometry of Glycoproteins: Methods and Protocols*, Methods in Molecular Biology, vol. 951, DOI 10.1007/978-1-62703-146-2_18, © Springer Science+Business Media, LLC 2013

LEVEL	REPRESENTATION EXAMPLE	NOTE
composition	9 Hex 2 HexNAc (or) $Hex_9HexNAc_2$ (or) 2 GlcNAc 9 Man	Indicates number and the type of monosaccharide Abbreviations Hex: hexose HexNAc: *N*-acetylhexosamine Man: mannose GlcNAc: *N*-acetylglucosamine
sequence	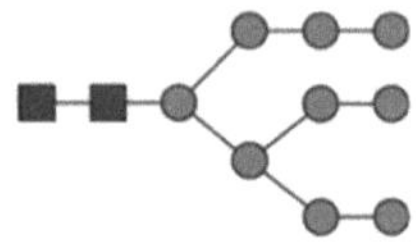	Called 'cartoon graphs'. Indicates monosaccharide and topology. Blue squares indicate GlcNAc and green circles indicate mannose.
sequence with linkage	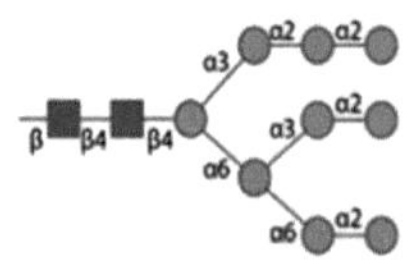	Cartoon graphs also indicating linkage information.

Fig. 1. *Three levels of representation for glycans.* The basic composition representation of a glycan gives us only the mass (or the monosaccharide composition) of the glycan. Sequence representations (or the cartoons) inform us the topology of the glycan including the branching structure. Addition of linkage types gives us the comprehensive view of the glycan. Additional information on symbol nomenclature and linkage types can be found in the textbook *Essentials of Glycobiology* (Varki, et al., 2009), which can be accessed on the NCBI bookshelf (http://www.ncbi.nlm.nih.gov/books/NBK1908). The glycans shown here are N-linked glycans taken from the CFG database (7). Note the core of these glycans with two GlcNAc and three Mannose residues.

two different ways. One is through database searching using glycan mass from a curated glycan database and the other is through de novo sequencing algorithms. These two types of methods have their own advantages and disadvantages. Database searching is commonly used for MALDI-based glycomics and provides results through a fast and precise search of the spectra, but cannot find any novel glycans that are not collected in the database. De novo sequencing methods typically involve fragmentation spectra. Although these algorithms are relatively slow and prone to errors, novel glycans can be discovered in this way. There are many tools that utilize both approaches for glycan detection and annotation. In this chapter, we describe three software tools that are used in N-linked glycomics.

Here, we introduce three software tools for high throughput glycan annotation and profiling in glycomics (see Note 1 regarding the comparative performance of the three tools). To gather comprehensive glycan annotation at the sequence and linkage level, we need to acquire more information from the different avenues. A preliminary glycan profile only gives us the mass of glycans but since some monosaccharides have exactly the same mass, e.g., mannose and galactose, GalNAc, and GlcNAc, etc., the precise monosaccharide composition cannot be deciphered from mass spectra alone. However, we can utilize tandem MS (MS^n) in combination with other software tools to elucidate both sequence and linkage information (4). Here we limit our discussion on the tools for N-glycan profiling. These methods can be extended to other types of glycosylations such as O-linked glycans. O-glycans have more diverse core structures, which means that the space of candidate of glycans is larger than that of N-glycans. As a result, it is harder to explore O-glycans using described software directly.

The lack of a comprehensive glycan database is a drawback since novel and rare structures cannot be identified by using database searching techniques. Discovery of novel glycan structures becomes more and more important in cases where diseases are related to rare glycosylation. We need to be more careful when examining the spectrum and always leave some tolerance for novel structure discovering.

2. Materials

In this section, we briefly introduce the methods and usage for the software tools. The software tools are listed in alphabetic order.

2.1. Cartoonist

Cartoonist (5) is an automated N-glycan profiling tool that can be used to annotate the spectrum. It begins with an archetype set of N-glycans, which it then expands using sets of predefined rules based on synthetic glycosylation pathways. By using these rules and an initial set of 300 archetypes, a total of 2,800 N-glycan candidates are derived.

Cartoonist assigns potential N-glycan to a peak based on mass, and uses the top 15 intense peaks to calibrate the result. The calibration simply takes the mass difference between predicted and observed mass value, and uses those pairs to fit with a linear model. Then this model is used to reevaluate the glycan assignments within each spectrum.

2.2. GlycoWorkbench

GlycoWorkbench (6) is a suite of programs for glycan profiling and interpretation. It not only supports various data formats, but also provides basic spectrum processing tools. In addition, it provides a user-friendly interface to draw glycan structures, which can be subsequently used to annotate the mass spectrum. Additionally, a simulated fragmentation mechanism also allows the user to view putative fragment peaks of the glycan in the spectrum.

In this manner, GlycoWorkbench can deal with both MS and tandem MS data. Users can load tandem MS data to obtain glycan structure and linkage information. Here, we focus on the glycan profiling functionality of the software from a single MS spectrum prospect. Users can either draw a specific structure or search certain glycans in public glycan database via GlycoWorkbench. It supports four different formats of public databases (CFG (7), Carbbank (8), GlycomeDB (9), and Glycosciences (10)). The software can achieve excellent interpretation of glycan structures, but consumes a lot of memory when profiling a whole spectrum.

2.3. MultiGlycan

MultiGlycan (11) uses N-glycan candidates derived from known N-glycan synthetic pathways (12). There are 328 N-glycans of different masses used in the analysis; but the users are allowed to specify specific N-Glycan compositions in this candidate list. Glycan annotation is then done at a composition level, and thus, glycan structures are not reported. It is worth mentioning that MultiGlycan not only directly uses mass as a feature for detection but also uses sophisticated mixture models to improve N-glycan annotation. MultiGlycan calculates the correlation between theoretical and experimental isotopic envelopes. It uses three different models for constructing theoretical isotopic envelopes for each glycan candidate. First, the glycan candidate mass is directly used to create an isotopic envelope. Second, a composite overlapping theoretical isotopic envelope comprising two glycan masses with a mass difference within a tolerance is constructed and matched to the observed isotopic envelope. Finally, a composite envelope is created based on the candidate glycan and an unknown compound. Using these three models, individual as well as overlapping glycan isotopic envelopes can be annotated, leading to an increased number of identified glycans. MultiGlycan also provides utilities that detect profile abundance variations across multiple samples. The correlation-based fit score is used to select confident glycan identifications. Similar to the gene-shaving technique (13), the software is equipped with a glycan shaving algorithm based on principal component analysis (PCA) to identify the top "n" (n is a user-specified number) glycan species that contribute most to the abundance variation (14). This is particularly useful for glycan biomarker discovery where abundance variations could be related to change of state between healthy and disease samples (1–3). MultiGlycan can also be used for O-glycan profiling and biomarker discovery if the user can input a predefined list of O-glycan compositions.

3. Methods

3.1. Cartoonist

Website: http://bio.parc.com/mass_spec. Supported spectrum format: msd. see Note 2.

1. File → Open, select msd format (see Note 3).
2. Click "Yes" in "Download cartoons" pop-up window.
3. The result will show in the window. An example is given in Fig. 2.

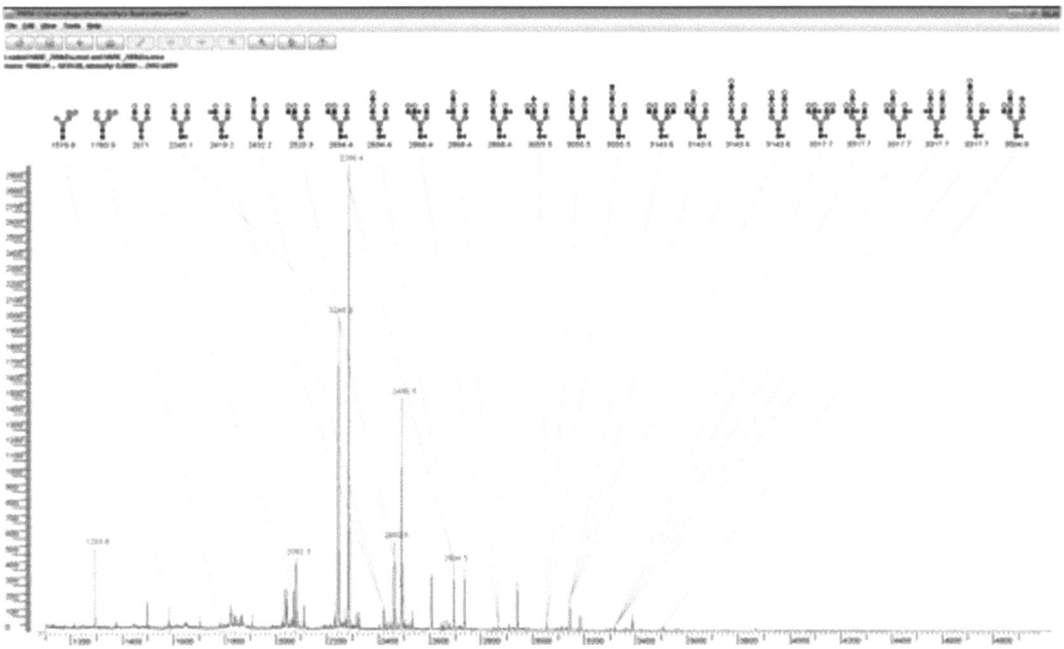

Fig. 2. *A snapshot of the PARC Mass Spectrum Viewer.* The glycan structures annotated in the input mass spectra is shown on the top.

3.2. GlycoWorkbench

Website: http://www.glycoworkbench.org. Supported mass spectrum data format: Plain text (Peak list), xml, mzData, mzXML, and t2d.

1. Load mass spectra file (in the formats of plain text, xml, mzXML, mzData, or t2d).
2. Click on the Profiler button in the Tools tab and click on "Annotate peaks with structures from database."
3. Choose one or multiple databases, derivatization, and reducing end.
4. Choice the fragment options in the pop-up window and click OK.
5. The result will be shown in the Search panel (example shown in Fig. 3).

3.3. MultiGlycan

Website: http://mendel.informatics.indiana.edu/~chuyu/MultiGlycan. Supported spectrum format Plain text (Peak list), mzXML, and RAW file (Thermo Scientific instruments).

1. Load spectrum file by click "Load" button in "Peak List Setting" panel.
2. Choose related options for the N-glycan profiling experiment, and load default or user-defined N-glycan composition file in "Glycan List Setting" Panel.
3. Click "Calculation."
4. Result will be shown in the lower table (see Fig. 4).

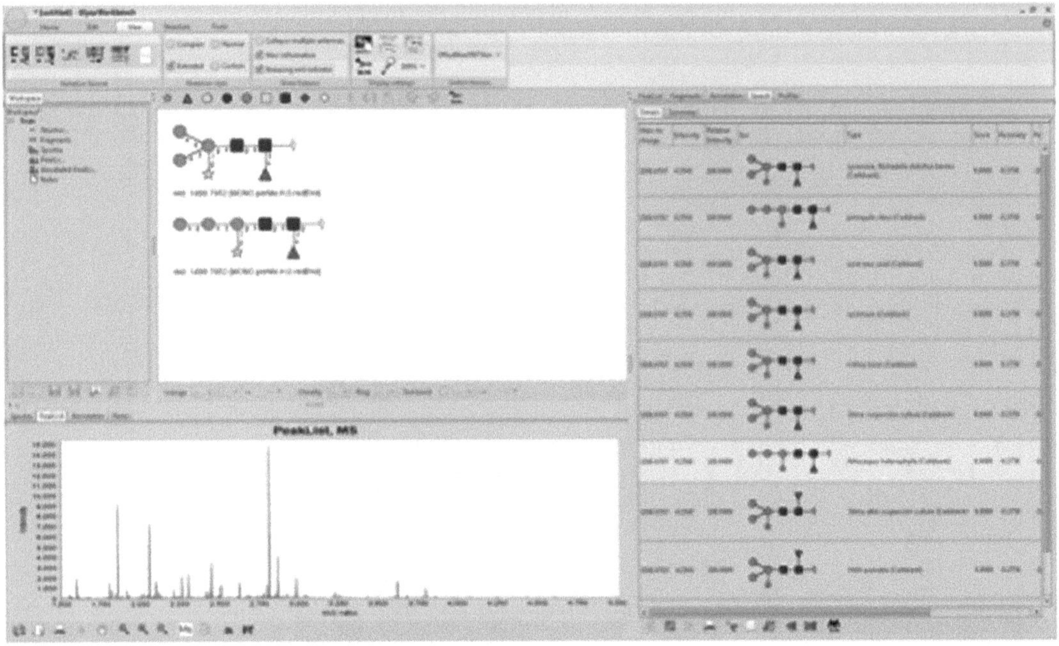

Fig. 3. *The user interface of GlycoWorkbench software.* There are four panels in GlycoWorkbench, workspace, canvas, spectrum view, and result list. The data can be loaded via right click on the workspace tree node. Canvas panel provides a GUI interface for user to draw glycan structures. Users also can view their raw spectrum in the spectrum view panel. Result list panel contains peaks, fragments, annotation, and profile list. All the results will be shown in this panel.

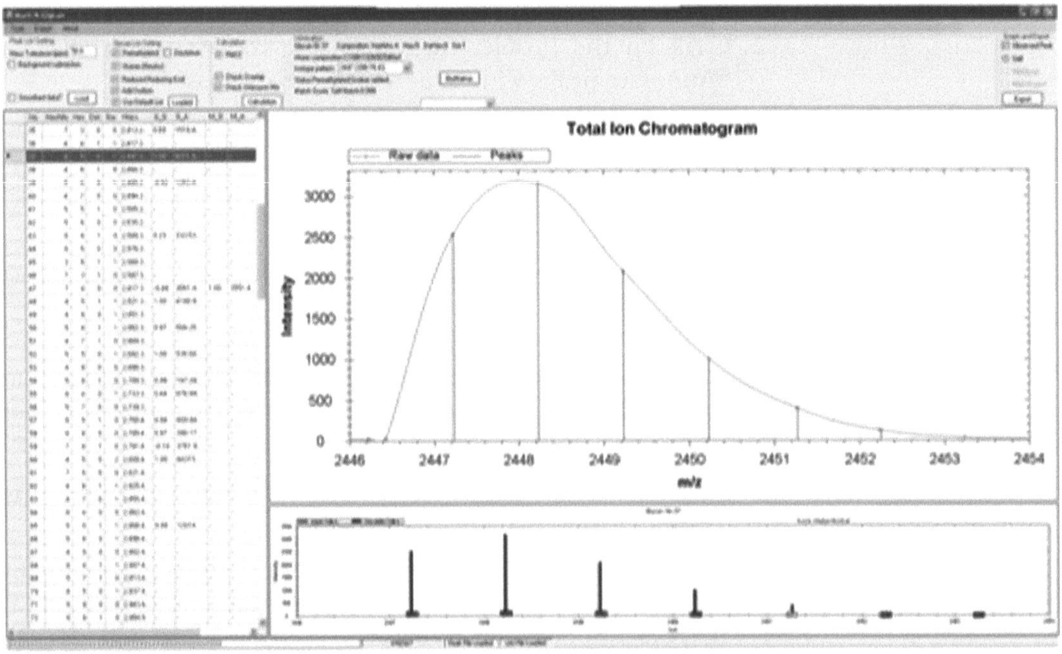

Fig 4. The user interface of MultiGlycan.

4. Notes

1. We used two MALDI TOF/TOF datasets from the glycomics analysis of human blood samples (15). (The input files can download from MultiGlycan website) One dataset is from hepatocellular carcinoma patients (HC-146) and the other is from healthy control (NC-33). All software tools were run on their default settings, and results were compared with each other. For ease, glycans that differ only in linkages are considered to be the same glycan. In Cartoonist result, we took the output file (msa) and counted the nonredundant peaks. In MultiGlycan result, we only took glycans that have the correlation scores above 0.7 in at least one of the three models. We note that before running GlycoWorkbench annotation, we loaded the spectrum (in peak list format) and conducted peak centroid via default setting. The total number of identified glycans is listed in Table 1. GlycoWorkbench has identified more than 600 glycans in both cases, much more than the number of glycan species we expect to observe in the human blood sample (15). Hence, a majority of them might be false positive identifications.
2. Cartoonist is integrated into the PARC Mass Spectrum Viewer (16) wrapped as an executable jar file. Users can directly execute or use a java command. These two execution methods will exhibit different behaviors in memory consumption and running time. Using the java command line (java –jar MassSpecViewer.jar) will consume more memory than the direct execution of the jar file, but the command line method runs faster. We ran a small sample data and list the performance both on running time and memory usage in Table 2.
3. Although PARC Mass Spectrum Viewer supports many formats of mass spectra, only the msd format can be used for Cartoonist annotation.

Table 1
Total number of identified glycans in two different datasets (HC-146, NC-33)

	Cartoonist	GlycoWorkbench	MultiGlycan
	HC-146		
Total identified glycan	48	691	120
	NC-33		
Total identified glycan	50	653	114

Table 2
Comparison for different invoking method of Cartoonist

	Running time (s)	Memory usage (MB)
Command line	50	1,800
Jar directly	215	300

Acknowledgments

This work is partially supported by National Science Foundation (DBI-0642897) and National Cancer Institute (1U01CA128535-01).

References

1. Martin-Rendon E, Blake DJ (2003) Protein glycosylation in disease: new insights into the congenital muscular dystrophies. Trends Pharmacol Sci 24(4):178–83
2. Scanlin TF, Glick MC (2000) Terminal glycosylation and disease: influence on cancer and cystic fibrosis. Glycoconj J 17(7–9):617–26
3. Turner GA (1992) N-glycosylation of serum proteins in disease and its investigation using lectins. Clin Chim Acta 208(3):149–71
4. Lapadula AJ et al (2005) Congruent strategies for carbohydrate sequencing. 3. OSCAR: an algorithm for assigning oligosaccharide topology from MSn data. Anal Chem 77(19):6271–9
5. Goldberg D et al (2005) Automatic annotation of matrix-assisted laser desorption/ionization N-glycan spectra. Proteomics 5(4):865–75
6. Ceroni A et al (2008) GlycoWorkbench: a tool for the computer-assisted annotation of mass spectra of glycans. J Proteome Res 7(4):1650–9
7. Raman R et al (2006) Advancing glycomics: implementation strategies at the consortium for functional glycomics. Glycobiology 16(5):82R–90R
8. Doubet S et al (1989) The complex carbohydrate structure database. Trends Biochem Sci 14(12):475–7
9. Ranzinger R et al (2008) GlycomeDB - integration of open-access carbohydrate structure databases. BMC Bioinformatics 9:384
10. Lutteke T et al (2006) GLYCOSCIENCES.de: an Internet portal to support glycomics and glycobiology research. Glycobiology 16(5):71R–81R
11. Yu C-Y et al (2010) Biomarker discovery by automatic annotation of N-glycan species in MALDI-TOF-TOF spectra, in 58th ASMS Conference. Salt Lake City
12. Krambeck FJ, Betenbaugh MJ (2005) A mathematical model of N-linked glycosylation. Biotechnol Bioeng 92(6):711–28
13. Hastie T et al (2000) 'Gene shaving' as a method for identifying distinct sets of genes with similar expression patterns. Genome Biol 1(2):RESEARCH0003
14. Kyselova Z et al (2007) Alterations in the serum glycome due to metastatic prostate cancer. J Proteome Res 6(5):1822–32
15. Tang Z et al (2010) Identification of N-glycan serum markers associated with hepatocellular carcinoma from mass spectrometry data. J Proteome Res 9(1):104–12
16. Marshall AG, Hendrickson CL, Jackson GS (1998) Fourier transform ion cyclotron resonance mass spectrometry: a primer. Mass Spectrom Rev 17(1):1–35

Part VI

Case Studies in Mass Spectrometry of Glycoproteins

Chapter 19

Quantitative Characterization of Glycoproteins in Neurodegenerative Disorders Using iTRAQ

Min Shi, Hyejin Hwang, and Jing Zhang

Abstract

Aberrant protein glycosylation has been recognized to be associated with many neurodegenerative disorders, including Alzheimer and Parkinson disease. Using mass spectrometry-based technologies to catalog and quantify glycoproteins in these diseases is expected to provide insight into not only the biochemical pathogenesis of neurodegeneration but also the biomarker discovery. This chapter describes a multidisciplinary approach to accomplish the goal of glycoprotein identification and quantification in human brain tissue and cerebrospinal fluid, which includes sample preparation, isobaric tag labeling of digested peptides, glycopeptide enrichment using hydrazide chemistry, protein/peptide identification and quantification by liquid chromatography-based high-resolution tandem mass spectrometry, as well as bioinformatic data processing.

Key words: Cerebrospinal fluid, Glycoprotein, Hydrazide chemistry, Isobaric tags for relative and absolute quantitation (iTRAQ), Mass spectrometry, Neurodegenerative disease, Proteomics

1. Introduction

Among various post-translational modifications of proteins, glycosylation represents the most common and complicated form. The protein sequence data suggests that more than half of all proteins are glycoproteins (1). The complexity arises primarily from the diversity of monosaccharide structure, sequence, and linkage. Additionally, on a single glycoprotein, there is normally a range of glycan structures associated with each potential glycosylation site (2). The two main types of glycosylation are N-linked and O-linked. N-linked glycosylation is particularly prevalent in proteins destined for the extracellular environment (3). The glycan moieties of these glycoproteins, by virtue of their diversity, play important roles not

Jennifer J. Kohler and Steven M. Patrie (eds.), *Mass Spectrometry of Glycoproteins: Methods and Protocols*, Methods in Molecular Biology, vol. 951, DOI 10.1007/978-1-62703-146-2_19, © Springer Science+Business Media, LLC 2013

only in modulating protein properties, such as stability and conformation, but also as key regulators of protein function, activity, localization, and interaction (4–6). Consequently, aberrant glycosylation has now been recognized as an attribute of many mammalian diseases, including hereditary disorders, immune deficiencies, cardiovascular diseases, cancer, and neurodegenerative diseases (4, 6, 7). In fact, many clinical biomarkers and therapeutic targets are glycoproteins (8–10).

In order to examine the disease-related glycosylation alteration, sensitive, fast, and robust analytical methods are required. Over the past few years, mass spectrometry (MS) has proven to be an important tool due to its high sensitivity, selectivity, and throughput. In neurodegenerative diseases such as Alzheimer disease (AD) and Parkinson disease (PD), early work has suggested that a systematic and detailed analysis of glycosylated proteins and their alterations in human brain tissues and body fluids, particularly cerebrospinal fluid (CSF), may provide pressingly needed biomarkers that can assist with clinical diagnosis, monitoring disease progression, and evaluating the effects of existing and future therapeutic drugs (7, 11–13). Additionally, unique disease biomarkers may reveal novel mechanisms underlying various neurodegenerative diseases, which are currently largely unknown, and may also provide new therapeutic targets.

The large dynamic range of protein concentrations in biological samples, particularly body fluids, is far beyond the analysis range of the current proteomic techniques (14, 15). Therefore, front-end enrichment and fractionation methods prior to MS analysis are necessary to enhance detection of low abundance glycoproteins, which often provide promising diagnostic and biological information (14, 15). Effective enrichment of glycosylated proteins is thus important to decrease sample complexity and to provide comprehensive glycoproteome coverage (16, 17). Glycoproteins can be enriched at the protein level and then digested into peptides (18, 19). However, this strategy suffers from solubility problems (particularly for those large membrane proteins) and steric hindrance when capturing proteins in their native forms. In the other probably more popular strategy, glycoproteins are digested first into peptides and then enriched using hydrazide chemistry (17, 20) or lectin affinity purification (21). While lectin affinity capture is easy to implement and multidimensional lectin chromatography could be effective, the binding selectivity of lectins to specific conformations of different carbohydrate moieties has limited the utility of lectin in global glycoprotein analysis (17, 22, 23). Hydrazide functionalized beads, on the other hand, appear to be a good matrix for trapping glycopeptides and glycoproteins by covalent bonding after oxidation with periodate (19, 23). Peptide moieties of the covalently captured N-linked glycopeptides can be released by treatment with peptide-N-glycosidase F (PNGase F) to allow peptide and glycosylation site identification.

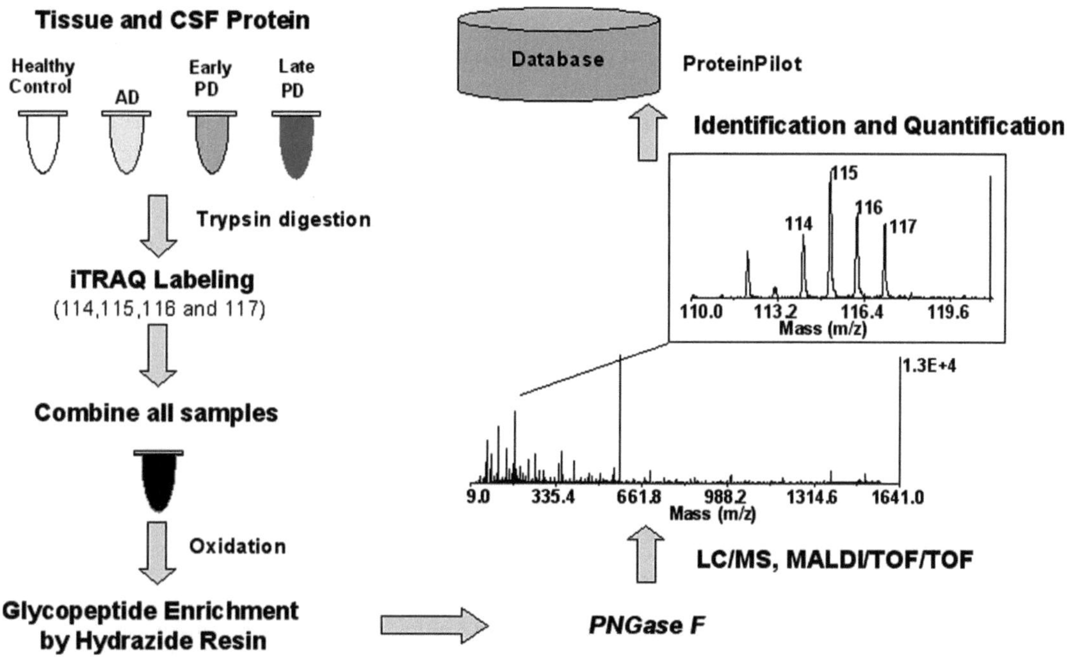

Fig. 1. Schematic diagram of a quantitative analysis of N-glycoproteins associated with neurodegenerative diseases. In this case, protein samples from healthy controls, patients with Alzheimer disease (AD), and early or late Parkinson disease (PD) are labeled with different iTRAQ reagents after trypsin digestion. All samples are mixed together before glycopeptides are enriched using the hydrazide resin. The N-glycopeptides are released by PNGase F and analyzed by tandem mass spectrometry (MS/MS). The fragmentation data of the peptides results in the identification of the labeled peptides and hence the corresponding proteins. Measurement of the intensity of the reporter ions generated from fragmentation of the tag attached to the peptides enables relative quantification of the peptides in each digest and hence the proteins from where they originate.

Characterizing glycoproteins in human brain tissue and CSF as extensively as possible is just the first step to define biomarkers unique to a neurodegenerative disease or disease progression. A more important process is to quantify the changes associated with a disease or a disease stage. Additionally, quantitative glycoproteomics can help to characterize the regulatory pathways and complex system networks by providing protein concentration information that corresponds to different cellular states (7). In the past several years, many MS-based quantitative proteomics methods have been developed (24). These methods include the use of chemical reactions to introduce isotopic tags at specific functional groups on peptides or proteins, such as isotope coded affinity tags (ICAT) (25) and isobaric tags for relative and absolute quantitation (iTRAQ) (26). ICAT quantification is restricted to cysteine-containing proteins, and it can only compare two conditions at a time. iTRAQ, on the other hand, is based on chemically tagging the *N*-terminus of peptides generated from protein digests that have been isolated from samples in, for example, different disease and control groups. The technique allows for the identification and quantification of all peptides as well as comparison of up to eight conditions simultaneously.

In this review, we will describe the detailed methods we have utilized in preparing human brain tissue and CSF samples before proteomics analysis, digesting protein samples with trypsin, followed by iTRAQ labeling and hydrazide bead capture, and performing quantitative MS analysis of N-linked glycoproteins unique to the disease and disease progression (see Note 1 for the option to characterize O-linked glycoproteins) (Fig. 1). A few recent methodology improvements over those used in the original publications (7, 12) are also incorporated.

2. Materials

2.1. Preparation of Tissue Sample

1. Glass-Teflon homogenizer (Wheaton, Millville, NJ).
2. 1 M 4-[2-hydroxyethyl]-1-piperazineethanesulfonic acid (HEPES), pH7.5.
3. Protease inhibitor cocktail (Cat.# P2714, Sigma-Aldrich, St. Louis, MO, or equivalent), dissolved in 10 mL of water as a 100× stock. Aliquot the stock and store at −20°C or lower.
4. Homogenization buffer: 7 M urea, 65 mM dithiothreitol (DTT), 2% 3-[(3-cholamidopropyl) dimethylamino]-1-propane sulfonate (CHAPS), 20 mM HEPES, pH7.5 (from 1 M stock), and protease inhibitor cocktail from Sigma-Aldrich or equivalent. Prepare freshly before use.
5. A probe (microtip) sonicator.

2.2. Preparation of CSF Sample

2.2.1. Collection of Human CSF

1. 24-gauge bullet-tip Sprotte spinal needle (Medecin Technik, Germany).
2. 1% lidocaine.
3. Protease inhibitor cocktail (see Subheading 2.1 above).

2.2.2. Hemoglobin Assay

1. Human Hemoglobin ELISA Quantitation Kit (Cat.# E80-135) from Bethyl, Montgomery, TX, including a sheep anti-human hemoglobin affinity purified antibody (1 mg/mL) as the coating/capturing antibody, a sheep HRP-conjugated anti-human hemoglobin antibody as the detecting antibody, and a human hemoglobin calibrator (2 mg/mL) at a working range of 6.25–400 ng/mL. Store at 4°C.
2. Polystyrene 96-well microplate (e.g., Cat.# 9017 EIA/RIA plate from Corning, Lowell, MA).
3. Coating buffer: 150 mM sodium carbonate, 350 mM sodium bicarbonate, 30 mM sodium azide. Adjust pH to 9.6 with HCl. Sterilize with a 0.22-μm filter and store at 4°C.

4. Wash solution: 50 mM Tris–HCl, pH 8.0, 140 mM NaCl, 0.05% Tween 20. Store at room temperature.
5. Blocking (Postcoat) solution: 50 mM Tris–HCl, pH 8.0, 140 mM NaCl, 1% BSA. Store at 4°C.
6. Sample/conjugate diluent: 50 mM Tris–HCl, pH 8.0, 140 mM NaCl, 0.05% Tween 20, 1% BSA. Store at 4°C.
7. Enzyme substrate solution: SureBlue TMB (3,3′,5,5′-tetramethylbenzidine) microwell peroxidase substrate (1-component) (Cat.# 52-00-01, Kirkegaard and Perry, Gaithersburg, MD; see Note 2). Store at 4°C.
8. Stopping solution: 2 M HCl (2 M H_2SO_4 can also be used). Store at room temperature.
9. A microplate spectrophotometer/reader for measuring absorbance at 450 nm.

2.3. TCA Precipitation and BCA Assay

1. Trichloroacetic acid (TCA), precooled at 4°C.
2. Acetone, precooled at −20°C.
3. Triethylammonium bicarbonate (TEAB) buffer, 1 M solution (Cat.# T7408, Sigma-Aldrich). Before use, add equal volume of H_2O to make the 0.5 M solution.
4. 8 M urea in 0.5 M TEAB buffer. Prepare immediately before use.
5. 2% SDS in 0.5 M TEAB buffer.
6. Triton X-100.
7. Bicinchoninic Acid Assay (BCA) protein assay kit from Pierce, Rockford, IL.
8. 96-Well polystyrene microplate, clear, flat bottom (e.g., Cat.# T-3016-5 from ISC Bioexpress, Kaysville, UT), for BCA assays.
9. A microplate spectrophotometer/reader.

2.4. Trypsin Digestion and iTRAQ Labeling

2.4.1. Trypsin Digestion

1. Sequencing grade modified trypsin (Cat.# V5111, Promega, Madison, WI). Store at −20°C or preferably −80°C.
2. Reducing Reagent, Cysteine Blocking Reagent from the iTRAQ Reagent Multi-Plex Kit (see Subheading 2.4.3 below).

2.4.2. SDS-PAGE and Silver Staining

1. 16.5% Tris-Tricine polyacrylamide gel (Cat.# 345-0063, Bio-rad, Hercules, CA, or equivalent).
2. Sample loading buffer, 2× (Cat.# 161-0739, Bio-rad, or equivalent). Add 20 μL β-mercaptoethanol to 980 μL sample buffer before use.
3. Electrophoresis buffer: 100 mM Tris, 100 mM Tricine, 0.1% SDS, pH 8.3. A 10× premixed solution could be obtained from Bio-rad (Cat.# 161-0744).

4. Fixing solution: 40% ethanol/10% acetic acid in deionized H_2O.
5. Wash solution: 30% ethanol in deionized H_2O.
6. Sensitizing solution: 0.02% sodium thiosulfate in deionized H_2O.
7. Silver solution: 0.2% silver nitrate, 0.02% formaldehyde (37%) in deionized H_2O.
8. Developing solution: 3% sodium carbonate, 0.05% formaldehyde (37%), 0.00005% sodium thiosulfate in deionized H_2O.
9. Stop solution: 0.5% glycine in deionized H_2O.

2.4.3. iTRAQ Labeling

1. iTRAQ Reagent Multi-Plex Kit (Applied Biosystem, Foster City, CA), including iTRAQ Reagents 114–117, iTRAQ Reagent Dissolution Buffer, Cysteine Blocking Reagent [200 mM methylmethane-thiosulfonate (MMTS) in isopropanol], Reducing Reagent [50 mM Tris(2-carboxyethyl)phosphine (TCEP) in H_2O], Denaturant (2% SDS), and ethanol. Store at −20°C.
2. 0.5 M TEAB buffer (see Subheading 2.3 above).
3. A SpeedVac® (Thermo Fisher, Waltham, MA) or similar vacuum concentrator.

2.5. Glycopeptide Enrichment

2.5.1. Desalting

1. Desalting column: Waters Oasis Sep-Pak® Vac C18 cartridge, 1 cc (Cat.# WAT023590, Waters, Milford, MA).
2. 5 M HCl.
3. 1% trifluoroacetic acid (TFA): for 5 mL, add 50 μL of TFA to 4.95 mL of HPLC-grade water.
4. Wetting/Eluting solution: 0.1% TFA in 50% acetonitrile (ACN). For 5 mL, add 500 μL of 1% TFA and 2.5 mL of ACN to 2 mL of HPLC-grade water.
5. Equilibration/Wash solution: 0.1% TFA (for 10 mL, add 1 mL of 1% TFA to 9 mL of HPLC-grade water).

2.5.2. Oxidation

1. 100 mM sodium periodate (10×).

2.5.3. Isolation of Glycoprotein by Hydrazide Resin

1. Hydrazide resin in isopropanol (Cat.# 153-6047, Bio-Rad).
2. Coupling buffer: 100 mM sodium acetate, 150 mM NaCl, pH 5.5.
3. 1.5 M NaCl.
4. 80% ACN.
5. 100% methanol.
6. 0.1 M NH_4HCO_3, pH 8.3.
7. PNGase F, 500,000 units/mL (Cat.# P0705L, New England BioLabs, Beverly, MA).
8. 0.5% TFA in HPLC-grade water (loading buffer for LC spotting).

2.6. LC-MS/MS Analysis of Glycopeptides

1. Reversed-phase (RP) nanocapillary LC system (LC Packings/ DIONEX Corporation, Sunnyvale, CA), or equivalent.
2. RP column: 15-cm × 100-μm-inner diameter Magic C18 capillary column, with 3-μm, 100-Å packing (Cat.# 161459, DIONEX Corporation), or equivalent.
3. Loading buffer: 0.5% TFA in HPLC-grade water.
4. Solvent A: 2% ACN, 0.1% TFA in HPLC-grade water.
5. Solvent B: 80% ACN, 0.08% TFA in HPLC-grade water.
6. 4,800 Plus MALDI TOF/TOF™ Analyzer (Applied Biosystems) with Data Explorer (Applied Biosystems) spectral analysis software, or equivalent.
7. Matrix solution: 7 mg/mL recrystallized α-cyano-4-hydroxycinnamic acid (CHCA), analytical-reagent grade (Cat.# 70990, Fluka, Ronokonkoma, NY; see Note 3 for recrystallization) in 60% ACN and 2.6% ammonium citrate prepared in HPLC-grade water.
8. Calibration standards (Cat.# 4333604, Applied Biosystems).
9. ProteinPilot™ software (Applied Biosystems) or equivalent.

3. Methods

3.1. Preparation of Tissue Sample: Estimated Timing, 1–2 h

1. Precool homogenizer, homogenization buffer, and sample tubes on ice. All the following steps should be performed on ice or at 4°C.
2. Quickly thaw ~100 mg of frozen tissue in 1 mL of homogenization buffer on ice. Cut tissue into small pieces with clean scissors.
3. Transfer sample into a glass-Teflon homogenizer and disrupt by using 20–30 up and down strokes. Transfer homogenized sample into a sample tube. Wash the homogenizer with ~0.4 mL homogenization buffer and transfer everything into the sample tube.
4. Insert the sample tube in an ice bucket and sonicate with a probe (microtip) sonicator for 10–20 pulses at a low setting. Chill the sample on ice for 10–30 s between each pulse (see Note 4).
5. Centrifuge at 14,000 × *g* for 15 min (4°C). Transfer supernatant into a clean tube. Proceed to "TCA precipitation and BCA assay."

3.2. Preparation of CSF Sample

3.2.1. Collection of Human CSF: Estimated Timing, 1–2 h

1. The procedure can only be performed by a medical personnel (usually a neurologist) and written informed consent should be obtained from the subject.
2. Place individuals in the lateral decubitus position and infiltrate the L4-5 interspace with 1% lidocaine to provide local anesthesia.
3. Perform a lumbar puncture atraumatically with a 24-gauge bullet-tip Sprotte spinal needle and draw CSF with sterile syringe(s).
4. Individuals must remain in bed for 1 h following lumbar puncture.
5. Collect all CSF samples in the morning after overnight fasting (see Note 5), and store at −80°C in small aliquots until further use.
6. (optional) Before use, add protease inhibitors to CSF samples (see Note 6). If the samples are to be reused, they should be further aliquoted into small volumes to minimize freeze-thawing.

3.2.2. Hemoglobin Assay (see Note 7): Estimated Timing, 5–6 h

1. Analyze standards, samples, blanks, and/or controls in triplicate. Dilute 1 μL of the coating antibody to 100 μL with coating buffer for each well to be coated in the plate. Incubate the plate for 60 min at room temperature with gentle shaking.
2. Aspirate the antibody solution from each well after incubation. Add 200 μL of wash solution to each well and then remove the solution by aspiration. Repeat for a total of three washes.
3. Add 200 μL of blocking solution to each well. Incubate for 30 min at room temperature with gentle shaking.
4. While blocking, prepare the standards (6.25, 12.5, 25, 50, 100, and 200 ng/mL) in the sample/conjugate diluent. Samples could be diluted 1:20 or more in the diluent. Store on ice before use.
5. After blocking, remove the solution and wash each well three times as in step 2.
6. Transfer 100 μL of the standard or sample to assigned wells. Incubate plate for 60 min at room temperature with gentle shaking.
7. Remove samples and standards, and wash each well five times as in step 2.
8. Dilute the HRP conjugate in conjugate diluent 1:10,000. Transfer 100 μL to each well. Incubate for 60 min at room temperature with gentle shaking.
9. Remove the HRP conjugate after incubation and then wash each well five times as in step 2.

10. Transfer 100 μL of enzyme substrate solution to each well. Incubate the plate for 20 min at room temperature. Watch the color changes to avoid overdeveloping.
11. To stop the TMB reaction, apply 100 μL of 2 M HCl to each well. If a different substrate is used, use the stop solution recommended by the manufacturer.
12. Using a microplate reader, read the plate at the wavelength that is appropriate for the substrate used (450 nm for TMB).

3.3. TCA Precipitation and BCA Assay: Estimated Timing, 3 h

1. Precool the sample and TCA at 4°C.
2. Add TCA into the sample tube to make a final TCA concentration of 20%. Mix by inverting the tube several times or vortexing gently.
3. Incubate at 4°C for 1 h.
4. Centrifuge at 15,000×*g* for 10 min (4°C). Remove supernatant.
5. Resuspend the pellet with 200 μL of cold acetone. Mix well by vortexing. If necessary, pipette up and down to break the large chunk of pellet.
6. Centrifuge at 15,000×*g* for 10 min (4°C). Remove supernatant.
7. Repeat steps 5–6.
8. Do not let the pellet dry. Redissolve the sample immediately as the following: add 5 μL of 2% SDS in 0.5 M TEAB buffer, mix well by vortexing; add 100 μL of 8 M urea in 0.5 M TEAB buffer, mix well by vortexing; add 2 μL of Triton X-100, and mix well by vortexing.
9. Measure protein concentrations with a BCA protein assay kit from Pierce following the manufacturer's microplate procedure. The samples need to be diluted with H_2O appropriately before assay. A standard curve ranging from 5 to 250 μg/mL is recommended.
10. Pool equal amount (protein) of individual samples in each comparing group to form group samples. Pooling at least five to ten individual samples in each group is desired. The pooled and remaining individual samples can be aliquoted and stored at −80°C.

3.4. Trypsin Digestion and iTRAQ Labeling

3.4.1. Trypsin Digestion: Estimated Timing, 18 h

1. Start with an equal amount (100–120 μg protein) of pooled samples from each comparing group (for example, healthy control, AD, early PD or late PD). Add the Reducing Reagent (50 mM TCEP) from the iTRAQ reagent kit to each sample to make a final concentration of 5 mM (see Note 8). Vortex to mix, and then spin.

2. Incubate at 37°C for 1 h. Spin to bring the sample to the bottom of the tube.
3. To each tube, add the Cysteine-Blocking Reagent (200 mM MMTS) from the iTRAQ reagent kit to make a final concentration of 10 mM (see Note 9). Vortex to mix, and then spin.
4. Incubate at room temperature for 10 min. Spin to bring the sample to the bottom of the tube.
5. Dilute the samples nine times with 0.5 M TEAB buffer (see Note 10). Save a small aliquot (1–3 μg; before digestion) for sodium dodecyl sulfate polyacrylamide gel electrophoresis (SDS-PAGE) and silver staining.
6. Dissolve trypsin at 1 μg/μL in Milli-Q H_2O or equivalent. Mix well by gentle vortexing. Keep on ice.
7. To each sample tube, add 5 μL of the trypsin solution. Vortex to mix thoroughly.
8. Incubate at 37°C for 2 h, preferably with shaking. Spin to bring the sample to the bottom of the tube.
9. To each sample tube, add another 5 μL of the trypsin solution. Vortex to mix thoroughly.
10. Incubate at 37°C overnight for a total of 16 h (including the first 2 h), preferably with shaking. Spin to bring the sample to the bottom of the tube. Save a small aliquot (1–3 μg; after digestion) for SDS-PAGE and silver staining. The digested samples could be stored at −20 or −80°C before iTRAQ labeling.

3.4.2. SDS-PAGE and Silver Staining (see Note 11): Estimated Timing, 3–4 h

1. Add at least an equal volume of sample loading buffer to the saved "before digestion" and "after digestion" samples. Run 16.5% Tris-Tricine SDS-PAGE, preferably with protein/peptide standards. If needed, the electrophoresis tank should be buried in ice to avoid overheating.
2. Rinse the gel with deionized water and incubate it in the fixing solution for at least 1 h (could be overnight).
3. Wash the gel twice with the wash solution for 20 min each, and then wash in deionized water for 20 min.
4. Incubate the gel in the sensitizing solution for 1 min, and then wash three times in deionized water for about 20 s each.
5. Incubate the gel in the silver solution for 20 min, and then wash three times in deionized water for about 20 s each.
6. Incubate the gel in the developing solution until clear bands appear (usually in 3–10 min). Watch to avoid overdeveloping.
7. Wash the gel in deionized water for about 20 s. Incubate in the stop solution for 5 min and then wash with deionized water three times for 5 min each. Figure 2 shows a typical expected result.

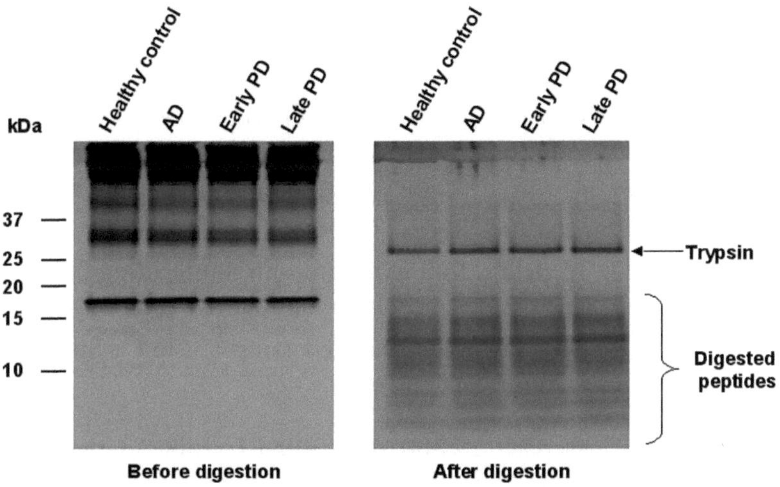

Fig. 2. Silver staining of protein samples before and after trypsin digestion. Proteins in human cerebrospinal fluid were extracted from healthy controls, patients with Alzheimer disease (AD), and early or late Parkinson disease (PD). Equal amounts of proteins from each group were digested by trypsin and small aliquots were analyzed by SDS-PAGE and silver staining. This was to check for equal loading and the degree of completion of tryptic digestion.

3.4.3. iTRAQ Labeling: Estimated Timing, 2 h

1. Dry down digested samples using a SpeedVac® or similar vacuum concentrator. Reconstitute each sample in ~30 μL of 0.5 M TEAB buffer.
2. Bring each vial of the iTRAQ Reagents that you need to room temperature for about 30 min.
3. Add 70 μL of ethanol to each vial. Vortex for 1 min to dissolve the reagents, and then spin down.
4. Transfer the contents of one iTRAQ Reagent vial to one sample tube so that each sample tube receives one iTRAQ reagent (for example, 114—healthy control, 115—AD, 116—early PD, and 117—late PD; see Note 12). Vortex to mix, and then spin.
5. Incubate the tubes at room temperature for 1 h.
6. Combine the contents of all iTRAQ-labeled sample tubes into a fresh tube. Vortex to mix well.

3.5. Glycopeptide Enrichment

3.5.1. Desalting: Estimated Timing, Up to 1 Day

1. To remove the free iTRAQ reagents, wet a C18 1 cc cartridge with 2 mL (1 mL×2) 0.1% TFA in 50% ACN. Next, equilibrate with 2 mL (1 mL×2) of 0.1% TFA.
2. Acidify the sample with 5 M HCl and 1% TFA in less than 100 μL (adjust to pH <3) and load slowly onto the cartridge. Reload the flow through slowly onto the cartridge (see Note 13).
3. Wash the sample cartridge with 3 mL (1 mL×3) of 0.1% TFA buffer.
4. Elute with 400 μL of the eluting buffer (0.1% TFA in 50% ACN). Repeat with another 400 μL of eluting buffer. Combine

the eluate and mix well. The sample could be stored at −20 or −80°C before oxidation.

3.5.2. Oxidation: Estimated Timing, Up to 1 Day

1. Split the eluate into two parts (~400 μL each). Add 100 mM sodium periodate to make a final concentration of 10 mM.
2. Incubate in the dark at room temperature for 1 h with gentle rotation.
3. To desalt, follow the steps in Subheading 3.5.1 but repeat the eluting one more time to collect a total of ~1.2 mL eluate.
4. Dry down the oxidized sample using a SpeedVac® or similar vacuum concentrator.

3.5.3. Isolation of Glycoprotein by Hydrazide Resin: Estimated Timing, 3 Day

1. Prepare 2 × 300 μL (300 μL/tube in two tubes) of moist hydrazide resin (50% slurry) per sample by spinning at 3,000 × *g* for 5 min and then removing the supernatants.
2. Wash the resin three times with deionized water and then three times with the coupling buffer. During each wash, resuspend the resin with three volumes (900 μL) of water or buffer, vortex to mix well, and then spin at 3,000 × *g* for 5 min to remove supernatant.
3. Resuspend the hydrazide resin with the coupling buffer (300 μL per tube) to make a 50% slurry.
4. Resuspend the oxidized sample from step 4 in Subheading 3.5.2 in 1.2 mL of the coupling buffer. Transfer 600 μL to each hydrazide resin tube.
5. Conjugate the glycopeptides to the hydrazide resin at room temperature by rotating for 24 h. Immobilized glycopeptides on hydrazide resin can be stored at 4°C for up to a month.
6. Collect the resin by centrifuging at 3,000 × *g* for 5 min. The supernatant, together with the supernatants from step 7, can be saved for non-glycopeptides identification.
7. Wash the resin three times with 1.5 M NaCl, three times with 80% ACN, three times with 100% methanol, three times with deionized water, and then three times with 0.1 M NH_4HCO_3, pH 8.3. During each wash, resuspend the resin with 1 mL of the wash solution, vortex to mix well, and then spin at 3,000 × *g* for 5 min to remove supernatant.
8. Resuspend the resin in each tube with 300 μL of 0.1 M NH_4CO_3 (pH 8.3) containing 1 μL of PNGase F. Incubate at 37°C for 12 h with shaking.
9. Add 1 μL of PNGase F. Incubate at 37°C for another 36 h with shaking.
10. Centrifuge at 3,000 × *g* for 5 min. Transfer the supernatant to a new tube.

11. Wash the resin twice with 0.1 M NH_4CO_3 (pH 8.3) and then twice with 80% ACN. During each wash, resuspend the resin with 200 μL of the wash solution, vortex to mix well, spin at 3,000 × *g* for 5 min, and then transfer the supernatant to a new tube.
12. Combine all the supernatants from steps 10 and 11. Dry down using a SpeedVac® or similar vacuum concentrator and then redissolve in 100 μL of 0.5% TFA.

3.6. LC-MS/MS Analysis of Glycopeptides: Estimated Timing, 3 Day Per Sample

1. Centrifuge the sample in the loading buffer (0.5% TFA) at 17,000 × *g* for 10 min (4°C) and then transfer supernatant into a glass vial.
2. Separate the sample using a C18 capillary column with a 2-solvent system (Solvent A, 2% ACN and 0.1% TFA; Solvent B, 80% ACN and 0.1% TFA) and a two-step elution program consisting of a linear gradient of 10–50% Solvent B in 155 min, and then a linear gradient of 50–80% Solvent B in 10 min. Use a flow rate of 0.4 μL/min and monitor the UV absorbance at 214 nm. The eluted gradient is mixed with the matrix solution and spotted onto a stainless steel MALDI plate to form a predefined 30 × 50 array (1,344 spots) using a Probot™ system (LC Packings/Dionex).
3. Run the mass spectrometric analysis using a 4,800 Plus MALDI TOF/TOF™ Analyzer with the reflector positive ion mode. Perform the default calibration before each run, with the mass accuracy calibrated to within 10 ppm using calibration standards. The setting parameters are the following: 500–5,000 *m/z* mass range with 1,000 shots per spectrum for MS analysis, a maximum of 50 precursors per spot with minimum signal/noise ratio of 20 for data-dependent MS/MS analysis, 2-kV collision energy for collision-induced dissociation (CID) with air as the collision gas, and 1925 acquisitions are accumulated for each MS/MS spectrum. Use the Data Explorer software for data acquisition and extraction of the monoisotopic masses.
4. Extract the MS/MS spectra (a typical spectrum is shown in Fig. 3) and search against the International Protein Index (IPI) human protein database from the European Bioinformatics Institute (EBI) with the ProteinPilot™ software and the Paragon™ method. The raw peptide identification results from the Paragon™ Algorithm searches are further processed with the Pro Group™ Algorithm within the ProteinPilot™ software before final display. The Pro Group Algorithm uses the peptide identification results to determine the minimal set of proteins that can be reported for a given protein confidence threshold. It is recommended to report all data based on 95% confidence for protein identification as determined by ProteinPilot (ProtScore ≥ 1.3) (see Note 14).

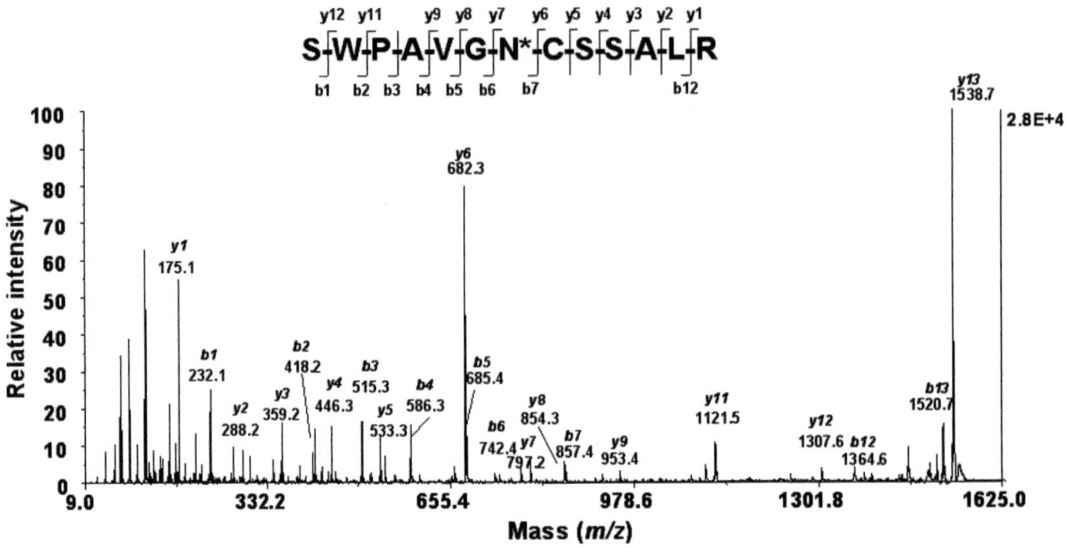

Fig. 3. Fragment mass spectrum of a tryptic N-glycosylated peptide. The sequence shown is SWPAVGNCSSALR from hemopexin, with y- and b-series ions interpreted.

5. After ProteinPilot data processing, the protein summary listing the identified proteins and their iTRAQ ratios can be exported as a tab delimited text file for further analysis. Identified proteins can be checked against the UniProtKB/Swiss-Prot database and the Institute for Systems Biology (ISB) database to determine whether they are glycoproteins with known glycosylation sites or probable/potential glycosylation sites.

4. Notes

1. For ease of analysis, we detected the N-linked glycopeptides only by using PNGase F in this approach. However, with a proper combination of O-glycosidase or chemical cleavage such as β-elimination, the O-glycopeptides can also be released from the solid support and analyzed by MS (27).
2. TMB (3,3′,5,5′-tetramethylbenzidine) is highly recommended but OPD (*o*-phenylenediamine dihydrochloride) or ABTS 2,2′-azino-bis-(3-ethylbenzthiazoline-6-sulfonic acid) can also be used. Wavelength should be 450 nm for TMB, 490 nm for OPD, and 405 nm for ABTS, respectively.
3. Recrystallization of matrix may significantly improve performance. This can be done by dissolving 1 g of CHCA in ~20 mL 100% ethanol (pre-warmed to 40–50°C), filtering immediately with a 0.2 μm hydrophilic polypropylene membrane, followed

by adding two volumes of cold HPLC-grade water, and then incubating at 4°C overnight. Filter the recrystallized matrix and then dry at room temperature overnight. The matrix should be stored in clean glass bottles and keep in dark.

4. The number of pulses and the sonicator setting need to be determined when different sonicators are used. Make sure lysis is complete (no more tissue chunks left) but the lysate does not foam excessively. It is better to use more pulses if needed rather than to increase the time of each pulse as heat is generated over time.
5. Human CSF is closely regulated via balanced secretion and absorption with an average circulating volume between 125 and 150 mL in an adult; as a result, the amount of CSF that can be obtained is usually limited to less than 25–30 mL. Additionally, there is a significant variation in CSF production during the day as well as a rostro-caudal gradient of protein concentration. Thus, it is critical to match CSF samples not only for the timing of CSF taps, but also for the fractions of CSF obtained. For a more detailed discussion, see a recent review article (15).
6. Protease inhibitor is not routinely included in a typical clinical CSF tap; we often add it into CSF as soon as the CSF is thawed for the first time, if the sample will be reused. However, it is still an unsettling issue whether it is necessary to add protease inhibitors before CSF samples are frozen. The controversy stems from inconsistent mass spectrometric results generated by various laboratories when protease inhibitors were supplemented. For example, aprotinin changes the sample MS spectrum for unknown reasons, and a number of small molecule inhibitors such as PMSF and AEBSF can form covalent bonds with proteins, thereby shifting their *m/z* spectra [see a review (15) for more details].
7. Blood contamination in CSF could happen during the lumbar puncture. Because the protein concentration in the CSF is relatively low compared to plasma (CSF/plasma: <1/200), and the protein profiles in CSF are similar to those in plasma (15), even a minor contamination of CSF with blood could significantly confound the interpretation of quantitative or qualitative proteomic analysis of CSF. Two criteria are commonly used in our lab to control for blood contamination: (1) CSF red blood cell (RBC) count, as determined by standard clinical chemistry laboratory, should be less than 10 RBC/μL, and (2) the hemoglobin level in CSF has to be less than 280 ng/mL, which is roughly equivalent to 1:500,000 dilution of human plasma (14 g/dL) or 10 RBC/μL (five million/μL).

8. TCEP is used to denature the sample so that thiols are not introduced to the sample. The presence of thiols that are typically introduced into sample mixtures by the addition of DTT or β-mercaptoethanol can interfere with the cysteine blocking step.
9. Typically, the cysteine blocking is done with iodoacetamide, but the iTRAQ kit includes MMTS. The advantage of MMTS is that it has fewer side reactions than does iodoacetamide (MMTS has fewer secondary alkylation targets). In addition, MMTS is a reversible blocker that can be useful should one decide to fractionate the sample by selectively isolating cysteine-containing peptides. However, it has been suggested that MMTS tends to alkylate cysteines less stoichiometrically than does iodoacetamide (MMTS leaves more unmodified cysteine residues). We have found the procedure employed by the iTRAQ kit to be easy to follow and to yield excellent labeling.
10. This is to reduce the urea concentration to be less than 1 M. High amounts of urea and some other detergents or denaturants can interfere with the trypsin digestion. The concentration limits of such detergents or denaturants at trypsin digestion are the following: SDS, 0.05%; OG (octyl β-d-glucopyranoside), 0.1%; NP-40, 0.1%; Triton X-100, 0.1%; Tween 20, 0.1%; CHAPS, 0.2%; and urea, <1 M. The other major concern is that the presence of primary amines, such as those in ammonium sulfate, -bicarbonate, -citrate, Tris buffer, etc., will interfere with the labeling process by competing for the iTRAQ reagents. If necessary, TCA or acetone precipitation can be performed either after reducing the sample and blocking the free cysteines (in the case of detergents that might be necessary for maintaining sample solubility), or just prior to tryptic digest (in the case of the presence of primary amines).
11. This is to check whether the trypsin digestion is complete and whether the protein/peptide amounts in each sample are still roughly the same. If digestion is not complete, add another batch of enzyme, and digest the samples for an additional 4 h at 37°C.
12. Usually, controls are labeled with the 114 reagent, but of course this is up to the investigator. It is critical to keep track of which sample received which label, as downstream quantification depends on the proper ratio of the disease or other group to the control group.
13. The sample can be loaded repeatedly to ensure binding. One can use a vacuum for wetting, equilibration, and washing. Supposedly MCX cartridges can be run dry without loss of recovery.

14. For each protein, Pro Group Algorithm reports two types of scores for each protein: unused ProtScore and total ProtScore. The total ProtScore is a measurement of all the peptide evidence for a protein, and is analogous to protein scores reported by other protein identification software. The unused ProtScore, however, is a measurement of all the peptide evidence for a protein that is not better explained by a higher ranking protein. In other words, the unused ProtScore is calculated with the unique peptides (peptides that are not used by the higher ranking protein), and it is a clearer indicator of protein evidence and assists in singling out members of a multiprotein family.

Acknowledgement

This work was supported by the National Institutes of Health grants AG033398, AG025327, NS057567, ES004696, and ES012703.

References

1. Apweiler R, Hermjakob H, Sharon N (1999) On the frequency of protein glycosylation, as deduced from analysis of the SWISS-PROT database. Biochim Biophys Acta 1473:4–8
2. Marino K, Bones J, Kattla JJ, Rudd PM (2010) A systematic approach to protein glycosylation analysis: a path through the maze. Nat Chem Biol 6:713–723
3. Roth J (2002) Protein N-glycosylation along the secretory pathway: relationship to organelle topography and function, protein quality control, and cell interactions. Chem Rev 102:285–303
4. Dwek RA (1996) Glycobiology: toward understanding the function of sugars. Chem Rev 96:683–720
5. Arnold JN, Wormald MR, Sim RB, Rudd PM, Dwek RA (2007) The impact of glycosylation on the biological function and structure of human immunoglobulins. Annu Rev Immunol 25:21–50
6. Lowe JB, Marth JD (2003) A genetic approach to Mammalian glycan function. Annu Rev Biochem 72:643–691
7. Hwang H, Zhang J, Chung KA, Leverenz JB, Zabetian CP, Peskind ER, Jankovic J, Su Z, Hancock AM, Pan C, Montine TJ, Pan S, Nutt J, Albin R, Gearing M, Beyer RP, Shi M, Zhang J (2010) Glycoproteomics in neurodegenerative diseases. Mass Spectrom Rev 29:79–125
8. Helzlsouer KJ, Bush TL, Alberg AJ, Bass KM, Zacur H, Comstock GW (1993) Prospective study of serum CA-125 levels as markers of ovarian cancer. JAMA 269:1123–1126
9. Lee R, Localio AR, Armstrong K, Malkowicz SB, Schwartz JS (2006) A meta-analysis of the performance characteristics of the free prostate-specific antigen test. Urology 67:762–768
10. Baselga J (2010) Treatment of HER2-overexpressing breast cancer. Ann Oncol 21(suppl 7):vii36–vii40
11. Sihlbom C, Davidsson P, Emmett MR, Marshall AG, Nilsson CL (2004) Glycoproteomics of cerebrospinal fluid in neurodegenerative disease. Int J Mass Spectrom 234:145–152
12. Pan S, Wang Y, Quinn JF, Peskind ER, Waichunas D, Wimberger JT, Jin J, Li JG, Zhu D, Pan C, Zhang J (2006) Identification of glycoproteins in human cerebrospinal fluid with a complementary proteomic approach. J Proteome Res 5:2769–2779
13. Sihlbom C, Davidsson P, Sjögren M, Wahlund L-O, Nilsson C (2008) Structural and quantitative comparison of cerebrospinal fluid glycoproteins in Alzheimer's disease patients and healthy individuals. Neurochem Res 33:1332–1340
14. Anderson NL, Anderson NG (2002) The human plasma proteome: history, character,

and diagnostic prospects. Mol Cell Proteomics 1:845–867

15. Zhang J (2007) Proteomics of human cerebrospinal fluid—the good, the bad, and the ugly. Proteomics Clin Appl 1:805–819
16. Novotny MV, Mechref Y (2005) New hyphenated methodologies in high-sensitivity glycoprotein analysis. J Sep Sci 28:1956–1968
17. Sun B, Ranish JA, Utleg AG, White JT, Yan X, Lin B, Hood L (2007) Shotgun glycopeptide capture approach coupled with mass spectrometry for comprehensive glycoproteomics. Mol Cell Proteomics 6:141–149
18. Kaji H, Saito H, Yamauchi Y, Shinkawa T, Taoka M, Hirabayashi J, Kasai K, Takahashi N, Isobe T (2003) Lectin affinity capture, isotope-coded tagging and mass spectrometry to identify N-linked glycoproteins. Nat Biotechnol 21:667–672
19. Zhang H, Li XJ, Martin DB, Aebersold R (2003) Identification and quantification of N-linked glycoproteins using hydrazide chemistry, stable isotope labeling and mass spectrometry. Nat Biotechnol 21:660–666
20. Tian Y, Zhou Y, Elliott S, Aebersold R, Zhang H (2007) Solid-phase extraction of N-linked glycopeptides. Nat Protoc 2:334–339
21. Kaji H, Yamauchi Y, Takahashi N, Isobe T (2006) Mass spectrometric identification of N-linked glycopeptides using lectin-mediated affinity capture and glycosylation site-specific stable isotope tagging. Nat Protoc 1:3019–3027
22. Nilsson CL (2003) Lectins: proteins that interpret the sugar code. Anal Chem 75:348A–353A
23. Wuhrer M, Catalina MI, Deelder AM, Hokke CH (2007) Glycoproteomics based on tandem mass spectrometry of glycopeptides. J Chromatogr B Analyt Technol Biomed Life Sci 849:115–128
24. Schulze WX, Usadel B (2010) Quantitation in mass-spectrometry-based proteomics. Annu Rev Plant Biol 61:491–516
25. Gygi SP, Rist B, Gerber SA, Turecek F, Gelb MH, Aebersold R (1999) Quantitative analysis of complex protein mixtures using isotope-coded affinity tags. Nat Biotechnol 17:994–999
26. Ross PL, Huang YN, Marchese JN, Williamson B, Parker K, Hattan S, Khainovski N, Pillai S, Dey S, Daniels S, Purkayastha S, Juhasz P, Martin S, Bartlet-Jones M, He F, Jacobson A, Pappin DJ (2004) Multiplexed protein quantitation in Saccharomyces cerevisiae using amine-reactive isobaric tagging reagents. Mol Cell Proteomics 3:1154–1169
27. Klement E, Lipinszki Z, Kupihar Z, Udvardy A, Medzihradszky KF (2010) Enrichment of O-GlcNAc modified proteins by the periodate oxidation-hydrazide resin capture approach. J Proteome Res 9:2200–2206

Chapter 20

Quantitative Proteomic Analysis of N-linked Glycoproteins in Human Tear Fluid

Lei Zhou and Roger W. Beuerman

Abstract

Human tear fluid is a complex biological fluid that is actually the extracellular fluid of the surface of the eye and can be considered a dilute protein solution. The quality of the tears are critical for vision as they provide the mirror smooth reflex for the surface of the eye and hence focusing of images on the retina. Changes in tear composition may reflect the health of the epithelial cells lining the ocular surface and they have been shown to be useful to discover biomarkers for eye diseases. Glycoproteins are potentially important biomarkers of disease and therapeutic targets and can also be found in human tears. In this chapter, we concentrated on technical details in quantitative proteomic analysis of N-linked glycoproteins in human tears by combining hydrazide chemistry enrichment of N-linked glycoproteins and iTRAQ technology.

Key words: Tear fluid, Tear proteomics, Glycoproteomics, Glycosylation, Hydrazide chemistry, iTRAQ, Quantitative proteomics

1. Introduction

Tear fluid covering the surface of the eye is an important component of the extracellular environment of the surface epithelial cell layer. Tear fluid contains electrolytes, proteins, lipids, mucins, some small organic molecules and metabolites. The function of the tear film includes lubrication, protection from disease, nutrition of the cornea, and a critical role in the optical properties of the eye (1). The tear protein composition is complex, both in the variety of proteins present as well as the dynamic concentration range which can span up to ten orders of magnitude (1). Analysis of tears is challenging due to the small sample size as the amount of tears collected for a single sample is only about 5 μL (2), which is minute compared to other body fluids such as serum, urine, and saliva.

Jennifer J. Kohler and Steven M. Patrie (eds.), *Mass Spectrometry of Glycoproteins: Methods and Protocols*, Methods in Molecular Biology, vol. 951, DOI 10.1007/978-1-62703-146-2_20,

Moreover, serial samples cannot be collected at short time intervals due to irritation effects.

Proteins are known to undergo post-translational modification (PTM) such as glycosylation and phosphorylation, which may change protein structure and function. Characterization of protein PTMs at the large scale is one of the important tasks in proteomics. Glycosylated proteins are of interest for their potential use as biomarkers and therapeutic targets (3–5). Two of the most common protein glycosylation patterns are O-linked and N-linked glycosylation. N-glycosylated proteins in particular have been the focus of many proteomics studies (6–8). Most N-linked glycoproteins are secreted and can be found in body fluids such as plasma (9, 10), urine (11), cerospinal fluid (12), tears (13), and saliva (14).

N-linked glycosylation occurs at the amide nitrogen of asparagine. The consensus motif for N-linked glycosylation is Asn-X-Thr/Ser (three-letter amino acid code), where X represents any amino acid except proline, though the less common motif Asn-X-Cys may also be found. These motifs allow the confirmation of the presence of an N-glycosylation site when analyzing peptide matches obtained from searches through protein databases.

Enrichment of glycoproteins can be achieved by a hydrazide-functionalized resin. The stable covalent hydrazone bonds can be formed between oxidized glycans and hydrazine functional groups. Subsequently, N-linked glycoproteins are selectively cleaved from the hydrazide resin using peptide-N-glycosidase F (PNGase F), an enzyme which specifically deglycosylates N-glycoproteins and not O-glycoproteins. Quantitative analysis of the changes of occupancy of N-glycosylation at specific sites (e.g., comparing disease samples with control samples) can be performed using iTRAQ reagents, which may provide disease biomarkers.

2. Materials

2.1. Collection of Human Tears

1. 10 μL calibrated glass microcapillary pipets, (Drummond Scientific Company, Broomall, PA, USA).
2. Protease inhibitor cocktail (Thermo Scientific Pierce, USA) 100× stock in DMSO, dilute ten times in water as working solution (10×).

2.2. Determination of Total Tear Protein Concentration Using DC Protein Assay

1. Use DC Protein Assay Kit (Bio-Rad Laboratories, Hercules, CA, USA).
2. DC Protein Assay Kit reagent contains REAGENT A (an alkaline copper tatrate solution), REAGENT B (a dilute Folin Reagent), and REAGENT S. Store at room temperature (25 ~ 30°C). Reagents A and B may also be stored at 4°C.

3. 96-well microplate.
4. Albumin standard: Bovine Serum Albumin (Thermo Scientific Pierce, USA) at 2.0 mg/mL in 0.9% saline and 0.05% sodium azide. Store at 4°C.

2.3. Desalting

1. Ethanol (100%, HPLC grade, Merck, Darmstadt, Germany).

2.4. Enrichment of Glycoproteins by Hydrazide Resin

1. Coupling Buffer (10×) (100 mM sodium acetate, 150 mM sodium chloride, glacial acetic acid for adjusting pH to 5.5, Bio-Rad Laboratories, Hercules, CA, USA).
2. Sodium periodate (Affi-gel® Oxidizer, Bio-Rad Laboratories, Hercules, CA, USA).
3. Glycerol, 200 mM in water (Sigma, St. Louis, MO, USA).
4. Hydrazide resin in isopropanol (Affi-gel® Hz Hydrazide Gel, Bio-Rad Laboratories, Hercules, CA, USA).

2.5. Tryptic Digestion

1. Urea buffer A: 8 M urea (Sigma, St. Louis, MO, USA), 200 mM 2-Amino-2-(hydroxymethyl)propane-1,3-diol (TRIS, from Bio-Rad Laboratories, Hercules, CA, USA), 0.05% SDS (Sigma, St. Louis, MO, USA), 5 mM EDTA (Sigma, St. Louis, MO, USA), pH 8.3.
2. Reducing solution: tris-(2-carboxyethyl)phosphine (TCEP, from Thermo Scientific Pierce, USA).
3. Methane methylthiosulfonate (MMTS, from iTRAQ™ Reagents Multiplex Kit).
4. Urea buffer B: 1 M urea (Sigma, St. Louis, MO, USA), 25 mM 2-Amino-2-(hydroxymethyl)propane-1,3-diol (TRIS, from Bio-Rad Laboratories, Hercules, CA, USA), pH 8.3.
5. Sequencing-grade trypsin (EMD Chemicals, Gibbstown, NJ, USA).
6. 1.5 M NaCl.
7. 80% acetonitrile (ACN)/0.1% TFA in water.
8. 100% Methanol.
9. 100 mM Ammonium bicarbonate.

2.6. Deglycosylation

1. Peptide-*N*-glycosidase F (PNGase F) (New England BioLabs, Ipswich, MA, USA).
2. 80% ACN.

2.7. iTRAQ Labeling (from iTRAQ™ Reagents Multiplex Kit)

1. Dissolution buffer.
2. Ethanol.
3. iTRAQ labeling reagents.

2.8. 2D nanoLC-ESI-MS/MS Analysis of Glycopeptides

1. SCX column: 300 mm (i.d.)×10 cm porosity 10 S SCX (Dionex, LC packings, USA).
2. Trap column cartridge: C18, 0.3×5 mm, from Dionex, LC Packings.
3. Reversed phase (RP) column: 10 cm×75 μm i.d. picofrit self-packed microcapillary LC column (New Objective, Woburn, MA, USA).
4. Packing material: Luna C18, 3 μm, 100A (Phenomenex Torrance, CA).
5. Solvent A: 0.1% formic acid in water.
6. Solvent B: 0.1% formic acid in ACN.

3. Methods

3.1. Collection of Human Tears

1. Obtain written informed consent from patients.
2. Collect tears from the patient's inferior cul-de-sac using fire-polished 10 μL calibrated glass microcapillary pipets, with special care taken not to touch the ocular surface (13, 15) (see Note 1).
3. Transfer tears from glass microcapillary pipets into an eppendorf tube, add 1 μL protease inhibitor cocktail and store at −80°C until further use.

3.2. Determination of Total Tear Protein Concentration Using DC Protein Assay

1. Prepare Reagent A' by adding 500 μL of Reagent A with 10 μL of Reagent S (ratio of 50:1).
2. Add 5 μL H_2O replicate into the 96-well plate as blank.
3. Pipette 5 μL of each BSA standards replicate (0.125, 0.250, 0.500, 0.750, 1.000, 1.500, and 2.000 mg/mL) into the 96-well plate.
4. Pipette 5 μL of unknown tear samples replicate into the 96-well plate.
5. Add 25 μL Reagent A' into each well.
6. Add 200 μL Reagent B into each well.
7. Incubate in the thermomixer at 25°C for 15 min.
8. Measure the absorbance at 720 nm on a Microplate Reader (see Note 2).

3.3. Desalting

1. Use 500 μg (total protein) sample of human tears.
2. Add 4 volumes cold ethanol to each sample.
3. Mix well and incubate overnight at −20°C.

4. Centrifuge at ≥10,000 × *g* for 15 min at 4°C and remove the supernatant.
5. Keep the pellet for the next step (see Note 3).

3.4. Enrichment of Glycoproteins by Hydrazide Resin

1. Resuspend the pellet in 1× Coupling Buffer (dilution of 10× Coupling Buffer, adjust pH to 5.5).
2. Add sodium periodate solution up to a final concentration of 15 mM and incubate in the dark for 1 h at room temperature.
3. Quench the excess sodium periodate by adding 200 mM glycerol to a final concentration of 20 mM and mixing for 15 min at room temperature.
4. Remove sodium periodate by ethanol precipitation as described earlier (Subheading 3.3).
5. Resuspend the pellet in 1× Coupling Buffer.
6. Equilibrate the hydrazide gel resin (250 μL slurry, supplied in isopropanol) by washing it four times in two volumes of 1× Coupling Buffer (100 mM sodium acetate, 150 mM sodium chloride, pH 5.5, diluted from 10× Coupling Buffer).
7. Resuspend the resin in 250 μL 1× Coupling Buffer.
8. Add oxidized glycoproteins to hydrazide-functionalized resins by incubating overnight at room temperature.

3.5. Trypsin Digestion

1. Allow gel to settle.
2. Transfer out supernatant.
3. Wash resin with urea buffer A for six times.
4. Reduce coupled proteins with 10 mM TCEP in urea buffer A for 45 min at room temperature.
5. Block cysteine residues with 200 mM MMTS (see Note 4).
6. Wash resin for six times using urea buffer B.
7. Resuspend resin in urea buffer B.
8. Add 3 μg of trypsin to 300 μg of total tear proteins to digest the glycoproteins by incubating overnight at 37°C.
9. Remove the trypsin-released peptides by washing the resin six times with three bed volumes of 1.5 M NaCl, followed by three times each with 80% ACN/0.1% TFA, 100% methanol and water, and six times with 100 mM NH_4HCO_3.

3.6. Deglycosylation

1. Resuspend resin in 100 mM NH_4HCO_3 to give a 50% gel slurry.
2. Release N-linked glycopeptides with 5 μL PNGase F by incubating overnight at 37°C.
3. Supernatant containing the deglycosylated N-glycopeptides is transferred into new tube.

4. Wash resin twice with 80% ACN.
5. Pool the washes with the supernatant and dry the deglycosylated glycopeptides via speedvac.

3.7. iTRAQ Labeling

1. Reconstitute lyophilized sample in 30 μL dissolution buffer.
2. Add 70 μL of ethanol (from iTRAQ Kit) to iTRAQ reagents 116/117. (First brought to room temperature), (see Note 5).
3. Transfer contents of individual iTRAQ vials into respective tubes containing sample solutions.
4. Incubate solutions at room temperature with mixing for 3 h.
5. Combine iTRAQ labeled fractions, mix and spin down prior to evaporate to dryness using speedvac.

3.8. 2D nanoLC-ESI-MS/MS Analysis of Glycopeptides

1. Analyze the glycopeptides using 2D nano-LC system (DIONEX, LC Packings, Sunnyvale, CA, USA) coupled to a nano-ESI Q-Star XL (Applied Biosystems, MDS Sciex, Concord, Ontario, Canada).
2. Directly couple the PicoFrit® microcapillary column with integrated spray tip to the Q-TOF mass spectrometer through a NanoSpray™ interface (Protana, Odense, Denmark).
3. Load the samples onto the first dimension, which is a strong cation exchange column (300 μm i.d. × 10 cm, porosity 10 S SCX, DIONEX, LC Packings) for 7 min at a flow rate of 30 μL/min.
4. Use ten steps of 20 μL-injection salt plug elutions (10, 20, 30, 40, 50, 75, 100, 250, 500, and 1,000 mM ammonium acetate).
5. Switch the system (Switchos, DIONEX, LC Packings) in-line to the C18 microcapillary column, which is the second dimension used in the 2D-LC analysis. A linear gradient of 0.1% formic acid (in ACN) from 5 to 60% over 135 min at a flow rate of ~300 nL/min should be delivered (UltiMate solvent delivery system, DIONEX, LC Packings).
6. The settings for the nano-ESI-MS/MS system are: ion spray voltage = 2,200 V, curtain gas = 20, declustering potential (DP) = 80 V, DP2 = 15 V, focusing potential = 265 V, collision gas setting = 5 for nitrogen gas. The Information-Dependent Acquisition (IDA) mode for the Analyst QS software (version 1.1, Applied Biosystems) was used to acquire the mass spectrometry data. The TOF-MS survey scan parameters used were as follows: 1 s TOF-MS survey scan was performed in the mass range of 300–1,200 Da, after which two product ion scans, each of 3 s, were carried out in the mass range of 100–1,800 Da. The switching criteria were set at ions with *m/z* greater than 350 and smaller than 1,200, a charge state of 2–4, and an abundance threshold of 8 counts/s, while former target ions were excluded for a total of 120 s.

3.9. Database Search and Data Analysis

1. ProteinPilot software (version 2.01, Applied Biosystems, MDS Sciex) was used to analyze the MS/MS.
2. Search the data against the IPI Human database (version 3.39), with a PSPEP setting of "Reversed Protein Sequences."
3. Choose emphasis on biological modifications, (see Note 6).
4. Threshold used for detected proteins and matched peptides was set at 1.3 (95% confidence), (see Note 7).
5. Key in the isotope correction factors for the iTRAQ™ labels according to those supplied with the reagent kit.

4. Notes

1. Normal tear volume is around 5–6 μL and the secretion rate is about 1.2 μL/min (16). The total protein concentration in tears is around 6–10 mg/mL (17). It may take several minutes to collect 5 μL of tears from patients. The glass microcapillary pipets used for tear collection are maintained in a sterile condition. The end of the tube will be put into the tear pool at the bottom, fornix, of the lower eyelid. The tears are collected by capillary action. This procedure is safe and painless. Special care must be taken not to touch the ocular surface and cause any discomfort, which may induce reflex tear. Another important tear collection method is to use the standard clinical Schirmer's strip (details can be found in (18)) and using the Schirmer's type I tear collection procedure.
2. Measurement of total protein concentration in tears: other protein assay kits, for example, BCA™ Protein Assay kit (Thermo Scientific Pierce, USA) and Coomassie Plus (Bradford) Protein Assay (Thermo Scientific Pierce, USA) can also be used.
3. Other protein precipitation methods may also be used, such as, acetone precipitation or TCA (trichloroacetic acid) precipitation.
4. Blocking reagent MMTS is typically used in iTRAQ experiments.
5. Figure 1 shows the overall workflow for identification and quantification of N-linked glycoproteins by combining hydrazide chemistry and iTRAQ. In this particular experiment, samples from control group were labeled with iTRAQ™ reagent 116, while those from the diseased group (a cornea disease called climatic droplet keratopathy, CDK) were labeled with iTRAQ™ reagent 117. CDK involves the spheroidal degeneration of the cornea, and yellowish deposits can be observed in the superficial corneal region (19). In this study (13), we identified a total of 43 unique N-glycoproteins, 19 of which have not previously been reported in tear fluid. By comparing

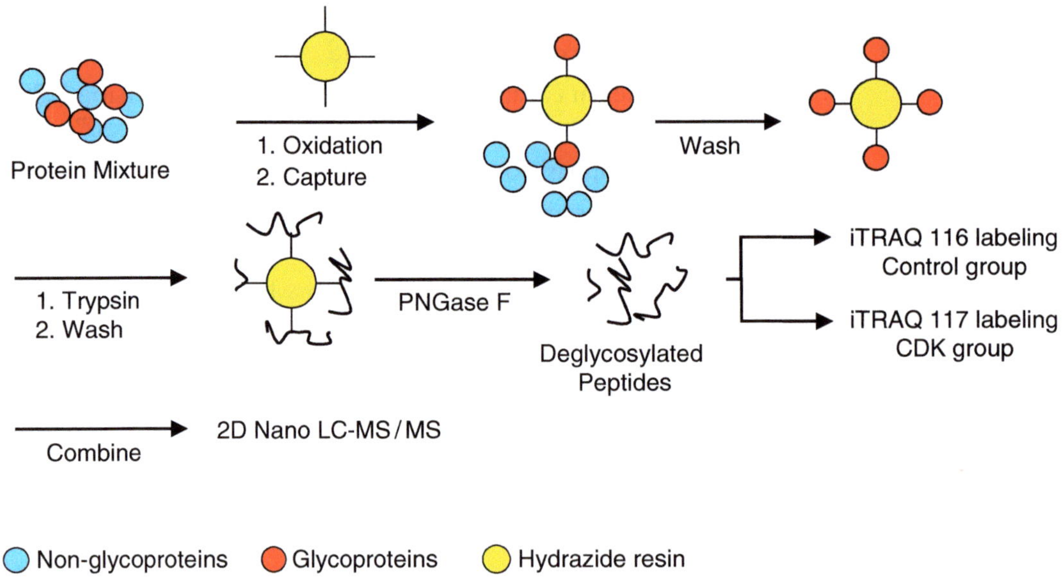

Fig. 1. Overall workflow for quantitative analysis of N-glycoproteins in tear fluid using glycoprotein capture (hydrazide resin), iTRAQ labeling and nanoLC-MS/MS. Control group was labeled with iTRAQ reagent 116 while diseased group was labeled with iTRAQ reagent 117 (reproduced from (13) with permission from ACS publications).

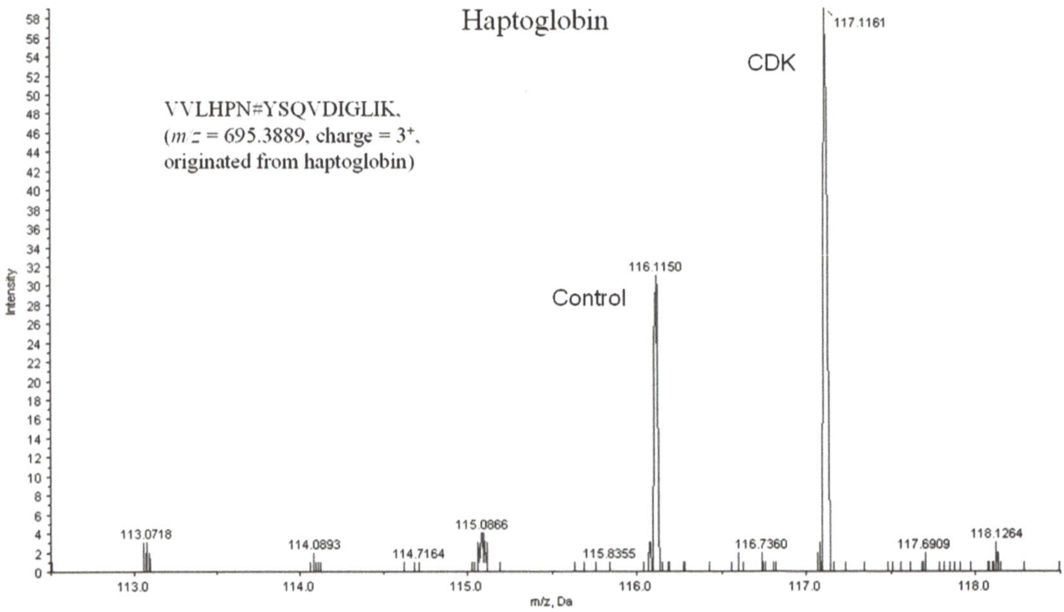

Fig. 2. Relative quantification for N-glycosylated VVLHPN#YSQVDIGLIK (a peptide fragment originated from haptoglobin, N# indicates the glycosylation site) at site 241 between tear samples from CDK and control using iTRAQ. Control group peptides were labeled with iTRAQ reagent 116, while those from CDK group were labeled with iTRAQ reagent 117 (reproduced from (13) with permission from ACS publications).

tears from control with CDK, increased N-glycosylation levels of four N-glycosylated proteins and decreased N-glycosylation level of one down-regulated N-glycosylated protein were observed in tears from CDK patients (Fig. 2 gives an example of relative quantitation).

6. N-glycosylation was confirmed by the presence of both the N-glycosylation motif (Asn-X-Thr/Ser) as well as a deamidated asparaginyl residue.
7. The threshold we used for identification of proteins is 95% confidence (or unused ProtScore > 1.3) in ProteinPilot software.

Acknowledgements

This research was supported by grants NMRC/0808/2003, NMRC/CPG002/2003, NMRC/0982/2005, and NMRC IBG from National Medical Research Council (NMRC), Singapore. Authors would like to thank Koh Siew Kwan and Chew Ai Ping for their technical support.

References

1. Zhou L, Beuerman RW, Foo Y, Liu S, Ang LP, Tan DT (2006) Characterisation of human tear proteins using high-resolution mass spectrometry. Ann Acad Med Singapore 35: 400–407
2. Zhou L, Huang LQ, Beuerman RW, Grigg ME, Li SF, Chew FT, Ang L, Stern ME, Tan D (2004) Proteomic analysis of human tears: Defensin expression after ocular surface surgery. J Proteome Res 3:410–416
3. Dube DH, Bertozzi CR (2005) Glycans in cancer and inflammation–potential for therapeutics and diagnostics. Nat Rev Drug Discov 4:477–488
4. Kobata A, Amano J (2005) Altered glycosylation of proteins produced by malignant cells, and application for the diagnosis and immunotherapy of tumours. Immunol Cell Biol 83: 429–439
5. Shepard HM, Lewis GD, Sarup JC, Fendly BM, Maneval D, Mordenti J, Figari I, Kotts CE, Palladino MA Jr, Ullrich A et al (1991) Monoclonal antibody therapy of human cancer: Taking the HER2 protooncogene to the clinic. J Clin Immunol 11:117–127
6. Lewandrowski U, Moebius J, Walter U, Sickmann A (2006) Elucidation of N-glycosylation sites on human platelet protein: a glycoproteomic approach. Mol Cell Proteomics 5:226–233
7. Zhang H, Li XJ, Martin DB, Aebersold R (2003) Identification and quantification of N-linked glycoproteins using hydrazide chemistry, stable isotope labeling and mass spectrometry. Nat Biotechnol 21:660–666
8. Zhang H, Yi EC, Li XJ, Mallick P, Kelly-Spratt KS, Masselon CD, Camp DG II, Smith RD, Kemp CJ, Aebersold R (2005) High throughput quantitative analysis of serum proteins using glycopeptide capture and liquid chromatography mass spectrometry. Mol Cell Proteomics 4:144–155
9. Liu T, Qian WJ, Gritsenko MA, Camp DG II, Monroe ME, Moore RJ, Smith RD (2005) Human plasma N-glycoproteome analysis by immunoaffinity subtraction, hydrazide chemistry, and mass spectrometry. J Proteome Res 4:2070–2080
10. Wang Y, Wu SL, Hancock WS (2006) Approaches to the study of N-linked glycoproteins in human plasma using lectin affinity chromatography and nano-HPLC coupled to electrospray linear ion trap–Fourier transform mass spectrometry. Glycobiology 16:514–523
11. Kreunin P, Zhao J, Rosser C, Urquidi V, Lubman DM, Goodison S (2007) Bladder cancer associated glycoprotein signatures revealed by urinary proteomic profiling. J Proteome Res 6:2631–2639

12. Pan S, Wang Y, Quinn JF, Peskind ER, Waichunas D, Wimberger JT, Jin J, Li JG, Zhu D, Pan C et al (2006) Identification of glycoproteins in human cerebrospinal fluid with a complementary proteomic approach. J Proteome Res 5:2769–2779
13. Zhou L, Beuerman RW, Chew AP, Koh SK, Cafaro TA, Urrets-Zavalia EA, Urrets-Zavalia JA, Li SF, Serra HM (2009) Quantitative analysis of N-linked glycoproteins in tear fluid of climatic droplet keratopathy by glycopeptide capture and iTRAQ. J Proteome Res 8:1992–2003
14. Ramachandran P, Boontheung P, Xie Y, Sondej M, Wong DT, Loo JA (2006) Identification of N-linked glycoproteins in human saliva by glycoprotein capture and mass spectrometry. J Proteome Res 5:1493–1503
15. Zhou L, Beuerman RW, Ang LP, Chan CM, Li SF, Chew FT, Tan DT (2009) Elevation of human α-defensins and S100 calcium-binding proteins A8 and A9 in tear fluid of patients with pterygium. Invest Ophthalmol Vis Sci 50:2077–2086
16. Mishima S, Gasset A, Klyce SD Jr, Baum JL (1966) Determination of tear volume and tear flow. Invest Ophthalmol 5:264–276
17. Gachon AM, Richard J, Dastugue B (1982) Human tears: Normal protein pattern and individual protein determinations in adults. Curr Eye Res 2:301–308
18. Zhou L, Beuerman RW, Chan CM, Zhao SZ, Li XR, Yang H, Tong L, Liu S, Stern ME, Tan D (2009) Identification of tear fluid biomarkers in dry eye syndrome using iTRAQ quantitative proteomics. J Proteome Res 8: 4889–4905
19. Gray RH, Johnson GJ, Freedman A (1992) Climatic droplet keratopathy. Surv Ophthalmol 36:241–253

Chapter 21

Elucidation of N-Glycosites Within Human Plasma Glycoproteins for Cancer Biomarker Discovery

Penelope Drake, Birgit Schilling, Brad Gibson, and Susan Fisher

Abstract

Glycans are an important class of post-translational modifications that decorate a wide array of protein substrates. These cell-type specific molecules, which are modulated during developmental and disease processes, are attractive biomarker candidates as biology regarding altered glycosylation can be used to guide the experimental design. The mass spectrometry (MS)-based workflow described here incorporates chromatography on affinity matrices formed from lectins, proteins that bind specific glycan motifs. The goal was to design a relatively simple method for the rapid analysis of small plasma volumes (e.g., clinical specimens). As increases in sialylation and fucosylation are prominent among cancer-associated modifications, we focused on *Sambucus nigra* agglutinin and AAL, which bind sialic acid- and fucose-containing structures, respectively. Positive controls (fucosylated and sialylated human lactoferrin glycopeptides), and negative controls (high-mannose glycopeptides from *Saccharomyces cerevisiae* invertase) were used to monitor the specificity of lectin capture and optimize the workflow. Multiple Affinity Removal System 14-depleted, trypsin-digested human plasma from healthy donors served as the target analyte. Samples were loaded onto the lectin columns and separated by high performance liquid chromatography (HPLC) into flow through and bound fractions, which were treated with PNGase F, an amidase that removes N-linked glycans and marks the underlying asparagine glycosite by a +1 Da mass shift. The deglycosylated peptide fractions were interrogated by HPLC ESI-MS/MS on a quadrupole time-of-flight mass spectrometer. The method allowed identification of 122 human plasma glycoproteins containing 247 unique glycosites. Notably, glycoproteins that circulate at ng/mL levels (e.g., cadherin-5 at 0.3–4.9 ng/mL, and neutrophil gelatinase-associated lipocalin which is present at ~2.5 ng/mL) were routinely observed, suggesting that this method enables the detection of low-abundance cancer-specific glycoproteins.

Key words: Lectin chromatography, Glycopeptide, Plasma, Cancer, Biomarker discovery, Mass spectrometry

Jennifer J. Kohler and Steven M. Patrie (eds.), *Mass Spectrometry of Glycoproteins: Methods and Protocols*, Methods in Molecular Biology, vol. 951, DOI 10.1007/978-1-62703-146-2_21, © Springer Science+Business Media, LLC 2013

1. Introduction

For more than 30 years aberrant glycosylation has been recognized as a hallmark of cancer (1). However, the complex nature of glycan structure and synthesis has constrained the pace of discoveries relating to their biological significance. Recent advances in carbohydrate chemistry, chemical biology, and mass spectrometry (MS) techniques have opened the door to rapid progress in correlating glycan structure and function (2–6). At the same time, the maturation of proteomics has put cancer biomarker discovery studies at the top of many to-do lists (7). The confluence of these two fields has led many investigators to the same conclusion: exploiting differences in glycosylation between malignant and healthy tissues likely affords excellent opportunities to identify sensitive and specific cancer biomarkers (8–14). Glycosylation machinery appears to be particularly sensitive to malignant transformation; as a result, the saccharide structures that are added to normal cellular proteins change, resulting in neoglycoforms that can be released from the cell through conventional secretory pathways, or as the result of enhanced proteinase activity. It is possible that a portion of these alternatively glycosylated molecules reach the bloodstream. As such, they could serve as early sentinels that enable cancer detection.

Investigators have developed a number of approaches to pursue these circulating biomarkers (15–20). The described workflow uses lectin chromatography performed on trypsin-digested samples, followed by desalting and deglycosylation steps prior to LC-MS/MS analyses. Accordingly, this method offers a sensitivity and specificity unmatched by similar commonly used approaches (e.g., lectin chromatography at the glycoprotein level, and hydrazide- or boronic acid-mediated chemical capture of glycoproteins/peptides). Specifically, it affords the ability to study oligosaccharide changes at the glycosite level, a powerful means to assess subtle disease-related modifications in carbohydrate structure and placement along the protein backbone.

2. Materials

2.1. Lectin-POROS Conjugation

1. POROS® AL 20 μm Self Pack® Media (Applied Biosystems, Carlsbad, CA, USA).
2. Unconjugated *Sambucus nigra* agglutinin (SNA) and *Aleuria aurantia* lectin (AAL) (Vector Laboratories, Burlingame, CA, USA).

3. Sodium cyanoborohydride (Sigma-Aldrich, St. Louis, MO, USA).
4. Quenching Buffer: 1 M Tris-Cl, pH 7.4: Weigh out 6.05 g Tris base and transfer to a 50 mL centrifuge tube. Add deionized water to 40 mL and vortex until the crystals have dissolved. Slowly introduce 38% hydrochloric acid (~3.7 mL) until the pH reaches 7.4; monitor with a pH meter. Add deionized water to reach a volume of 50 mL.
5. 1 M NaCl: Weigh out 2.92 g sodium chloride and transfer to a 50 mL centrifuge tube. Add deionized water to 50 mL and vortex until the salt has dissolved.

2.2. Lectin Affinity Column Packing

1. Column components (Upchurch Scientific, Oak Harbor, WA, USA): 1 Omega Column 2×50 mm (catalog number OC-201A); 2 Frits PEEK 0.5 μm (OC-205); 2 Omega End Fittings (OC-411B); 2 Omega Column couplers (OC-412B); 1 Omega Column 4.6×50 mm (OC401-A); 1 column connector (no catalog number, this part is custom made by Upchurch).
2. Phosphate buffered saline (PBS; Invitrogen, Carlsbad, CA, USA).
3. PBS containing 0.1% sodium azide: To 500 mL PBS, add 0.5 g sodium azide and a stir bar. Mix well until dissolved.

2.3. Lectin Affinity Chromatography

1. Buffer A: 25 mM Tris buffer, pH 7.4, 50 mM sodium chloride, 10 mM calcium chloride, 10 mM magnesium chloride. In a glass flask, combine 450 mL high performance liquid chromatography (HPLC) water, 1.51 g Tris base, 1.46 g NaCl, 0.5 g $CaCl_2$, and 1.0 g $MgCl_2$. With a stir bar, mix well. Once the solids have dissolved, pH the sample to 7.4 by adding 38% HCl (~700–900 μL); monitor with a pH meter. Add water to a final volume of 500 mL; filter through a 0.2 μm membrane; store at 4°C.
2. Buffer B: 0.5 M acetic acid. Add 14 mL glacial acetic acid to 486 mL deionized water. Filter through a 0.2 μm membrane; store at room temperature.
3. Glycopeptide standards from Sigma-Aldrich: human lactoferrin (catalog number L0520) and yeast invertase (I0408). Standards should be trypsin-digested and desalted as previously described (21, 22).

2.4. Sample Preparation

1. Deplete the 14 most abundant proteins from plasma using a Multiple Affinity Removal System (MARS)-Hu-14 HPLC column, according to the manufacturer's instructions (Agilent; Santa Clara, CA).
2. Depleted plasma should be trypsin-digested and desalted as previously described (21, 22).

2.5. Sample Desalting I

1. Oasis HLB SPE cartridges (catalog number WAT094225), 1 cc volume, and Extraction Manifold (e.g., catalog number WAT200609); both from Waters (Milford, MA, USA).

2.6. Sample Deglycosylation

1. Peptide: *N*-glycosidase F (PNGase F, glycerol-free; New England Biolabs, Ipswich, MA, USA).

2.7. Sample Desalting II (Optional)

1. MicroSpin™ columns (5–200 µL elution volume, 5–60 µg capacity; The Nest Group, Southboro, MA, USA).

3. Methods

3.1. Prepare Lectin-Conjugated POROS

1. Put on a mask to protect against inhalation of POROS-AL beads during steps 1 and 2. Weigh out the desired amount of POROS beads (100 mg beads = ~300 µL final volume, see Note 1) and transfer into a clean Eppendorf tube (see Note 2).
2. Wash the beads with addition of 1 mL PBS. Pellet beads by centrifugation at the highest speed for 3 min. Remove the supernatant and repeat.
3. Weigh out the desired amount of unconjugated lectin (3–12 mg/300 µL beads) and transfer to a clean Eppendorf tube (see Note 3). Add PBS to yield a 5–20 mg/mL solution; the volume of the lectin solution should be twice that of the bead volume. Remove an aliquot of this pre-conjugation solution and note the remaining volume.
4. Transfer the remaining lectin solution to the POROS beads. Add sodium cyanoborohydride to a final concentration of 50 mM. Place the tube on a rocker and react overnight at room temperature. (Sodium cyanoborohydride is toxic and must be handled in a fume hood. Contaminated waste must be disposed of appropriately.)
5. Pellet the POROS beads as in step 1. Remove the supernatant and save as the post-conjugation solution.
6. Wash the beads with 1 mL Quenching Buffer. Pellet the beads and discard the supernatant.
7. Block the remaining reactive sites on the POROS beads with 1 mL Quenching Buffer. Add sodium cyanoborohydride to a final concentration of 50 mM. Place the tube on a rocker and incubate at room temperature for 30 min.
8. Pellet the beads and discard the supernatant.
9. Wash the beads with 1 mL 1 M NaCl. Pellet the beads and discard the supernatant. Repeat four times for a total of five washes.

10. Resuspend the POROS beads in desired volume of PBS. If packing one column (~200 mL bed volume), resuspend in 400 μL. The beads are now ready to pack (see Note 4).
11. Determine the amount of protein that was conjugated to the POROS beads by measuring the protein concentration (e.g., using a bicinchoninic acid assay) of the pre-conjugated and post-conjugated lectin solutions. The difference in concentrations is the amount of protein that was conjugated to the beads. The amount of protein conjugated/volume of beads equals the concentration of lectin on the beads (see Note 5).

3.2. Pack the Lectin-Conjugated POROS Beads into a PEEK Column

1. Assemble the packing system as shown in Fig. 1 and support it on a metal ring stand.
2. Transfer conjugated POROS beads into the upper column (reservoir). If needed, add PBS to the reservoir until the buffer reaches the top of the column. Gently place the end fitting

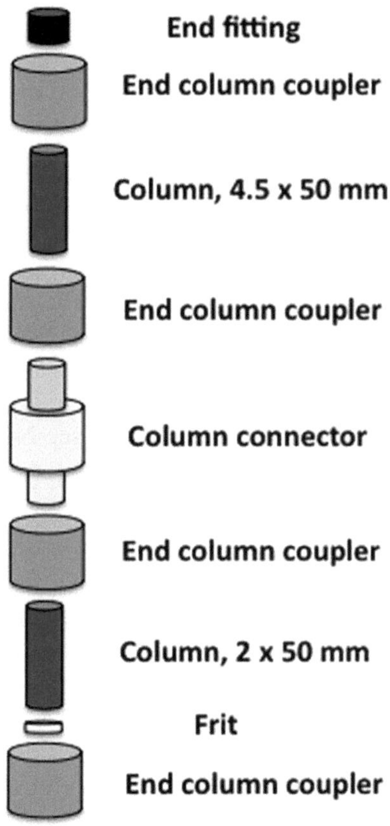

Fig. 1. Column packing assembly. The packing system consists of, from *bottom* to *top*: end column coupler, frit, column (2 × 50 mm), end column coupler, column connector, end column coupler, column (4.5 × 50 mm), end column coupler. The upper column serves as a reservoir for the packing material.

onto the top of the column; avoid trapping air bubbles. Connect the end fitting to the HPLC system.

3. Pack the column by flowing PBS through the system. Start with a flow rate of 250 μL/min. Increase the flow rate by 250 μL/min each minute until either 4,000 μL/min or the maximum pressure (3,000 psi) has been reached. Continue at the maximum flow rate until 35 mL of PBS have passed through the column (see Note 6).
4. Turn off the pumps and allow the pressure on the column to drop to <20 psi.
5. Gently disassemble the packing system starting from the top. When the top end column coupler of the packed (bottom) column is reached, remove carefully. Some packed material may be extruding from the column. With a razor blade or similar sharp edge, gently wipe away the excess beads, leaving a packed surface that is even with the top of the column. Do not apply pressure to the beads.
6. Disengage the packed column from the ring stand. Place a new frit into a new end coupler capped with an end fitting. Hold this with the frit facing up; turn the packed column over and connect the open end to the frit/end coupler.
7. Label the column appropriately (see Note 7). It is now ready for use. When not in use, the lectin column must be stored at 4°C in PBS containing 0.1% sodium azide.

3.3. Establish Chromatography Conditions

3.3.1. Program the Lectin Enrichment Method

1. The details of programming an HPLC will vary according to the specifics of the manufacturer's software. We use a Michrom Paradigm MG4 HPLC. On this machine, methods are built and accessed under the "Setup" tab at the top of the screen. Program the gradient method shown in Table 1. Injection volume: 100 μL (see Note 8).

   Buffer A: 25 mM Tris buffer, pH 7.4, 50 mM sodium chloride, 10 mM calcium chloride, 10 mM magnesium chloride. Buffer B: 0.5 M acetic acid.

2. If an autosampler is available, program the following schedule for fraction collection. Otherwise, collect the following fractions by hand (times indicated are minutes after injection): flow through 2.5 to 9; bound 9.75 to 15. Example traces are shown in Fig. 2.

3.3.2. Validate the Lectin Column Performance

1. Before using a newly prepared column to lectin-enrich samples, column function should be verified using the glycopeptide standards lactoferrin and invertase (Sigma L0520 and I0408, respectively). Digested standards (50 μg each) should be applied to the column and fractions collected as described in Subheading 3.3.1. Fractions should be desalted as described in

Table1
Gradient program for lectin affinity chromatography

Time	Flow rate (μL/min)	% A	% B
0:00	80	100	0
9:00	80	100	0
9:01	500	0	100
13:50	500	0	100
13:51	3,000	100	0
19:50	3,000	100	0

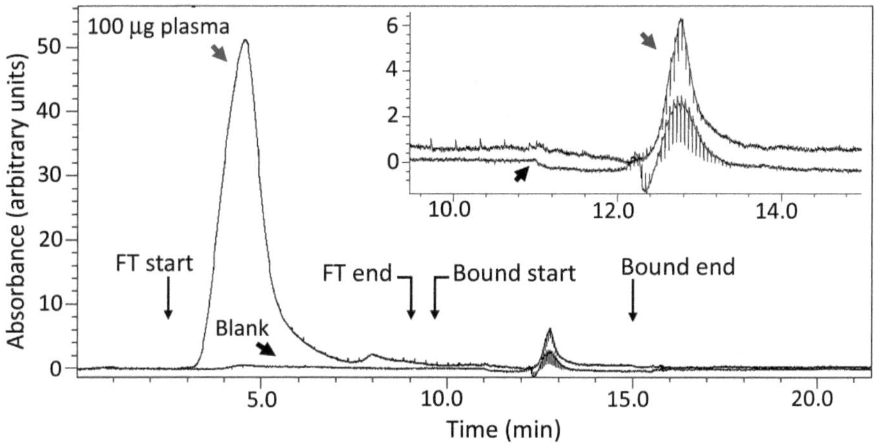

Fig. 2. Lectin chromatography separates the vast majority of unglycosylated peptides from specifically modified glycopeptides. The elution of trypsin-digested peptides and glycopeptides from MARS 14-depleted human plasma (100 μg) was monitored by absorbance at 280 nm (*grey arrowhead*). *Arrows* indicate the start and end time points for the collection of flow through (FT) and bound fractions. The background absorbance is indicated by the trace obtained after injecting Buffer A alone (*Blank, black arrowhead*). Note that a very small fraction of total sample binds the column and elutes in the bound fraction (see inset; *grey* and *black arrowheads* designate plasma and Buffer A loads, respectively).

Subheading 3.5, and then analyzed by MALDI-MS to monitor the presence of intact glycopeptides in the fractions (21). As shown in Fig. 3, invertase glycopeptides should bind weakly or not at all, and the majority of lactoferrin glycopeptides should appear in the bound fraction (see Note 9).

3.4. Prepare Samples for Lectin Affinity HPLC and Perform Chromatography

1. Prior to use in this protocol, human plasma should be depleted of the 14 most abundant proteins using a MARS Hu-14 column. Samples (e.g., depleted plasma, cell line conditioned media, tumor lysates) should be individually trypsin-digested and desalted as previously described (21, 22).

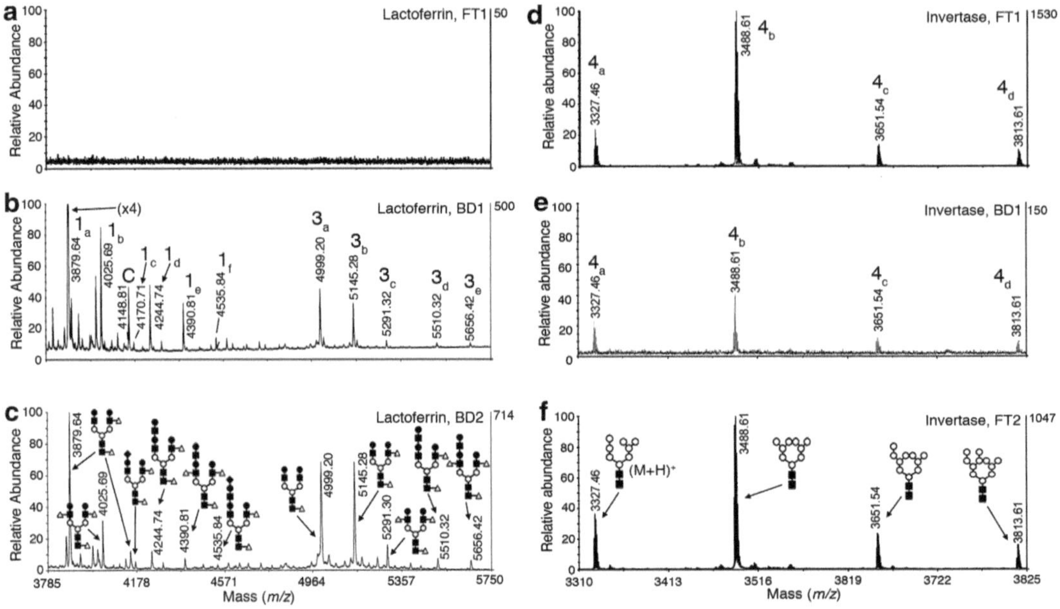

Fig. 3. Glycopeptide standards bound to AAL, which interacts with fucosylated oligosaccharides. Glycopeptides from human lactoferrin, which carries fucosylated and sialylated structures, served as positive controls; glycopeptides from invertase, which carries high-mannose structures, served as negative controls. Lactoferrin (**a**–**c**) and invertase (**d**–**f**) glycopeptides were affinity selected on AAL-conjugated POROS. Flow-though (FT) and bound (BD) fractions were collected and analyzed by MALDI-MS or MALDI-MS/MS. (**a**) MS spectrum of the lactoferrin FT1 fraction. (**b**) MS spectrum of the BD1 fraction. Peaks corresponding to various fucosylated glycoforms of three related peptides were observed (1) TAGWNVPIGTLRPFLNWTGPPEPIEAAVAR, (2) PFLNWTGPPEPIEAAVAR, and (3) LRPFLNWTGPPEPIEAAVAR; where **N** corresponded to glycosylated Asn with NWT as the consensus glycosylation sequence. (**c**) The BD1 fraction was rechromatographed and subjected to MS analysis; consistent with the initial results, glycopeptides were detected in the BD2 but not in the FT2 fraction (data not shown). The peptide backbone for these structures was confirmed by MS/MS. Monosaccharide composition and, in some cases, partial sequence information was also obtained for the carbohydrate structures. These data (and the depicted glycans) were consistent with previously identified lactoferrin carbohydrate structures, and the fact that this molecule is highly fucosylated (25, 26). (**d**) MS spectrum of the invertase FT1 fraction. Peaks corresponded to high-mannose structures appended to peptide "4" (AEPILNISNAGPWSR). (**e**) MS spectrum of the BD1 fraction. The glycopeptides were also observed but at approximately tenfold less intensity. (**f**) The FT1 fraction was rechromatographed and subjected to MS analysis. The four molecular ions, which were also observed in the FT1 fraction (compare panels (**d**, **f**)), were interrogated by MS/MS and the peptide backbone and monosaccharide composition were confirmed. These data (and the depicted glycans) are consistent with the fact that invertase presents high-mannose structures (27). No invertase glycopeptides were detected in the rechromatographed BD2 fraction (data not shown) (■, *N*-acetylglucosamine; ●, galactose; ○, mannose; △, fucose; ◆, sialic acid).

2. Determine the amount of sample to be chromatographed (in μg; see Note 10). Dilute the sample to a final volume of 110 μL. The injection volume is 100 μL (see Note 11).
3. Use the lectin enrichment HPLC method described in Subheading 3.3.1. The column and buffers are at room temperature while in use.
4. Attach either the AAL or the SNA lectin column to the HPLC and run a blank method (injecting only Buffer A).
5. Chromatograph samples, including blank methods between analytes of different origin.

3.5. Desalt Collected Fractions

For each fraction use one 1 cc Waters Oasis HLB SPE cartridge and Waters manifold as follows:

1. Attach the number of required cartridges to vacuum manifold.
2. Wash each cartridge three times with 1.5 mL 80% acetonitrile in 1% formic acid. Vacuum gauge on manifold should read 5–20 inches Hg.
3. Equilibrate cartridges with 1.5 mL 0.1% formic acid. Vacuum gauge on manifold should read 5–20 inches Hg.
4. Slowly load entire volume of one sample onto one cartridge. Vacuum gauge on manifold should read 2–2.5 inches Hg, and flow rate should not exceed 1 mL/min.
5. Wash cartridges with 3 × 1 mL 0.1% formic acid. Vacuum gauge on manifold should read 5–20 inches Hg.
6. Slowly elute peptides/glycopeptides into clean, labeled 2 mL Eppendorf tubes with 1.5 mL 80% acetonitrile in 0.1% formic acid. Vacuum gauge on manifold should read 2–2.5 inches Hg, and flow rate should not exceed 1 mL/min. One collection tube should be used for each cartridge.
7. Neutralize eluate by adding 75 μL 0.5 M ammonium bicarbonate to each collection tube. Target neutralized pH is 7.0–8.0, test with pH indicator paper.
8. Using a centrifugal concentrator, reduce sample volume to 50–100 μL (~2 h @ 35°C).

3.6. PNGase F-Digestion of Glycopeptides

1. Test sample pH to ensure it is between 7.0 and 8.0. If needed, add 0.5 M ammonium bicarbonate to increase pH.
2. Add 0.5 μL (250 U) glycerol-free PNGase F (New England Biolabs) to each sample tube and incubate at 37°C overnight (see Note 12).

3.7. Desalt Samples Prior to LC-MS/MS (Optional)

1. If desired (see Note 13), desalt each sample using a MicroSpin™ column as follows.
2. First, flick the column with a fingertip several times to settle the beads. Then, slide the adaptor collar onto the column and place in a 2 mL Eppendorf tube.
3. Pipette 100 μL 100% acetonitrile to the top of the column and centrifuge at 200 × *g* for 1 min.
4. Repeat step 3 using 80% acetonitrile in 0.1% formic acid.
5. Repeat step 3 twice using 0.1% formic acid to equilibrate the column.
6. Place the column into a new 2 mL Eppendorf tube. Pipette the sample onto the top of the column and centrifuge at 200 × *g* for 1 min.
7. Wash the column by pipetting 100 μL 0.1% formic acid to the top of the column and centrifuging at 200 × *g* for 1 min.

8. Repeat step 7.
9. Elute the sample into a clean Eppendorf tube by pipetting 100 μL 80% acetonitrile, 0.1% formic acid to the top of the column and centrifuging at 200 × *g* for 1 min.
10. Add an additional 100 μL 80% acetonitrile, 0.1% formic acid to the top of the column and centrifuge at 200 × *g* for 1 min.
11. Reduce the sample to a volume of <2 μL by centrifugal concentration.
12. Resuspend the sample in 20 μL 0.1% formic acid. It is now ready for analysis by LC-MS/MS.

3.8. Interpret LC-MS/MS Results

1. Perform LC-MS/MS analyses (see Note 14) and search the data (see Note 15).
2. Assign glycopeptides (see Note 16) using the following criteria: (a) Presence of an N-linked consensus sequence (NXS/T, where X is any amino acid except proline); (b) The +1 Da mass shift located at the asparagine residue within the consensus sequence (observed as the conversion of asparagine to aspartic acid). A representative spectrum is shown in Fig. 4, depicting the identification of a glycopeptide from cadherin-5, a glycoprotein that typically circulates at 0.3–4.9 ng/mL (23). To

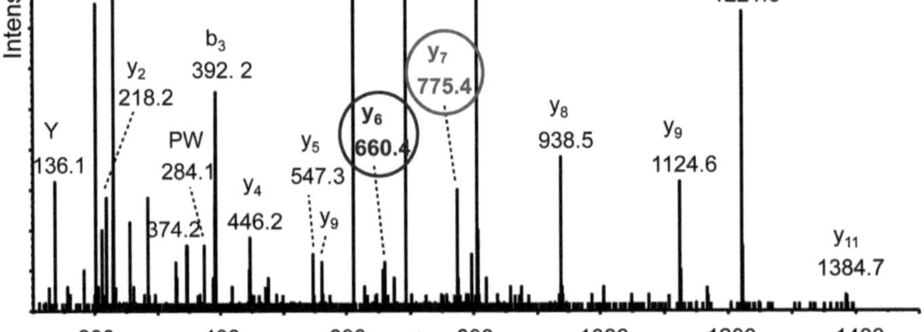

Fig. 4. High-confidence identification of a glycopeptide from cadherin-5, a glycoprotein that typically circulates in human blood at ng/mL levels. ESI-MS/MS spectrum resulting from CID on the molecular ion MH_2^{2+}, selected at *m/z* 806.91 $^{2+}$ (*M* = 1611.80). The spectrum was searched using Protein Pilot and peptide EVYPWYN*LTVEAK was identified (*N** = deamidated *N*) with a peptide confidence of 99. Fragment ions y6 and y7 clearly show an increment mass difference of Δ*M* = 115 Da, localizing the deamidation to N442, the glycosite.

ensure inclusion of glycosites containing lysine and/or arginine in the X position (e.g., NKT), which are likely to have been cleaved by trypsin, the amino-acid residue following the carboxy-terminal cleavage site should also be considered. Peptides containing the motif NGS or NGT should be interpreted with caution due to the fact that asparagine residues in that sequence are prone to chemical deamidation during overnight trypsin digestion (24).

3. For an extensive description of acquisition methods, database searching, and representative data obtained from pooled plasma of healthy donors, please refer to Drake et al. (21).

4. Notes

1. The PEEK columns used in this protocol are ~200 μL in volume. The method calls for preparation of ~300 μL of beads for each packed column to allow for losses during bead conjugation and column preparation. Typically, extra beads remain after the packing step. These may be stored in PBS with 0.5% sodium azide at 4°C until needed. From time to time after extended use, the upstream frit degrades and beads are lost. Extra beads can then be added (either with the described packing method, or, if <1 mm^3 beads were displaced, a spatula may be used to plaster a centrifuged bead pellet into the void) and one can continue to use the lectin column. In our hands, well-maintained columns retain functionality for up to 6 months and ≥50 lectin enrichment HPLC cycles.
2. Use round bottom, 2 mL Eppendorf tubes. The round bottom allows for the beads to flow freely during the mixing steps. By contrast, beads get trapped in the narrow bottom end of 1.5 mL tubes and mix poorly.
3. This recommended amount of lectin for conjugation allows for discretion according to reagent costs.
4. This is a potential stopping point. The beads may be stored in an Eppendorf tube at 4°C in PBS with 0.5% sodium azide until they are packed into the PEEK column.
5. The optimal on-bead concentration probably varies for each lectin. However, good results can be obtained using columns with 2–20 mg/mL of SNA or AAL. The conjugation efficiency is typically ~80–85%.
6. Place a 50 mL centrifuge tube under the bottom end of the column prior to packing, and use the markings on the tube to monitor the volume of PBS that elutes from the column.

7. The column is directional, meaning that buffers should flow through from upstream to downstream in the same direction in which the column was packed. Indicate the intended direction of flow with an arrow on the column body.
8. The method was designed for a 200 μL autosampler loop and 100 μL injection syringe. Larger or smaller autosampler loops and syringes can be used with corresponding adjustments to the injection volume and fraction collection.
9. Although specific instruments and analytical conditions will vary between laboratories, we used the following MS/MS methods: Mass spectra were acquired by matrix-assisted laser desorption ionization–time-of-flight (MALDI-TOF) MS using a 4,800 MALDI TOF-TOF Proteomics Analyzer (Applied Biosystems, Foster City, CA; USA/MDS Sciex, Concord, ON, Canada) equipped with TOF/TOF™ ion optics, a 200 Hz Nd:YAG laser, controlled by the 4,000 Series Explorer Software V3.5.28193. The positive-ion MS spectra in the *m/z* range of 800–6,000 Da were manually acquired by accumulating 1,500–2,000 shots at a fixed laser intensity (4,000–4,200). The accelerating voltage was 20 kV. Other instrumental settings were optimized to afford a resolution of >15,000 at *m/z* = 2093.89 Da. MS/MS spectra were generated by employing collision-induced dissociation (CID) using the following settings: collision cell floated at 1 kV; resolution of precursor ion selector = 300 or 400 full-width half-maximum (FWHM); metastable suppressor: on; total shots per spectrum = 4,000; and fixed laser intensity = 5,000–5,500. No collision gas was used. Manufacturer-supplied Plate Model and Default MS Calibration algorithms were employed to generate external calibration of the MS mode using monoisotopic molecular ions from a combination of six peptide standards (des-Arg^{1}-bradykinin, angiotensin I, [Glu^{1}]-fibrinopeptide B, and three ACTH peptides: 1–17, 18–39, and 7–38). Acceptance criteria thresholds for a generation of MS calibration files required at least four standard molecular ions with S/N of ≥300, mass tolerance of 0.5 Da, and maximum outlier error of 25 ppm. Default calibration of the MS/MS spectra used a minimum of five matched fragment ions of [Glu^{1}]-fibrinopeptide B that were detected with a minimum S/N of 120, within a mass tolerance window of ±1 Da and a maximum outlier error of 20 ppm. Data were processed using Data Explorer Software Version 4.9 (Applied Biosystems). Samples were desalted and concentrated using C18 ZipTips® (Millipore). Then aliquots of 0.5 μL (~15 nM–3 μM) in 0.1% TFA were mixed on a MALDI target with the matrix (α-cyano-4-hydroxycinnamic acid, 5 mg/mL in 80% ACN/0.1% TFA/10 mM dibasic ammonium phosphate) at a 1:1 ratio (v/v). All data were manually processed and interpreted.

Note that LC-ESI-MS/MS may also be used to monitor elution of the glycopeptide standards; in this case the flow through and bound fractions should be processed separately, and analyzed using the described sample workflow. The presence of invertase glycopeptides (negative control) in the flow through, and lactoferrin glycopeptides (positive control) in the bound fraction should then be assessed.

10. There is wide latitude in the sample load for this method, from ~10 μg to 1 mg (21). As a general rule, the percent enrichment (# of glycopeptides observed/# of total peptides observed) decreases with increasing sample load. However, the total number of glycopeptides identified increases until the column capacity is reached. The column capacity varies with the lectin and sample type (e.g., plasma or conditioned medium; (21) and Drake unpublished observations). Typically, when using clinical samples, the peptide load will be determined by the amount of available material (i.e., sample volume is the limiting factor). Robust data may be obtained with as little as 50 μg of MARS 14-depleted plasma.
11. Buffer A, which contains divalent cations that are critical for optimal lectin binding, should be added to the sample (prior to injection onto the lectin column) to comprise ≥25% of the final volume.
12. A wide range of enzyme:substrate ratios and reaction volumes was tested to determine the optimal PNGase F-digestion conditions for this workflow. Over the range of enzyme concentrations tested, 1–20 U/μg of peptide, only a modest increase (max. $\Delta = 2\%$) in product formation was observed. Accordingly, the method employs the highest enzyme concentration that was practical given cost considerations (2.5 U/μg).
13. There is no need to desalt the sample if there is a trap column in line with the LC-MS/MS system.
14. As with any LC-MS/MS experiment, inclusion of multiple technical replicates will increase the number of peptides sampled for MS/MS. A minimum of two replicates per lectin fraction is recommended. Although specific instruments and analytical conditions will vary between laboratories, we used the following HPLC and MS methods: Peptides were analyzed by reverse-phase nano-HPLC-ESI-MS/MS using an Eksigent nano-LC 2D HPLC system (Eksigent, Dublin, CA) which was directly connected to a quadrupole time-of-flight (QqTOF) QSTAR Elite mass spectrometer (MDS SCIEX, Concord, CAN).

    Peptide mixtures were loaded onto a guard column (C18 Acclaim PepMap100, 300 μm I.D. × 5 mm, 5 μm particle size, 100 Å pore size, P/N 160454, Dionex, Sunnyvale, CA) and washed with the loading solvent (0.1% formic acid, flow rate:

20 μL/min) for 10 min. Subsequently, samples were transferred onto the analytical C18-nanocapillary HPLC column (C18 Acclaim PepMap100, 75 μm I.D. × 15 cm, 3 μm particle size, 100 Å pore size, P/N 160321, Dionex, Sunnyvale, CA) that was directly connected to a New Objective PicoTip Emitter (FS-360-20-10-N-20-C12DOM, tip ID = 10 μm, Woburn, MA). Peptides were eluted at a flow rate of 300 nL/min using the following gradients: Gradient 1 (~3 h): at 2% solvent B in A for 5 min, 2–40% solvent B in A (from 5 to 125 min), 40–90% solvent B in A (from 125 to 140 min), and at 90% solvent B in A (from 140 to 149 min), with a total runtime of 194 min (including mobile phase equilibration), or the shorter Gradient 2 (~2 h): 2–40% solvent B in A (from 0 to 60 min), 40–90% solvent B in A (from 60 to 75 min), and at 90% solvent B in A (from 75 to 85 min), with a total runtime of 120 min (including mobile phase equilibration). Solvent A consisted of 0.1% formic acid in 98% H_2O/2% acetonitrile and solvent B consisted of 0.1% formic acid in 98% acetonitrile/2% H_2O. Mass spectra (ESI-MS) and tandem mass spectra (ESI-MS/MS) were recorded in positive-ion mode with a resolution of 12,000-15,000 FWHM. The nanospray needle voltage was typically 2,300 V in HPLC-MS mode. Spectra were calibrated in static nanospray mode using MS/MS fragment ions of a Glu-Fibrinogen B peptide standard (y_1 fragment ion with m/z at 175.1195, and y_{11} fragment ion with m/z at 1285.5444). For collision-induced dissociation tandem mass spectrometry (CID-MS/MS), the mass window for precursor ion selection of the quadrupole mass analyzer was set to ±1 *m/z*. The precursor ions were fragmented in a collision cell using nitrogen as the collision gas. Advanced information dependent acquisition (IDA) was used for MS/MS collection, including QSTAR Elite (Analyst QS 2.0) specific features, such as "Smart Collision" and "Smart Exit" (fragment intensity multiplier set to 2.0 and maximum accumulation time at 2 s) to obtain MS/MS spectra for the six most abundant parent ions following each survey scan. Dynamic exclusion features were based on value M not *m/z* and were set to exclusion mass width 50 mDa and exclusion duration of 120 s.

15. We search with a number of bioinformatics search engines, including Protein Pilot (Applied Biosystems), Mascot (Matrix Science), and Global Proteome Machine (http://www.thegpm.org). Different search engines identify complementary unique glycopeptides from the same dataset; therefore, searching the same files with multiple software tools increases the number of glycopeptides observed.

16. The availability of PNGase F allows for the deglycosylation and facile analysis of N-linked glycopeptides. Due to technical

constraints (i.e., the lack of a similar enzyme to globally remove O-linked species), lectin enrichment of O-linked glycopeptides was not addressed in these studies, and all further references to glycopeptides pertain to N-linked species unless otherwise stated.

Acknowledgement

This work was supported by the Clinical Proteomic Technologies for Cancer initiative, 5U24CA126477-04.

References

1. Meezan E, Wu HC, Black PH, Robbins PW (1969) Comparative studies on the carbohydrate-containing membrane components of normal and virus-transformed mouse fibroblasts. II. Separation of glycoproteins and glycopeptides by sephadex chromatography. Biochemistry 8:2518–2524
2. Barkauskas DA, An HJ, Kronewitter SR, de Leoz ML, Chew HK, de Vere White RW, Leiserowitz GS, Miyamoto S, Lebrilla CB, Rocke DM (2009) Detecting glycan cancer biomarkers in serum samples using MALDI FT-ICR mass spectrometry data. Bioinformatics 25:251–257
3. Goldberg D, Bern M, Parry S, Sutton-Smith M, Panico M, Morris HR, Dell A (2007) Automated N-glycopeptide identification using a combination of single- and tandem-MS. J Proteome Res 6:3995–4005
4. Harvey DJ (2005) Structural determination of N-linked glycans by matrix-assisted laser desorption/ionization and electrospray ionization mass spectrometry. Proteomics 5: 1774–1786
5. North SJ, Hitchen PG, Haslam SM, Dell A (2009) Mass spectrometry in the analysis of N-linked and O-linked glycans. Curr Opin Struct Biol 19:498–506
6. Royle L, Campbell MP, Radcliffe CM, White DM, Harvey DJ, Abrahams JL, Kim YG, Henry GW, Shadick NA, Weinblatt ME, Lee DM, Rudd PM, Dwek RA (2008) HPLC-based analysis of serum N-glycans on a 96-well plate platform with dedicated database software. Anal Biochem 376:1–12
7. Wong SC, Chan CM, Ma BB, Lam MY, Choi GC, Au TC, Chan AS, Chan AT (2009) Advanced proteomic technologies for cancer biomarker discovery. Expert Rev Proteomics 6:123–134
8. Barrabes S, Pages-Pons L, Radcliffe CM, Tabares G, Fort E, Royle L, Harvey DJ, Moenner M, Dwek RA, Rudd PM, De Llorens R, Peracaula R (2007) Glycosylation of serum ribonuclease 1 indicates a major endothelial origin and reveals an increase in core fucosylation in pancreatic cancer. Glycobiology 17:388–400
9. Kuzmanov U, Jiang N, Smith CR, Soosaipillai A, Diamandis EP (2009) Differential N-glycosylation of kallikrein 6 derived from ovarian cancer cells or the central nervous system. Mol Cell Proteomics 8:791–798
10. Meany DL, Zhang Z, Sokoll LJ, Zhang H, Chan DW (2009) Glycoproteomics for prostate cancer detection: changes in serum PSA glycosylation patterns. J Proteome Res 8:613–619
11. Misonou Y, Shida K, Korekane H, Seki Y, Noura S, Ohue M, Miyamoto Y (2009) Comprehensive clinico-glycomic study of 16 colorectal cancer specimens: elucidation of aberrant glycosylation and its mechanistic causes in colorectal cancer cells. J Proteome Res 8:2990–3005
12. Mizuochi T, Nishimura R, Derappe C, Taniguchi T, Hamamoto T, Mochizuki M, Kobata A (1983) Structures of the asparagine-linked sugar chains of human chorionic gonadotropin produced in choriocarcinoma. Appearance of triantennary sugar chains and unique biantennary sugar chains. J Biol Chem 258:14126–14129
13. Ohyama C, Hosono M, Nitta K, Oh-eda M, Yoshikawa K, Habuchi T, Arai Y, Fukuda M (2004) Carbohydrate structure and differential binding of prostate specific antigen to Maackia amurensis lectin between prostate cancer and benign prostate hypertrophy. Glycobiology 14:671–679

14. Taylor AD, Hancock WS, Hincapie M, Taniguchi N, Hanash SM (2009) Towards an integrated proteomic and glycomic approach to finding cancer biomarkers. Genome Med 1:57
15. Cho W, Jung K, Regnier FE (2008) Use of glycan targeting antibodies to identify cancer-associated glycoproteins in plasma of breast cancer patients. Anal Chem 80:5286–5292
16. Heo SH, Lee SJ, Ryoo HM, Park JY, Cho JY (2007) Identification of putative serum glycoprotein biomarkers for human lung adenocarcinoma by multilectin affinity chromatography and LC-MS/MS. Proteomics 7:4292–4302
17. Jung K, Cho W, Regnier FE (2009) Glycoproteomics of plasma based on narrow selectivity lectin affinity chromatography. J Proteome Res 8:643–650
18. Kaji H, Saito H, Yamauchi Y, Shinkawa T, Taoka M, Hirabayashi J, Kasai K, Takahashi N, Isobe T (2003) Lectin affinity capture, isotope-coded tagging and mass spectrometry to identify N-linked glycoproteins. Nat Biotechnol 21:667–672
19. Plavina T, Wakshull E, Hancock WS, Hincapie M (2007) Combination of abundant protein depletion and multi-lectin affinity chromatography (M-LAC) for plasma protein biomarker discovery. J Proteome Res 6:662–671
20. Zhang H, Li XJ, Martin DB, Aebersold R (2003) Identification and quantification of N-linked glycoproteins using hydrazide chemistry, stable isotope labeling and mass spectrometry. Nat Biotechnol 21:660–666
21. Drake PM, Schilling B, Niles RK, Braten M, Johansen E, Liu H, Lerch M, Sorensen DJ, Li B, Allen S, Hall SC, Witkowska HE, Regnier FE, Gibson BW, Fisher SJ (2011) A lectin affinity workflow targeting glycosite-specific, cancer-related carbohydrate structures in trypsin-digested human plasma. Anal Biochem 408:71–85
22. Keshishian H, Addona T, Burgess M, Kuhn E, Carr SA (2007) Quantitative, multiplexed assays for low abundance proteins in plasma by targeted mass spectrometry and stable isotope dilution. Mol Cell Proteomics 6:2212–2229
23. Soeki T, Tamura Y, Shinohara H, Sakabe K, Onose Y, Fukuda N (2004) Elevated concentration of soluble vascular endothelial cadherin is associated with coronary atherosclerosis. Circ J 68:1–5
24. Krokhin OV, Antonovici M, Ens W, Wilkins JA, Standing KG (2006) Deamidation of -Asn-Gly- sequences during sample preparation for proteomics: consequences for MALDI and HPLC-MALDI analysis. Anal Chem 78:6645–6650
25. Matsumoto A, Yoshima H, Takasaki S, Kobata A (1982) Structural study of the sugar chains of human lactoferrin: finding of four novel complex-type asparagine-linked sugar chains. J Biochem 91:143–155
26. Spik G, Strecker G, Fournet B, Bouquelet S, Montreuil J, Dorland L, van Halbeek H, Vliegenthart JF (1982) Primary structure of the glycans from human lactotransferrin. Eur J Biochem 121:413–419
27. Trimble RB, Atkinson PH (1986) Structure of yeast external invertase Man8-14GlcNAc processing intermediates by 500-megahertz 1H NMR spectroscopy. J Biol Chem 261:9815–9824

Chapter 22

Characterizing the Glycosylation State of Therapeutic Recombinant Glycoproteins

Nicole Samuels, David Kates, Jun Liu, and Joanne Severs

Abstract

As an increasing number of recombinant therapeutic glycoproteins are manufactured and investigated, the importance of their attached glycans is becoming more widely reported and understood. Regulatory agencies expect detailed "extended characterization" of the glycoprotein as well as routine, well-controlled "release assays" with specifications to be employed for quality control of each manufactured lot. In this chapter we will briefly discuss relevant glycan issues in the area of therapeutic recombinant glycoprotein manufacture and describe in detail two assays that are employed in the development of, for example, recombinant Factor VIII for the treatment of hemophilia.

Key words: Mass spectrometry, FVIII, Recombinant, Glycoprotein

1. Introduction

The discovery and development of recombinant DNA technology has been the cornerstone of the Biotechnology industry. The majority of recombinant therapeutic proteins are produced in mammalian cells, mainly due to their ability to generate high quality proteins with biochemical properties that are similar to their human counterparts. Recombinant glycoproteins constitute an ever increasing portion of clinical and marketed therapeutic proteins. The most prevalent glycosylation sites in therapeutic glycoproteins are at asparagine (Asn-X-Thr/Ser consensus sequence), serine and threonine residues (1). N-linked oligosaccharides contain a common pentasaccharide core and may be classified as a high mannose, hybrid type, or complex type structure. The N-glycan core structure can be further elongated and adorned with additional

Jennifer J. Kohler and Steven M. Patrie (eds.), *Mass Spectrometry of Glycoproteins: Methods and Protocols*, Methods in Molecular Biology, vol. 951, DOI 10.1007/978-1-62703-146-2_22, © Springer Science+Business Media, LLC 2013

monosaccharides such as fucose and sialic acid. Conversely, O-linked glycans have up to eight different core structures, which also can be further extended. In addition to heterogeneity in the structure of the glycan located at a particular site, heterogeneity in glycosylation occupancy with respect to the site of glycan attachment is also observed in glycoproteins. As a result, heterogeneity in glycosylation can potentially give rise to great structural diversity and pose significant challenges for the characterization of recombinant glycoprotein-based drugs.

Glycan structure and composition can influence the biological function and circulation of therapeutic glycoproteins in the body. Micro-heterogeneity of human IgG glycans has been reported to have an effect on biological functions such as complement dependent cytotoxicity (CDC), antibody-dependent cytotoxicity (ADCC), binding to Fc receptors and binding to the CIq protein (2). The presence of bisecting GlcNAc has been shown to lead to an increase in affinity for FcγRIII with an accompanying increase in ADCC for an IgG (3). Protein sialylation can affect absorption, serum half-life, and clearance from the serum, as well as the physical, chemical, and immunogenic properties of some glycoproteins (4). For example, N-linked oligosaccharide structures on the B-domain of Factor VIII (FVIII) recognize the carbohydrate recognition domains of the asialoglycoprotein receptor and play a role in the clearance of FVIII from the circulation. The treatment of FVIII with neuraminidase has been shown to enhance its binding to the asialoglycoprotein receptor approximately 100-fold (5). Due to the impact glycan structure and composition can have on the function and half-life of therapeutic glycoproteins, manufacturers have sought to maintain optimal and homogeneous glycosylation from batch-to-batch. Protein sialylation, glycoform profiles, and glycan site occupancy, are just a few of the protein quality attributes impacted by cellular, media, and process effects. As glycosylation is sensitive to the particular cells and conditions in which a protein is produced, a key to consistently producing the desired glycoform profile for a therapeutic glycoprotein is gaining an understanding the impact of manufacturing parameters on product quality (1). Knowledge of the relationship between manufacturing variables, including the choice of a production cell line, the manufacturing mode bioreactor control parameters and media components and product quality can be exploited to optimize protein glycosylation, and potentially enhance the efficacy of therapeutic glycoproteins (1, 6).

The most commonly used cell lines in cell culture processes for therapeutic glycoproteins are derived from hamster and murine sources and have the potential to modify surface glycans with epitopes that are not normally found, or found in low levels on human glycoproteins (7). The presentation of these nonhuman glycan structures, which could potentially affect the safety profile of a glycoprotein therapeutic, is another area that needs consideration on a case-by-case basis. *N*-glycolylneuraminic acid (NGNA) is formed from the hydroxylation of sialic acid and is a potential component of therapeutic glycoproteins expressed in mammalian cell lines. Most humans have an intake of this species through the digestion of animal products in their diet. Additionally, terminal sialic acids, which can be either 2,3 or 2,6-linked in human proteins are reported to be only 2,3-linked in glycoproteins produced by Chinese hamster ovary (CHO) cell lines (4). Galactose-α 1,3-galactose is normally expressed in non-primate mammals and is sometimes found on glycoproteins expressed in mammalian cell lines (7). Most notably, this oligosaccharide is present on the F_{ab} portion of the monoclonal antibody Cetuximab and was linked to severe hypersensitivity reactions in patients (8). As a result, care must be taken to monitor and characterize glycoprotein therapeutics during manufacture for nonhuman epitopes that may initiate adverse reactions in patients.

As discussed above, glycan structure and composition can modulate the pharmacological properties of therapeutic glycoproteins to different extents. Therefore, methods for monitoring parameters such as glycan heterogeneity, monosaccharide composition, and the degree of sialylation are just a few examples of critical assays to have in the analytical toolbox (9). Routinely used methods are often employed for release testing of glycoprotein therapeutics for quality control. These methods are used to demonstrate consistent glycosylation during the manufacture of therapeutic glycoproteins from batch-to-batch and are usually validated late in the drug development cycle. Additionally, more specialized methods may be reserved for detailed "Extended Characterization" of clinical and reference materials. In general, a combination of enzymatic digestions (of either the glycan or protein backbone) along with chromatographic and, for extended characterization, mass spectrometric analyses are often employed. Factors to be demonstrated include: degree of glycosylation site occupancy, oligosaccharide structure/abundance, sialic acid content/type/linkage, degree of sialylation of terminal galactose residues, site specific structure, linkage determination, and α-1,3 linked galactose percentage. The agencies responsible for reviewing clinical and license applications for each country will request to see evidence of well-controlled test methods as well as the results for each manufactured product lot and reference standards prior to human treatment.

FVIII is an essential cofactor in blood coagulation. Recombinant FVIII is administered in the treatment of patients with hemophilia A, where the endogenous FVIII is deficient or defective (10). FVIII consists of heavy and light chains, which are cleaved by thrombin (Factor II) to yield an active trimer. Full length FVIII is comprised of 2,332 amino acid residues and contains up to 25 potential N-glycan sites, as well as O-glycans and tyrosine sulfation (11).

Recombinant FVIII is expressed in mammalian cell lines and is arguably the most complex therapeutic glycoprotein manufactured. Herein we present a HPLC-based method for the analysis of 2-anthranilic acid (AA) labeled N-linked oligosaccharides following enzymatic release from FVIII immobilized on nylon membranes. This method can also be applied to solution digestions. We also present a liquid chromatography mass spectrometry (LC-MS) method to analyze FVIII glycoforms following digestion with thrombin. Additional issues to be considered in developing methods for the analysis of glycoprotein drugs are the formulation components that are added for stabilizing purposes. Nonionic surfactants present an analytical challenge for some mass spectrometers and we present a method for in-line removal of Tween 80 prior to the analysis of the FVIII.

The first method describes a qualitative "extended characterization" analysis by which purified rFVIII preparations are digested with human thrombin, and then analyzed by reversed phase high pressure liquid chromatography (RP-HPLC) and electrospray ionization mass spectrometry to confirm integrity of the chains and post-translational modifications. In vitro treatment of full length rFVIII preparations with thrombin yields A3C1C2, A1, A2, a1, a3, and B fragments (10). Application of the digested sample to a RP-HPLC column results in the elution of rFVIII fragments in an order of increasing hydrophobicity. Interfacing RP-HPLC with electrospray ionization mass spectrometry (ESI-MS) provides intact mass assignments for the rFVIII fragments. The B-domain of full length rFVIII is excluded from this mass spectral analysis due to the large mass and heterogeneity imparted by as many as 19 potential N-glycans. The specific conditions stated here were developed to overcome the high Tween 80 levels that are often added to formulation buffers to ensure product shelf-life stability. For many electrospray ionization mass spectrometry systems, the levels of Tween 80 sufficient to stabilize proteins often interfere with the analysis and necessitates the removal of this nonionic surfactant from the protein sample. Figures 1 and 2 show examples of the deconvoluted mass spectra acquired for the glycoforms of the A3C1C2 and A1 domains of the protein.

The second method describes an assessment of the glycosylation state of rFVIII by measuring the relative ratios of N-linked oligosaccharides. In this method, the N-linked oligosaccharides are released from samples of recombinant protein bound to nylon spin filters, using the enzyme PNGase F. The released N-glycans are then labeled with anthranilic acid (2-AA) using reductive amination with sodium cyanoborohydride as reductant (12). The 2-AA labeled components of recombinant protein preparations are separated by HPLC in normal phase/ion exchange mixed-mode and using fluorescence emission at 425 nm after excitation at 360 nm. This procedure separates the labeled components primarily

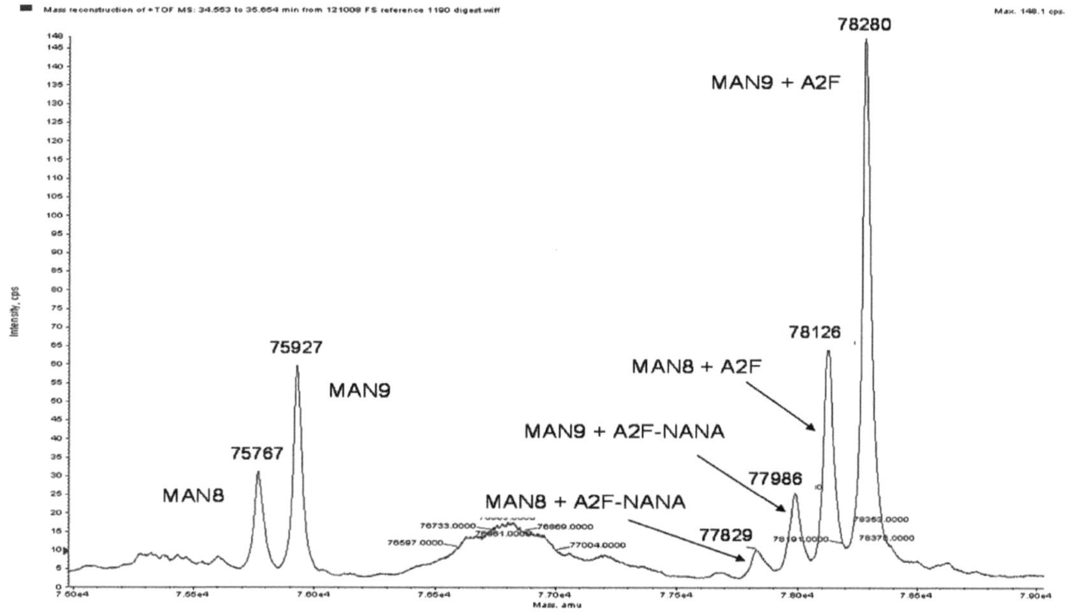

Fig. 1. Deconvoluted mass spectrum of the A3C1C2 domain of rFVIII. MAN8 = oligomannose-8, MAN9 = oligomannose-9, A2F = di-sialylated, biantennary, core fucosylated, and A2F-NANA = mono-sialylated, biantennary, core fucosylated.

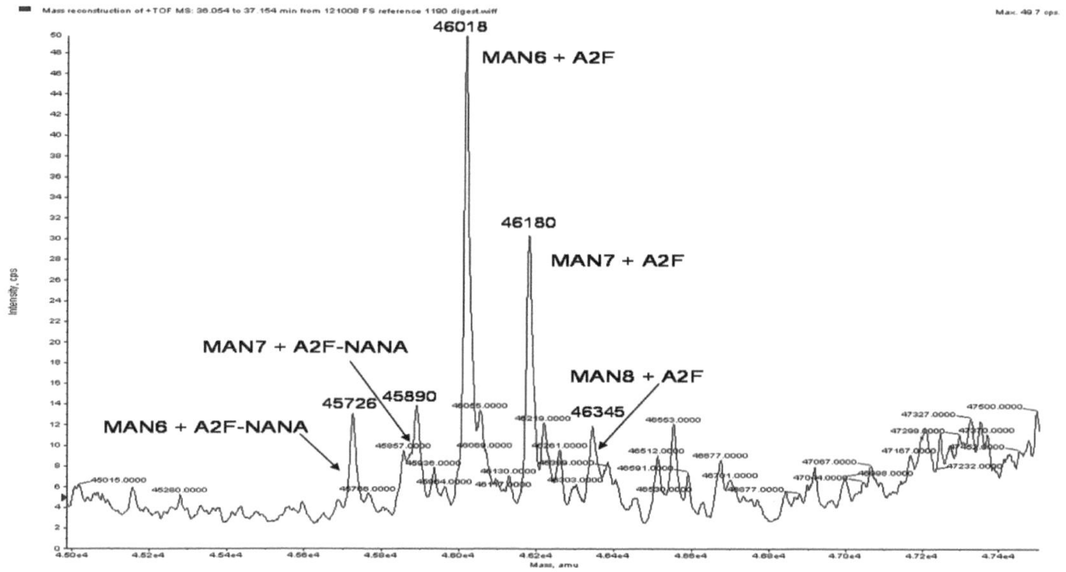

Fig. 2. Deconvoluted mass spectrum of the A1 domain of rFVIII. MAN6 = oligomannose-6, MAN7 = oligomannose-7, MAN8 = oligomannose-8, A2F = di-sialylated, biantennary, core fucosylated, and A2F-NANA = mono-sialylated, biantennary.

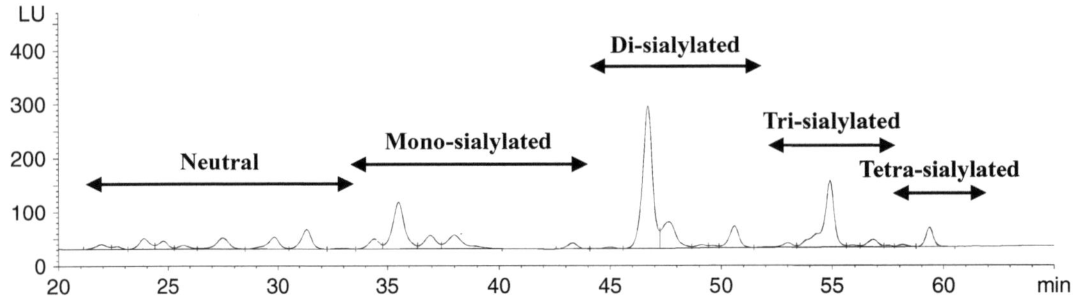

Fig. 3. Oligosaccharide map of rFVIII.

according to their ionic charge, which is dependent on the number of sialic acid groups covalently bound to the oligosaccharides, and also by their size/structure (Fig. 3).

2. Materials

2.1. LC-MS Analysis of a Thrombin Digest of rFVIII Formulated Product

1. rFVIII reference standard and test samples (100 μg/mL in concentration and formulated with Tween 80).
2. Human alpha thrombin (1 U/μL).
3. PPACK dihydrochloride (D-Phe-Pro-Arg chloromethylketone). PPACK is an irreversible inhibitor of thrombin. Reconstitute PPACK with water to a concentration of 0.5 mg/mL. Reconstituted PPACK can be aliquoted and stored at −70 °C.
4. Acetonitrile and formic acid (HPLC grade).
5. Trifluoroacetic acid (TFA) (sequencing grade).
6. Reversed phase column: C4, 5 μm, 300A, 1.0 × 150 mm.
7. Capillary HPLC system (see Note 1).
8. Electrospray ionization-mass spectrometer.

2.2. Solid Phase Oligosaccharide Release, 2-AA Labeling and Map Analysis of rFVIII

1. Analytical Column; Shodex, Asahi Pak 5 μm NH2P-50 2D (2 × 150 mm).
2. Water (≥18 MΩ).
3. 1× PBS solution.
4. 1.0% Polyvinylpyrrolidone-360 (PVP) blocking solution (in water).
5. PNGase F.
6. 10 mM Tris-acetate digest buffer solution (pH 8.3).

7. Anthranilic acid.
8. Sodium cyanoborohydride.
9. 95% acetonitrile and 20% acetonitrile.
10. Mobile Phase A (2% acetic acid, 1% tetrahydrofuran, in acetonitrile).
11. Mobile Phase B (5% acetic acid, 1% tetrahydrofuran, 3% triethylamine in water).
12. 2% sodium acetate, 4% boric acid in MeOH.
13. 2-AA Labeled glycan standards.
14. Nylon spin filter tubes (Corning/Costar).
15. Microfuge tubes, 2.0 mL.
16. Autosampler vials, vial caps (pre-slit septa).
17. 12 tube vacuum manifold with fitting for spin-tube filters attached to vacuum.
18. Benchtop microcentrifuge.
19. Incubation oven, 37°C.
20. Heat block, 80°C.
21. pH meter.
22. Analytical balance.
23. HPLC: Equipped with eluent degas module, gradient pump module, column heater, autosampler, and fluorescence detector.

3. Methods

3.1. LC-MS Analysis of a Thrombin Digest of rFVIII Formulated Product

3.1.1. Thrombin Digestion of rFVIII

1. Transfer 100 μL of the rFVIII reference standard and test samples (100 μg/mL protein) into separate microcentrifuge tubes.
2. Thaw an aliquot of thrombin and dilute tenfold in water just before use (mix 10 μL of 1 U/μL thrombin + 90 μL water). Immediately after the dilution, add 2 μL of diluted thrombin to each tube (ratio of 0.02 Units of thrombin activity per 1 μg of rFVIII). Vortex gently until well mixed and centrifuge briefly in a low speed microcentrifuge. Incubate all samples at 37°C for 30 min.
3. Thaw PPACK right before use and stop thrombin digestion with the addition of 1 μL of 0.5 mg/mL PPACK to each digested rFVIII sample. Vortex and microcentrifuge tubes briefly.

3.1.2. Column Conditioning Step (Using Non Acid-Containing Solvents)

1. Set column temperature to 65°C.
2. Set auto sampler sample tray temperature to 4°C.

3. Set injection volume to 0 μL.
4. Set flow rate to 100 μL/min.
5. Set Solvent A to be 100% water and Solvent B to be 100% acetonitrile.
6. Run Gradient 1:
   5% B at 0 min.
   5% B at 5.0 min.
   70% B at 5.1 min.
   70% B at 25.0 min.
   90% at 25.1 min.
   90% at 30.0 min.
   5% B at 30.1 min.
   5% B at 40.0 min.

3.1.3. Tween 80 Wash Step (Using Non Acid-Containing Solvents)

1. Set column temperature to 65°C.
2. Set auto sampler sample tray temperature to 4°C.
3. Set injection volume to 40 μL thrombin digested rFVIII.
4. Set flow rate to 100 μL/min.
5. Set Solvent A to be 100% water and Solvent B to be 100% acetonitrile.
6. Run Gradient 2:
   5% B at 0 min.
   5% B at 5.0 min.
   70% B at 5.1 min.
   70% B at 25.0 min.
   90% at 25.1 min.
   90% at 30.0 min.
   5% B at 30.1 min.
   5% B at 40.0 min.

3.1.4. rFVIII Elution Step (Acid-Containing Solvents)

1. Set column temperature to 65°C.
2. Set auto sampler sample tray temperature to 4°C.
3. Set injection volume to 0 μL.
4. Set HPLC flow rate to 50 μL/min.
5. Set Solvent A to be 90% water, 10% acetonitrile, 0.5% formic acid, 0.005% TFA.
6. Set Solvent B to be 90% acetonitrile, 0.5% formic acid, 0.005% TFA.

7. Run Gradient 3:
   5% B at 0 min.
   5% B at 20.0 min.
   45% B at 30.0 min.
   80% B at 50.0 min.
   95% B at 51.0 min.
   95% B at 55.0 min.
   5% B at 56.0 min.
   5% B at 70.0 min.
8. Data is acquired, deconvoluted, and analyzed with a calibrated mass spectrometer (see Notes 1 and 2).

3.2. Solid Phase Oligosaccharide Release, 2-AA Labeling, and Map Analysis of rFVIII

3.2.1. Protein Capture

1. Preparation of System Suitability standard: Reconstitute a 100 pmol vial of 2-AA labeled A1F and A2 glycan standards with 100 μL of water. Combine equal amounts of A1F and A2 glycans, then dilute 50-fold with water. For use, add 100 μL of 95% acetonitrile and transfer to autosampler vial.
2. Set incubator oven to 37°C.
3. Set heat block to 80°C.
4. Remove nylon spin filter inserts from the centrifuge tubes and place on the vacuum manifold device. Prepare one filter for each sample.
5. Rinse each nylon filter two times with 100 μL of water.
6. Rinse each filter with 100 μL of 1× PBS.
7. Load protein solution samples onto filter and draw through by vacuum.
8. Wash each filter with 100 μL of 1× PBS. In some cases, reduction and alkylation of the protein may be required prior to blocking with PVP-360.
9. Remove the vacuum source from the vacuum manifold device.
10. Add 100 μL of 1.0% PVP-360 blocking solution to each filter and incubate for 10 min at room temperature.
11. Reconnect the vacuum source and draw the blocking solution through the filter.

3.2.2. Enzymatic Release

1. Rinse each filter three times with 100 μL of E-pure water.
2. For the final rinse, remove the filters from the vacuum manifold and transfer to labeled microcentrifuge tubes.
3. Spin 1 min at 10,000 RPM to remove excess water.

4. Transfer filter to new, labeled microcentrifuge tube for enzymatic release.
5. Prepare enzyme digest solution.
6. Add 50 μL of digest solution to each of the filters, ensuring that the filter is completely covered with liquid then cap each tube.
7. Incubate the tubes for 2 h at 37°C.
8. Remove samples and allow them to cool at least 2 min.
9. Spin samples approximately 1 min at 10,000 RPM to collect released glycans.

3.2.3. Label Glycans with 2-AA

1. Weigh anthranilic acid (0.045 g) and sodium cyanoborohydride (0.045 g) in separate microcentrifuge tubes.
2. Add 1.5 mL of sodium acetate/boric acid/methanol solution to the anthranilic acid and mix.
3. Transfer the solution to the sodium cyanoborohydride tube and mix to dissolve.
4. Add 100 μL of 2-AA labeling stock solution to each of the samples, close cap and mix well.
5. Incubate the tubes for 60 min at 80°C.
6. Remove samples and allow them to cool at least 3 min.

3.2.4. Post-Reaction Sample Clean-Up

1. Dilute each sample with 1 mL, 95% acetonitrile.
2. For each labeling reaction, rinse one new nylon spin filter with 2 × 200 μL, 95% acetonitrile using the vacuum manifold.
3. Load the diluted samples onto the filters and draw through by vacuum to bind the labeled glycans.
4. Wash each filter with 2 × 200 μL, 95% acetonitrile.
5. Remove the filters from the vacuum manifold and spin for 1 min at 10,000 RPM to remove excess 95% acetonitrile.
6. Transfer the filters into clean, labeled microcentrifuge tubes.
7. Elute the labeled glycans with 50–400 μL of 20% acetonitrile. Spin in microcentrifuge to collect eluted samples.
8. Dilute 50 μL of each sample with 200 μL of 95% acetonitrile for HPLC run.

3.2.5. HPLC Analysis

1. Set HPLC flow rate to 0.2 mL/min.
2. Set excitation wavelength to 360 nm and emission wavelength to 425 nm.
3. Set injection volume to 100 μL.
4. Set column temperature to 50°C.

5. Run the following gradient:
   30% B at 0 min.
   30% B at 5 min.
   35% B at 10 min.
   50% B at 34 min.
   65% B at 46 min.
   80% B at 52 min.
   95% B at 55 min.
   95% B at 60 min.
6. The relative % areas of the major peaks (neutral, mono-sialylated, di-sialylated, tri-sialylated, and tetra-sialylated species) are determined by integration of the HPLC fluorescence chromatogram.
7. The 2-AA labeled N-glycans have been analyzed by in-line LC-MS in the negative ion mode to confirm the identities of the chromatographic peaks.

4. Notes

1. When using a binary HPLC system, the three gradients for the thrombin digest LC-MS method are run consecutively for each sample. However, the protein digest is only injected during the second gradient and mass spectra data are acquired during the third gradient. Although the LC-MS thrombin map method for rFVIII was developed using a binary HPLC system, a quaternary system is preferred as the three gradients can be easily consolidated into a single method. A constant calculated amount of each rFVIII thrombin digest should be injected.
2. Mass deconvolution software programs should be reviewed to assure that relative quantitation is maintained.

References

1. Hossler P, Khattak SF, Li ZJ (2009) Optimal and consistent protein glycosylation in mammalian cell culture. Glycobiology 19:936–949
2. Raju TS (2003) Glycosylation variations with expression systems and their impact on biological activity of therapeutic immunoglobulins. Bioprocess Int 1:44–53
3. Davies J, Jiang L, Pan LZ, LaBarre MJ, Anderson D, Reff M (2001) Expression of GnTIII in a recombinant anti-CD20 CHO production cell line: expression of antibodies with altered glycoforms leads to an increase in ADCC through higher affinity for FC gamma RIII. Biotechnol Bioeng 74:288–294
4. Bork K, Horstkorte R, Weidemann W (2009) Increasing the sialylation of therapeutic glycoproteins: the potential of the sialic acid biosynthetic pathway J. Pharm Sci.98:3499–3508
5. Bovenschen N, Rijken DC, Havekes LM, Van Vlijmen BJ, Mertens M (2005) The B domain

of coagulation factor VIII interacts with the asialoglycoprotein receptor. J Thromb Haemost 3:1257–1265

6. Solá R, Griebenow K (2010) Glycosylation of therapeutic proteins. An effective strategy to optimize efficacy. BioDrugs 24:9–21
7. Purcell RT, Lockey RF (2008) Immunologic responses to therapeutic biologic agents. J Investig Allergol Clin Immunol 18:335–343
8. Chung CH, Mirakhur B, Chan E, Le QT, Berlin J, Morse M, Murphy BA, Satinover SM, Hosen J, Mauro D, Slebos RJ, Zhou Q, Gold D, Hatley T, Hicklin DJ, Platts-Mills TA (2008) Cetuximab-induced anaphylaxis and IgE specific for galactose-alpha-1,3-galactose. N Engl J Med 358:1109–1117
9. Higgins E (2010) Carbohydrate analysis throughout the development of a protein therapeutic. Glycoconj J 27:211–225
10. Fang H, Wang L, Wang H (2007) The protein structure and effect of factor VIII. Thromb Res 119:1–13
11. White G, Pickens E, Liles D, Roberts H (1998) Mammalian recombinant coagulation proteins: structure and function. Transfus Sci 19:177–189
12. Anumula KR, Dhume ST (1998) High resolution and high sensitivity methods for oligosaccharide mapping and characterization by normal phase high performance liquid chromatography following derivatization with highly fluorescent anthranilic acid. Glycobiology 8:685–694

INDEX

Jennifer J. Kohler and Steven M. Patrie (eds.), *Mass Spectrometry of Glycoproteins: Methods and Protocols*, Methods in Molecular Biology, vol. 951, DOI 10.1007/978-1-62703-146-2, © Springer Science+Business Media, LLC 2013

I

K

L

M

N

O

P

Q

R

S

MIX
Papier aus verantwortungsvollen Quellen
Paper from responsible sources
FSC® C105338

If you have any concerns about our products,
you can contact us on
ProductSafety@springernature.com

In case Publisher is established outside the EU,
the EU authorized representative is:
Springer Nature Customer Service Center GmbH
Europaplatz 3, 69115 Heidelberg, Germany

Printed by Libri Plureos GmbH
in Hamburg, Germany